PERSONS AND THEIR BODIES:
RIGHTS, RESPONSIBILITIES, RELATIONSHIPS

Philosophy and Medicine

VOLUME 60

Editors

The titles published in this series are listed at the end of this volume.

PERSONS AND THEIR BODIES:
RIGHTS, RESPONSIBILITIES, RELATIONSHIPS

Edited by

MARK J. CHERRY

Saint Edward's University
Austin, Texas, U.S.A.

KLUWER ACADEMIC PUBLISHERS
DORDRECHT / BOSTON / LONDON

A C.I.P Catalogue record for this book is available from the Library of Congress.

ISBN 0-7923-5701-9

Published by Kluwer Academic Publishers,
P.O. Box 17, 3300 AA Dordrecht, The Netherlands

Sold and distributed in North, Central and South America
by Kluwer Academic Publishers,
P.O. Box 358, Accord Station, Hingham, MA 02018-0358, U.S.A.

In all other countries, sold and distributed
by Kluwer Academic Publishers, Distribution Center,
P.O. Box 322, 3300 AH Dordrecht, The Netherlands

Printed on acid-free paper

Printed and bound in Great Britain by MPG Books Ltd., Bodmin, Cornwall.

TABLE OF CONTENTS

SECTION FOUR / THE BODY FOR PROFIT: ORGAN SALES AND MORAL THEORY

SECTION FIVE / PERSONS AND THEIR BODIES: KEY ARGUMENTS AND CONTEMPORARY CRITIQUES

PREFACE

This volume is a contribution to the literature of the philosophy of medicine and bioethics which takes as its heuristic focus the sale of human organs. The volume seeks insights from the history of philosophy, including, ancient philosophy, natural law theory, traditional Christianity, contemporary legal analysis and post-modern critique of moral theory. The volume's historical conception was a core set of papers written for a Liberty Fund Colloquium held in Houston, Texas, directed by Baruch A. Brody. This conference focused on the freedoms and responsibilities persons have towards their bodies. What began as the subject for a single colloquium, however, developed into a sustained philosophical dialogue.

I want to express my thanks to all of those who made the initial colloquium possible, the Liberty Fund, Inc., Baruch A. Brody, as well those participants, who though not represented in this volume, contributed to the intellectual discussion which framed the original set of essays and thereby indirectly impacted the framing of this volume. I could not have produced the final version without the support and kindness of the Center for Medical Ethics and Health Policy. I wish to thank in particular the *Philosophy and Medicine* series editors, H. Tristram Engelhardt, Jr. and Stuart F. Spicker, for their generosity in supporting the publication of this volume.

This project has benefitted through the kind efforts of many. I am in the particular debt of Thomas J. Bole, III, who labored with the project for some time, finding authors and focusing commentaries so that contributors would address critical perspectives from the history of philosophy to contemporary concerns in bioethics and health policy. The project's final form emerged from discussions with the contributors as well as with my colleagues H. Tristram Engelhardt, Jr., Ruiping Fan, and George Khushf. I am much in their debt. Discussions with Laurence McCullough, Stephen Wear, Nicholas Capaldi, and Joseph Boyle also helped to shape the character of this volume's contribution to the debates surrounding organ sales and health care policy. I wish to acknowledge also my debt to the authors in this volume for their patient kindness in revising and rewriting their essays. I have learned a great deal.

M.J. Cherry (ed.), Persons and their Bodies: Rights, Responsibilities, Relationships, vii–viii.
© 1999 *Kluwer Academic Publishers. Printed in Great Britain.*

After numerous revisions a sustained set of analyses has been forged. During its period of development much has been altered and many additions have been made. The issues range from ontological questions regarding mind and body, to the ethical underpinnings of governmental policy regarding the selling of internal organs. Through the entire volume, though, the essays focus on the moral permissibility of selling human organs for transplantation. The volume makes no claim to have encompassed all of the relevant issues or even all of the relevant philosophical perspectives. However, it is my hope that the papers gathered here will assist in clarifying the issues at stake in the medical and public policy fields, and serve to spur further dialogue. It is a great pleasure to present this international discussion as part of the *Philosophy and Medicine* book series.

Mark J. Cherry
Saint Edward's University
Austin, Texas   *August 1999*

MARK J. CHERRY

PERSONS AND THEIR BODIES:
RIGHTS, RESPONSIBILITIES, AND THE SALE OF
ORGANS

I. INTRODUCTION:
TAKING THE PHILOSOPHICAL BACKGROUND SERIOUSLY

Contemporary debates about organ sales are largely innocent of the history of thought on the matter. This volume seeks to remedy this short coming. Contemporary positions for or against a market in human organs are nested within moral intuitions, ontological or political theoretical premises, or understandings of special moral concerns, such as permissible uses of the body and its parts, which have a long history of analysis. The essays compass the views of Plato and Aristotle (James Hankinson), Thomas Aquinas (Thomas Bole), John Locke (Eric Mack and George Khushf), Immanuel Kant (Tom Powers), G.W.F. Hegel (Thomas Bole), John Stuart Mill (Wendy Donner), and Christianity (Allyne Smith and Gerald McKenny). Attention is also given to particular methodological approaches, such as the phenomenology of the body (Drew Leder and S. Kay Toombs), natural law theory (Joseph Boyle), legal theory (Christian Byk, Stephen Wear, and Donna Kline), and libertarian critique of ethical and legal theory (H. Tristram Engelhardt, Jr.).

These discussions cluster, therefore, a number of conceptually independent philosophical concerns: (1) What is the appropriate under-standing of the relationship between persons and their bodies? (2) What does it mean to "own" an organ? (3) Do governments have moral authority to regulate how persons use their own body parts? (4) What are the costs and benefits of a market in human organs? Such questions are related by an urgent public health challenge: the considerable disparity between the number of patients who could significantly benefit from organ transplantation and the number of human organs available for transplant. In 1996, for example, in the United States, 72,386 patients waited on the United Network for Organ Sharing lists for transplants (UNOS, 1998). By August of 1998 the waiting lists included 57,839 patients (UNOS, 1998). Yet, in 1997 only 19,998 organ transplants of all

M.J. Cherry (ed.), Persons and their Bodies: Rights, Responsibilities, Relationships, 1–32.
© 1999 *Kluwer Academic Publishers. Printed in Great Britain.*

types were performed (UNOS, 1998). Organ availability is not expected to increase significantly in the near future. Proposals to address this crisis, include national educational programs on the benefits of transplantation and the pressing need to donate organs. Enacted and proposed policies range from requiring physicians to ask families to donate organs from recently deceased relatives, making organ retrieval upon death mandatory, rewarding those who do donate, to creating for-profit commercial markets in human organs.[1] This volume explores the theoretical, normative, and historical foundations of such alternative policies.

II. THE PUBLIC HEALTH CARE CHALLENGE: WHEN ALTRUISM IS INSUFFICIENT

The challenges are significant. In 1996, 4022 patients died waiting for suitable organs (UNOS, 1998). Many others endured temporary life sustaining measures, such as kidney dialysis, while waiting on the organ que. While there were 40,634 registrants for kidney transplants in August of 1998, only 11,409 were performed in 1997. Similarly, there were 11,115 registrants for a liver transplant, but only 4166 transplants during 1997. The data regarding pancreas and heart transplants is similar: there were 416 patients registered for a pancreas and 4118 for a heart, with 206 pancreas transplants and 2292 heart transplants performed in 1997. The following chart summarizes this comparison with regard to various organs (UNOS, 1998).

Type of Organ	Number of Registrants[2] (August 1998)	Number of Transplants performed (1997)
Kidney	40,634	11,409
Liver	11,115	4,166
Pancreas	416	206
Kidney-Pancreas	1,765	854
Intestine	90	67
Heart	4,118	2,292
Heart-Lung	240	62
Lung	3,006	942
	61,489 Total Registrants (57,839 patients)	19,998 Total

Circumstances are apparently more serious in other countries. For example, surgeons in India report that 80,000 persons per annum present with end stage renal failure. There are relatively few resources for kidney dialysis (a total of 613 hemodialysis machines in 1988). Without organs, many patients will die (Reddy *et al.*, 1990, pp. 910-911; Thiagarajan *et al.*, 1990, pp. 912-914; Dossetor *et al.*, 'Discussion', 1990, p. 935; Johny *et al.*, 1990, pp. 915-917).[3]

Waiting times are being exacerbated by the growing pool of potential transplant candidates. The median waiting time for a kidney transplant was 400 days in 1988 compared to 824 days in 1994, with blood type B patients waiting 1,329 days. Patients listed for repeat kidney transplants waited 1,550 days, in 1993. Generally, patients with panel reactive antibodies (PRAs) of 20-79% waited much longer (1,325 days in 1994) than patients with PRAs under 20% (619 days in 1994). Median liver waiting time increased from 33 days in 1988 to 246 days in 1995, and again to 366 days in 1996. Patients with blood type O experienced the longest wait for livers of 501 days in 1996. Increasing wait times are also being experienced for other organs (UNOS, 1998).[4]

Given such circumstances, attention has turned to the potential of a commercial market to alleviate organ shortages and to decrease the time patients spend on waiting lists. For example, a futures market or a market in cadaver organs might significantly address heart or pancreas needs. Indeed, a commercial market in human organs may be most feasible where need is the greatest: kidney and liver transplants. Approximately 3,628 kidneys transplants from living donors were performed in 1997 (UNOS, 1998). Perioperative mortality for nephrectomy is very low, approximately 0.03%, with other major complications occurring in less than 2% of cases. Moreover, provided that healthy liver donors retain a substantial portion of their organ, its overall regeneration is possible after the donation of a segment (see, for example, Singer *et al.*, 1989; 1990; Siegler, 1992). Most centers with surgeons who have considerable experience in partial hepatechtomy report very low mortality and morbidity for such operations (Iwatsuki *et al.*, 1983; Nagao *et al.*, 1985; Hardy *et al.*, 1998).[5]

Despite such possibilities, a for-profit commercial market in human organs is often denounced as inappropriately commodifying the human body. Selling human organs for a profit is held to be exploitative and degrading, morally analogous to slavery, as well as incompatible with basic human values, such as human dignity, and important social goals, such as equality and a spirit of altruism. The human body, it is argued, should not be treated as property (Scott, 1981; 1990). Financial incentives are believed to coerce the poor into selling parts of their bodies. For example, in 1970 the Committee on Morals and Ethics of the Transplantation Society held that "the sale of organs by donors living or dead is indefensible under any circumstance" (1970) and the World Health Organization's (WHO) "Guiding Principles on Human Organ Transplantation" prohibits giving and receiving money for organs. Moreover, the WHO urges member states to legislate forbidding the commercial trafficking of human organs (WHO, 1991). Currently, at least thirty-nine countries proscribe the purchase of human organs for transplantation.[6] For example, in the United States, Title III (Section 301) of the federal "National Organ Transplant Act" makes it unlawful for any person knowingly to acquire, receive or otherwise transfer human organs for valuable consideration for use in transplantation.

III. THE EMERGING CONSENSUS AGAINST ORGAN SALES

The emerging consensus against a for-profit market in human organs is marked by a view that organs should not be treated as commodities. Organs are to be viewed as a social resource to be distributed according to medical necessity. They are to support public interests, rather than to be sold for private commercial gain.[7] They are to be a "gift of extraordinary magnitude" which transplantation surgeons "hold ... in trust for society" (Transplantation Society, 1985, p. 462). It is not just that organs are to be donated in the spirit of altruism. They are to be nationalized so as to constitute a "national resource to be used for the public good ... to best serve the public interest" (U.S. Task Force, p. 9). Since the "... physicians who select the recipient of a donated organ are making decisions about how a scarce public resource should be used" (U.S. Task Force, p. 86), organs are to be allocated on the basis of acceptable medical criteria and social goals, rather than patient financial status. These foundational background assumptions, in particular that available human organs are a public resource which states are morally in authority to allocate, colors much of the debate regarding the permissibility of an organ market (Gorovitz, 1987).[8]

Various policy statements have been broadly influential in shaping the emerging consensus against a market. For example, the Transplantation

Society (1985), the World Medical Association (1985; 1987), UNESCO (1989), the World Health Organization (WHO)(1991), the Nuffield Council on Bioethics (1995) and the U.S. Task Force on Organ Transplantation (1986) each condemns the creation of a for-profit market in human organs. The position is supported by arguments which evaluate the moral and political theoretical parameters as well as the costs and benefits of a market in human organs. Such arguments purport to show that, unlike altruistically motivated donation, (1) offering financial incentives undermines consent, coercing the poor into selling their organs, and (2) that a market in human organs exploits the poor, violates human dignity, and is morally repugnant. Moreover, opponents argue that such a market would lead to (3) greater inequality between the rich and the poor as well as (4) worse health care outcomes than the current system of donation.

A. Consent to Organ Donation: Commercialism vs. Altruism

For consent to organ donation to be morally effective, it must be voluntary. While it is often difficult to insulate patients and family members from institutional and social pressures, it is argued that potential organ donors "should be free of any undue influence and pressure and sufficiently informed to be able to understand and weigh the risks, benefits and consequences of consent" (WHO, 1991, p. 472). As the U.S. Office of Technology Assessment Report points out, psychological, emotional and medical needs, as well as a desire to please others may influence one to donate organs (1987, p. 96). Family members may agree to donate organs to avoid confrontations or to satisfy personal, family or social objectives. Nevertheless, opponents of an organ market argue that unlike such motivations, the prospect of financial gain inappropriately influences a subject's decision to give consent.

Consider, for example, the Committee on Morals and Ethics of the Transplantation Society's description of money's coercive potential:

> In a South American country ... advertisements from desperate individuals have appeared in newspapers offering a kidney or even an eye ... for money. ... It does not seem unlikely that a few of the unscrupulous will acquiesce to the profit motive ... (pp. 461-462).

Financial gain is considered an influence which overwhelms and subjugates the voluntary consent of impoverished potential donors (Sells

1994; 1992; Daar and Sells 1990). The choice to sell a kidney in order perhaps to better the economic status of one's family is considered a decision without scruples.

Numerous professional statements endorse this assessment. For example, the U.S. Task Force recommended that transplanting kidneys from living unrelated donors be prohibited whenever financial gain rather than altruism is the motivating factor (1986, p. 10). It called for federal and state prohibition (p. 10). Similarly, the WHO's guiding principles on organ transplantation proscribe advertising for organs with an intent to offer or seek payment (1991, p. 472). The Transplantation Society simply condemns advertising by transplant surgeons for donors or recipients (p. 463), and the American Medical Association prohibits the purchasing of organs from living donors for transplantation (1994/95, p. 26).[9]

In contrast, altruism is believed to support individual freedom by fostering personal choice. As the Transplantation Society argued with regard to living unrelated donors: "altruism on the part of the donor may be a real motivating factor ... the wish to donate an organ need not be a sign of mental instability" (p. 466). Moreover, they proposed guidelines for the distribution and use of organs from cadaver sources and living donors which specifically require that donors be altruistically motivated.

> It must be established by the patient and transplant team alike that the motives of the donor are altruistic and in the best interests of the recipient and not self-serving or for-profit ... especially in the exceptional case where the emotionally related donor is not a spouse or a second degree relative, the donor advocate would ensure and document that the donation was one of altruism and not self-serving or for-profit (1985, para 2).

Unlike for-profit commercial transfers, altruistic donation purportedly binds persons to their families and communities.

The market is viewed as corrosive of "gift-of-life" sentiments, which have often characterized organ procurement. After all, altruistic donation has powerful psychological and social repercussions which are of value to society and interpersonal relationships. Altruistic donation is seen as a free expression of important human values as well as of communal commitments. Fox and Swazey encapsulate this view as they lament that many assert that organ transplantation is analogous to a commercial industry and product. Rather than thought of as parts of living persons, "offered in life or death to sustain known or unknown others, that resonate with the symbolic meaning of our relation to our bodies, our selves, and to each other", Fox and Swazey express concern that organs are becoming mere things, i.e., as "just organs"(1992, p. 207). A market in human organs, they argue, would ignore and undermine the gift exchange dimensions of organ donation, with its obligations of giving, receiving, and repaying. It would undermine our willingness to address common problems with collective resolve (see also Gorovitz, 1984, p. 12). As Leder argues, the importance of such gifts to individual recipients as well as for community solidarity is believed to be overlooked by commodification (1999).

Altruistic donation is regarded as central to maintaining both public trust and organ availability. The argument is that when organs are donated to the community, they carry "the hopes of the donor's family ... that the organ will be used to the best possible advantage" (Sells, 1994, p. 1017). As Mary Ellen McNally, coordinator for the New England Organ Bank in 1990 emphasized, trust that organs are not commodities to be sold underlies public willingness to donate: "the sale of those organs would diminish the goodness of that gift" (Dossetor *et al.*, 'Discussion', 1990, p. 933).

In contrast, others advocate a system of death benefit payments to motivate families of potential organ donors on the grounds that "Our concerns must focus not on some philosophic imperative such as altruism but on our collective responsibility for maximizing life saving organ recovery" (Peters, 1991, p. 1305). Yet others, such as Edmund Pellegrino, seek to rebut these proposals as "logically, ethically, and practically flawed" (Pellegrino, 1991, p. 1305). Since such proposals appear to reject the centrality of the gift relationship to organ donation, they are criticized as endorsing values which are antithetical to the altruistic, community oriented, culture of donation.[10]

While the purchased organ is viewed as a commodity, coerced by poverty from the seller, altruistically donated organs are understood as binding persons to their community and family, as well as fostering "warm satisfaction" in the minds of donors or their bereaved families (Sells, 1994, p. 1016; Sadler and Sadler, 1984). While couched in somewhat rhetorical terms, the process of organ donation is regarded as giving something intrinsically valuable to the donor or his family. In short, opponents of organ sales argue that commercialism treats persons as individuals in isolation from family and community, coercing some through the offer of money into parting with their organs. However, altruistic donation is held to foster the expression of important personal values as well as family commitments and social solidarity. Thus, donation is consistent with free voluntary consent (Beauchamp and Childress, 1994; Simmons, 1981).

B. Organ Markets are Morally Repugnant, Exploit the Poor, and Violate Human Dignity

Buying organs from living vendors for transplantation arouses in many feelings of gruesome horror. Commercial schemes are viewed as intrinsically exploitative. They are seen to be incompatible with human dignity and morally repugnant. The historical precedent for such markets, according to Russell Scott, is chattel slavery in which the human body becomes mere property (1990, p. 1003). At the very least, this view does not take seriously the involuntary character of hereditary slavery. For Childress, condemnation of organ markets is grounded in basic principles, such as "the dignity of the individual" and "respect for persons" (1989, p. 88; see also OTA, 1983, pp. 130-132). For Ramsey, behind the repugnance lies a real danger that such practices "will only erode still more our apprehension that man is a sacredness in the biological order ... and our respect for men of flesh who are only to be found within the ambience of bodily existence" (1970, p. 209; see also Scorsone, 1990).[11] Others view an organ market as representing an extreme in human greed (Abouna et al., 1990).

For many the moral repugnance has a straightforward basis. For example, the WHO resolution 'Development of guiding principles for human organ transplantation,' affirms that commerce in human organs is "inconsistent with the most basic human values and contravenes the Universal Declaration of Human Rights ..." (WHA 40.13; 1994, p. 467; see also WHO 1991).[12] The WHO resolution 'Preventing the purchase and sale of human organs,' similarly asserts that the commercial sale of organs is exploitative and incompatible with human dignity (WHA 42.5). This resolution argues that prohibition is necessary to

> prevent the exploitation of human distress, particularly in children and other vulnerable groups, and to further the recognition of the ethical principles which condemn the buying and selling of organs for purpose of transplantation ... (p. 467).

Organ selling purportedly exploits the distress of those in need.

As an example of a commercial venture, consider H. Berry Jacobs, M.D., founder of International Kidney Exchange Inc. Jacobs asked 7,500 hospitals if they would participate in his plan to broker human kidneys. He proposed to offer the poor in Third World countries and the United States whatever price would induce them to sell a kidney, and then negotiate acquisition, for a profit, by Americans able to pay for the organs. Jacobs' plan aroused cries of moral indignation in the United States Congress. It was denounced by the National Kidney Foundation, The Transplantation Society, and the American Society of Transplantation Surgeons. Professional organizations resolved to expel members involved in such transactions (Fox and Swazey, 1992, p. 65). As the U.S. Task Force summarized the objection: "We are alarmed that ... certain transplant centers are reportedly brokering kidneys from living unrelated donors. We find this practice to be unethical and to raise serious questions about the exploitation and coercion of people, especially the poor" (p. 98). The contention is that the rich exploit the poverty of sellers, who given better circumstances would not have sold their organs (Abouna *et al.*, 1990, p. 919; see also Dossetor and Stiller, 1990).

The potential for exploitation and the violation of human dignity, which is perceived to underlie commerce in human organs, is held to trump the possibility of increasing life sustaining transplants. As the Nuffield Council Report argues, certain uses of human tissue are morally unacceptable because they "fail to respect others or to accord them dignity ... they injure human beings by treating them as things, as less than human, as objects for use" (1995, para 6.7). Uses which are morally acceptable protect "... a central element of the undefined, yet widely endorsed, demand for respect for the human body and for respect for human dignity" (para 6.4). This distinction is grounded in a judgment regarding whether or not such uses properly respect human dignity (Meyer, 1995; Grubb, 1995; Walsh, 1995). As Keyserlingk argues, even though such requirements may make kidney procurement inefficient, this does not provide a sufficient reason to override society's commitment to preventing exploitation and preserving human dignity:

> Assuming and adopting as we do the Kantian injunction that persons should always be treated as autonomous ends and not merely as means to the ends of others, a morally justifiable policy will be one which is likely to provide the largest number of ... organs ... without violating the human dignity of potential donors. As for a policy permitting the sale and purchase of human organs, it may well be in some respects the

most efficient approach, at least for those in a position to buy and sell, and it can provide direct and full control to buyer and seller; *but because it so badly fails the other tests, it should be rejected* (1990, p. 1005, emphasis added).

Efficient, effective organ procurement and transplantation should not be understood as a good worth pursing at the price of human dignity.

Human organs, Keyserlingk argues, have a special status due to their intimate relation with persons. Furthermore, in life the body as a whole is a good, because it is the center and means of personal awareness and vehicle of communication. It is this link between self and body which grounds bodily inviolability. These characteristics, he concludes, make the body and its parts properly the subject of altruism and gift giving rather than commercial sale (Keyserlingk, 1990, p. 1005). To respect these characteristics of persons is to respect human dignity. Organ sales in contrast, treat the body as a collection of parts,[13] independently of the body's intimate relationship to the life of a persons. Organ sales thereby violate human dignity. As Bob Brecher summarizes:

> however much the Turkish peasant who sold a kidney may have needed the money he was paid; however genuinely he may have wished to exercise his autonomy in this enterprising venture ... however sincere his wish to benefit his family with the proceeds, and however great their need; nevertheless what he did was wrong (1994, pp. 1001-1002).

Insofar as the commercial vending of human organs violates human dignity, is deemed exploitative, and morally repugnant, that more individuals will suffer and die than on a for-profit scheme of organ vending is judged to be an insufficient reason to justify the creation of such a market.

C. Equality and Social Justice

Building on the assumption that organs are a resource to be utilized for the public good with an independent assumption of egalitarianism, the argument levied against a market in human organs is that equal access to organ transplants is the only distribution consonant with the equality of all candidates. Social justice, it is contended, requires equality of opportunity through redistributing property and forbidding the private sale and purchase of better basic health care. As Childress notes "the individual's personal and transcendent dignity ... can be protected and witnessed to by a recognition of his equal right to be saved" (1970; see also, 1989, p. 88; 1986, p. 4).[14] The Transplantation Society contends, for example, that allowing the purchase of organs for transplant would lead to inequitable health care outcomes. If

> wealthy individuals from other countries are placed on transplant lists ... they compete with local patients for scarce cadaver kidneys ... private hospitals in Europe now perform kidney transplants for foreigners who can afford the substantial fees ... The unacceptable consequence of this is that kidneys go only to patients who can pay (1985, p. 460).

The concern is that a market in human organs will lead to a situation in which the poor sacrifice their bodies for the health of the rich, while the rich gain unequal, and therefore unfair, access to a scarce medical resource (Dickens, 1990; Jonsen, 1987; Blumstein, 1990; 1992).

This conclusion resonates with the President's Commission for the Study of Ethical Problems in Medicine and Biomedical and Behavioral Research, when it held that "In light of the special importance of health care, the largely undeserved character of the differences in health status, and the uneven distribution and unpredictability of health care needs, society has a moral obligation to ensure adequate care for all" (1983). From this perspective, society has a duty to sustain the existence of each individual that extends not only to providing basic health care, but also, if necessary, to organ transplantation (Ignatieff, 1984; Daniels, 1985).[15]

The U.S. Task Force similarly concluded that availability of transplantation should not be limited by patient financial status. They argued that "All transplant procedures recognized as medically effective, should be made available through reimbursement mechanisms for the care of patients who have no other source of funds ..." (1986, p. xxi, see also pp. 9-11). The Task Force equated equitable access to transplantation with the prohibition of organ sales: "... implementation of equitable access prohibits any elements of commercialization in the distribution of organs" (p. xxi). Insofar as there exists a social commitment to provide all with equal access to adequate health care, including organ transplants, equitable access implies access independent of wealth (Task Force, p. 105).

As the Massachusetts Task Force on Organ Transplantation elaborates this point, rationing human organs based on the ability to pay sends the message that

> we do not believe in equality and that a price can and should be placed on human life and that it should be paid by the individual whose life is at stake. Neither belief is tolerable in a society in which income is inequitably distributed (1984, p. 233).

Each policy stresses equality as a moral constraint on public health care policy (Townsend and Davidson, 1982).

Rather than recognizing the guarantee of property rights and free collaboration as core to human respect and dignity, despite inequalities, as Engelhardt argues, opponents of an organ market reflect a vision of social justice which brings into question the good fortunes of those who have more opportunities, wealth, and resources to purchase more extensive health care, such as organ transplantation (Engelhardt, 1999; see also 1987, pp. 339-353; 1984). The argument in part concerns which values ought to weigh more in the calculation of the costs and benefits of public health care policy. However, it is also a debate over which values may be sacrificed for others (Cantarovich, 1990, p. 927; Pauly, 1987) in that allocation and rationing decisions are inevitable (Evans, 1987a; 1987b).

For example, one might argue that losses in equal access to expensive health care can be compensated at the level of social policy by gains in individual liberty as well as by greater organ availability, and thus less total human suffering. Many assess the value considerations otherwise. As Annas argues, a market in human organs places a very high value on individual rights and a very low value on equality and fairness (1987, p. 332; see also, Annas, 1983; Basson, 1979). For example, Caplan argues that

> Allocating lifesaving organs by the ability of those in need to pay what the market will bear is blatantly unfair to the poor ... To argue that the sale of organs as a business is motivated by humanitarian concerns ... simply flies in the face of the fact that selling organs can only increase the cost for those who now receive them for free (1983, p. 23; see also Caplan 1984; 1992).

In short, equality is viewed as an essential element of social justice and as able to trump important liberty interests. Moreover, the additional costs of providing more organs to save the lives of more people is considered unjustifiable.

If one assumes that organ donation is a gift to society, and that the state has the moral authority to regulate procurement and distribution, it is plausible to raise concerns on grounds of equality that "It seems unfair and even exploitative for society to ask people to donate organs if these organs will be distributed on the basis of the ability to pay" (Task Force, p. 104). If poor and rich alike donate, but only the rich receive organ transplants, a significant social class of those who donate will not benefit from transplantation. Thus, opponents conclude, organs should be distributed to medically eligible recipients regardless of their ability to pay for the transplant:[16] on these grounds a for-profit market is argued to lead to unequal health care outcomes, and therefore, as being unfair and unjust.[17] Medical suitability and severity of illness are viewed as consisting of more objective data appropriately weighed in allocation of services. They appear to satisfy the requirement that "Criteria for organ placement must be objective, medically sound, and publically (sic) stated" (U.S. Task Force, p. xxi). However, no one seemed to notice that the argument as structured was less one against a market in organs as one against (1) selling organs which were dontated or (2) charging for health care services using donated organs.

D. Health Care Consequences

An additional concern is that a market in human organs would harm the health of donors and recipients. Consider, for example, the Transplantation Society's assessment:

> Normal safeguards which protect family donors and recipients are threatened by brokerage arrangements, when otherwise unacceptable donors may become acceptable if the price is right. Similarly, one can argue that less than ideal kidneys could be sold more cheaply than organs of good quality. Even if motivation of both donor and recipient is correct, the sale of organs essentially forces the donor to have an operation ... Indeed, as a practical example, it could be argued that the buying and selling of blood for transfusion ... has led to a less safe and more expensive service (p. 462).

The Transplantation Society identifies three important special health care costs: first, that a market in human organs will lead to procuring lesser quality organs, second, that this will lead to lower life expectancies for recipients, and third, that vendors will incur inappropriate surgical harms, including possibly death.

Many share the first concern. As Caplan argues, medicine cannot morally allow "itself to be used by those who would risk their own health out of greed, desperation or ignorance" (1983, p. 23). Similarly, Frier and Mavrodes conclude that the profit motive would likely reduce care and caution in the selection of suitable organs (1980; see also Jonsen 1997; Kennedy, 1979). So too Abouna *et al.* argue that an organ market would lead to inferior quality medical care with higher complication rates (1990, p. 918). The concern is that, if profit is the primary objective then "normal standards of medical screening may well not be exerted; postoperative deaths from HIV transmission at the time of transplantation have been reported" (Sells, 1990, p. 931; see also Fox and Swazey, 1992, p. 208). Moreover, family members may not be willing to donate genetically well matched organs, if it is possible to purchase the needed organ.

The second criticism resonates with general anxiety regarding graft survival. The long term survival of transplants may be superior when organs are procured from living relatives rather than from unrelated donors (Abouna *et al.*, 1990 p. 919; see also Cook and Terasaki, 1987). In one study, the patient survival at one year for transplants utilizing purchased kidneys from unrelated donors was 81.5%, compared to 97.8% using living related donors (Daar *et al.*, 1990). However, other studies rebut this conclusion. One study in India suggested that graft survival from live unrelated donors at one year appeared to be moderately superior to those from live related transplants (84% verses 81% respectively). They acknowledged, though, that they did not view the difference as statistically significant (Thiagarajan *et al.*, 1990, p. 913; see also Reddy, 1990). If poor transplant results lead to an increased number of transplants per patient, this will be an inefficient and ineffective use of scarce health care resources, presuming that the market does not sufficiently increase the supply of organs to more than compensate for any such inefficiencies.

The third criticism identified by the Transplantation Society is often phrased in terms of the Hippocratic injunction not to do harm, *primum non nocere* (*Epidemics*, book 1, chapter XI; see Edelstein, 1967; Jonsen, 1977). Organ vendors receive operations for which there is no medical indication. "Operations should be performed for therapeutic reasons: a financial reward does not represent a therapeutic indication for surgery" (Sells, 1990, p. 931; see also Cantarovich, 1990, p. 927; Kilbrandon, 1968). Moreover, if the vendor is poor, once his organ is removed and he is discharged from the hospital, there may be no guarantee of adequate follow up. The donor may ill afford any future hospitalization for possible complications of the operation (Abouna et al., 1990). While some have argued that there is no morally relevant difference between selling one's time and selling one's own redundant organs, this is argued to be a minority view (Jonsen, 1997, p. 242; see also Blumstein and Sloan, 1989).

Additional criticisms are that legitimating a market in human organs would indirectly[18] lead to a decrease in the number of kidneys available for transplant as well as an unacceptable rise in the financial costs of procuring organs. The first concern is tied to the view that "Making organs available only to those who can afford them will only feed preexisting suspicion and paranoia" (Youngner, 1990, p. 1015). There are three primary possibilities. First, if organ sales are permissible this may so inflame public response as to turn the public against the organ transplant system. As a consequence fewer people would be inclined to donate.[19] Second, if patients purchase organs there is less incentive to create and promote local donation programs. Third, an organ market may also impact living donor related transplantation. For example, in Kuwait "Several well-matched relatives of potential recipients ... withdrew their offer of donation after they learned that their relatives [could] go to India and buy a kidney in the market place" (Abouna et al., 1990 p. 919; see also Dossetor et al., 'Discussion', 1990, pp. 933-934). Insofar as a decrease in altruism becomes widespread, and if the number of organs available for purchase does not increase to meet the shortfall, there may be fewer organs available (Skelly, 1987).

The Office of Technology Assessment Report identifies two types of additional financial costs which would be incurred if donors are financially compensated for their organs: the actual cost of compensation and the costs of administering the exchange, i.e., additional transaction costs (1987, p. 116). Transaction costs would likely include advertising for potential vendors, screening for tissue compatibility, as well as the cost of negotiating between patients and vendors over contractual conditions for the transfer of property rights in the organs. If a market would add significant financial barriers to transplantation, this would further limit the ability of many to gain access to such health care, thereby producing worse health care outcomes than the current system of donation (Baily, 1988).

In short, it is argued that a market in organs will lead to harms to organ vendors, inefficient and ineffective use of scarce health care resources, decrease the number of available organs, raise health care costs and procure lower quality organs for transplantation. Such health care costs if realized may outweigh the potential benefits a market in human organs may possess.

IV. CONTROVERSIES AND CRITICISM

While the focus of debate framing the global consensus has been on the ways in which a market in human organs may undermine consent, exploit the poor, and violate human dignity, create greater inequality between rich and poor, and lead to worse health care outcomes than the current system of donation, more detailed analysis is required adequately to evaluate such concerns. Such a consensus, while pervasive, fails to be sufficiently justified. Consider, for example, the following argument, addressed by surgeons in India to the International Transplantation Society at a conference on the permissibility of a market in human organs. Thiagarajan *et al.* argue against

> those who have allocated to themselves the right to sit in judgment, based on their own environment and prejudices, and to exclude from scientific discussion those observations that come to alternative and controversial yet acceptable, practices prevailing in other less fortunate areas of the world (1990, p. 914).

While pointing to the success of their transplantation program in terms of graft survival, lack of donor mortality, return of patients to productive lives, and assistance in resolving the great financial need of donors, Thiagarajan *et al.* lament that the greatest value lost is professional and academic respectability.

Similarly, Reddy *et al.* ask themselves "Do we buy or let die?" (p. 911). They acknowledge that in India money often changes hands even among related living donors and recipients. Moreover, they rebut the purported moral repugnance which appears to be foundational to the "consensus". Organ selling is not mere utilitarian spare parts medicine, they argue, it focuses on basic human values, community connectedness and fundamental social goals. It conveys important moral benefits.

To dismiss the idea of paid donors as the ethics of expediency is to deny these patients the right to live. We serve only the corrupt and the unscrupulous if we deny the patient benefit of a transplant that is medically indicated because of our fear that the paid donation process is too complex to be regulated (p. 911).

Purchasing human organs for transplantation, they conclude, is in keeping with ethical values, as well as understandings of fundamental commitments to medical care.

The response of the Transplantation Society to such criticisms invoked the authority of superior information and expertise. Nicholas Tilney stated, for example, that "We felt that we had the knowledge and expertise to create these guidelines, knowing some of the facts and innuendoes ..." (Tilney in Dossetor *et al.,* 'Discussion', 1990, p. 935). The criticism of the Indian surgeons, though, remains unaddressed: more careful analysis is required before one should pass judgment against a commercial market in human organs. Before ratifying a global consensus, more voices ought to be heard. As Johny *et al.* point out, while the buying and selling of organs is condemned internationally, this doctrine is largely based on sensational press reports concerning the involvement of unscrupulous middlemen, as well as the concern to protect desperate recipients and needy poor from exploitation (1990, p. 917). Insofar as such concerns can be appropriately dealt with, legitimate objections to a market in human organs will be diminished.

V. PHILOSOPHICAL ARGUMENTS AND BACKGROUND ASSUMPTIONS

In the following chapters, the authors explore the moral, ontological, and political theoretical concerns at stake in such a bioethical and public policy debate. For example, the first cluster of essays addresses the relation of ontological presuppositions to moral considerations. Consider, for example, the contrast between the ancient Greeks and both contemporary phenomenological and traditional Christian approaches. On the one hand, Greek philosophy, Hankinson argues, often stigmatized bodily sensual life as bestial. The association of the soul with the physical body distorted the person from his true purpose: contemplation. The body was understood as a burden to bear rather than a blessing to celebrate. On the other hand, Toombs explores the vibrancy and complexity of the person as the living body, while Smith reflects on the traditional Christian view of the person as a unity of body and soul together, which participates in the Divine image. In each case the body is affirmed as good.

As Toombs notes, every particular ontology of person and body has moral implications. She argues that the lived body is distinct from all other types of material objects and that it cannot be conceptualized adequately as owned property. To claim that my body is "mine" does not in itself express an ownership relation. It rather betokens an experiential awareness of the living body, an intimate being-in-the-world. Consequently, "the recognition that I AM my body focuses one's attention on the role of personal participation in the prevention of illness and treatment of disease" (p. 87). Embodiment, she argues, grounds duties to promote one's health and bodily integrity.

While Smith also denies that "ownership" appropriately conceptualizes the relationship between person and body, he identifies the foundation of duties to the body not in the self, but in God. For the Christian, the rubric for understanding the relation is not ownership but stewardship, i.e., the Christian is responsible for the proper uses of himself, because his owner (God) will hold him to account for his stewardship. As Smith notes many Orthodox view such duties as ruling out a market in human organs. Selling organs is seen as commercializing body parts and therefore as unworthy; such a market would be prohibited by concerns for the protection of human life and its dignity.

According to Donner, it is likely that John Stuart Mill would have agreed with Smith's assessment that an organ market ought to be proscribed, albeit for secular reasons. While Mill rejected the notion of duties to oneself, he held that individuals should stimulate each other to develop the higher faculties and the general promotion of the good. It is plausible, Donner argues, to use Mill's theory of personal self development to conceptualize activities which commodify the human body as degrading. Involvement would have "... a degree of folly, and a degree of what might be called ... lowness or depravation of taste, which though it cannot justify doing harm to the person who manifests it, renders him necessarily and properly a subject of distaste, or in extreme cases, even of contempt ..." (pp. 66-67). However, as a utilitarian, if a market in human organs would maximize the good, in terms of quality and quantity of life, health benefits, and social outcomes, Mill may, in principle, be committed to accept its licitness.

The second set of essays grapple with the intersection of individual rights and political authority. Insofar as persons have rights over their body parts or personal responsibility to take care of their health, one must assess, first, whether such rights can be alienated, and second, whether satisfaction of particular duties to one's body should be coerced through force of law. According to Boyle, the natural law tradition views persons as in authority over themselves. Generally, governments do not have moral authority to interfere in decisions that adults make concerning their own health. Good health and bodily integrity, however, are the means through which one carries out commitments to others. Therefore, "The most common and plausible limitations on individual freedom ... are those in which bodily intangibility and patient autonomy are not violated or are violated only in minimal ways; for example, activities which cause various forms of pollution, spread disease, or put others' life and health at risk" (p. 128). Insofar as sufficient care is taken to preserve health and functional integrity, though, natural law considerations of respect and appropriate uses of one's body would not necessarily proscribe for-profit transactions in human organs.

Khushf and Mack each consider the implications of Locke's account of natural rights and the limited state. Khushf notes that Locke bases his arguments against suicide and slavery in theological premises: persons are not their own property, rather they belong to God. In particular, only God has absolute dominion over human life. Such a position, Khushf argues, provides the foundation for inalienable rights to life which proscribe individual freedoms to commit suicide, or to cede absolute authority over oneself to another. However, insofar as procurement of redundant internal organs is accomplished by an experienced surgeon in a suitably sterile environment, organ sales are not analogous to either slavery or suicide.

In contrast, Mack argues that Locke views natural rights as defining for each person a basic sphere of moral jurisdiction within which each right holder may do as he sees fit. Such natural rights act as side constraints over against others, but do not, according to Mack, stand as barriers to choices about one's own body. For example, an individual may alienate his rights and, thereby, open himself up to treatment which otherwise would be impermissible: e.g., to the surgeon's opening of the thoracic cavity or the oncologist's injection of life-threatening chemicals. Natural rights are, therefore, alienable. Given such an account, a market in human organs would likely be permissible.

The third cluster of essays is less inclined to view a market in human organs as morally licit. For example, Powers argues that Kant's account of personhood as the infinitely valuable source of moral consciousness and reasoning grounds deontological limits to the permissible uses of the body. Kant forbids both suicide and the selling of body parts. According to Powers, Kant's conception of human dignity and of the centrality of persons to the moral project, blunt the consequentialist considerations which lie behind organ procurement. One's body parts, insofar as they are integral to one's ability to comply with the moral law, may not be removed and sold at will. Powers argues that Kant would view such actions as only serving to debase the human person.

Similarly, Leder holds that the quest for control over the body leads to human degradation. Organ transplantation technology, he argues, treats the body as a *res extensa*, a mere machine to be refined by well trained mechanics, using inter-changeable parts. As Byk points out, the body is thereby reduced to the sum of its parts; its character as the embodiment of a person is recast as property. Such inappropriate depersonalization, Leder argues, is exemplified in the for-profit market in human organs.

Even if the market would lead to depersonalization, or harms to human dignity, one must still consider whether the modern secular state possesses moral authority to forbid such consented-to exchanges of body parts. Engelhardt, Kline, and Bole, each consider the limits of state authority to govern free consensual uses of body parts. While Kline contends that there exists moral authority for state intervention regarding such practices, Engelhardt and Bole each retort that there is no general secular moral basis to sustain such a claim. According to Engelhardt, reason alone cannot establish a unique canonical account of moral intuitions, middle-level principles, consequences, or virtues to establish such governmental moral authority without begging the question of which ranking of values or principles should guide. In the absence of such an account, moral authority must be sought in the consent of persons. Kline contends, though, that Engelhardt has proven too much: given his arguments, he is unable even to justify the authority of mutual agreement as an acceptable ground of common moral action.

However, as Bole argues, while reason cannot by sound argument establish a particular content-full moral view, it can show that absent a rationally required objective content, persons themselves must be taken as the ground of moral authority. Therefore, absent actual agreement, persons have general secular rights to be left alone in their interaction with consenting others. Furthermore, as Bole argues, Hegel demonstrated, perhaps inadvertently, that the market provides a framework in which persons can respect the freedom of others while fashioning common agreements through peaceable negotiation. The free-market is the arena in which moral strangers, i.e., persons who do not share a common moral framework or a common moral authority, can negotiate morally authoritative agreements, which bind a society of moral strangers into a peaceable whole. In such terms, market arrangements for the sale of human organs can be morally congenial. At the very least, as Bole and Engelhardt argue, legal prohibition is in general secular terms morally unjustifiable.

Beyond considerations of embodiment, ownership, and morally justified political authority, there are various special moral concerns which bear on a market in human organs. As Wear *et al.* point out, the crafting of public policy must consider moral concerns regarding social and personal welfare, often captured under the general rubric of exploitation. Market exploitation often works by offering inducements to those who are vulnerable, such that intrinsically unattractive options become, all things considered, the best available choice. The concern is that financial incentives will tip the balance of interests inducing the poor to sell their organs to the rich. Given conditions of poverty, the poor will have significant financial incentives to undergo the risks of surgery. On the other hand, if the rich are dying of organ failure, they will similarly be under significant pressure to purchase a life saving organ. However, one must further assess whether the market would exploit persons in greatest need, or whether forbidding organ markets is itself exploitative. By forbidding organ sales, are the financially secure and able-bodied exploiting the poor to assuage their own feelings of guilt, so that they can sleep at night thinking that they have saved the poor from themselves? That is, the rich and able-bodied under this circumstance would be exploiting the poor so as to be able to have the poor not challenge their view of proper moral conduct.

As McKenny observes, the historical philosophical and religious lexicon which contemporary bioethical discussion typically engages to address particular concerns, such as the moral permissibility of organ selling has been remarkably thin in comparison with the robust corpus that the West inherited from its Aristotelian, Christian, Jewish, and classical liberal background. As the essays in this volume illustrate, attention to this literature discloses the rich counter-balancing interests and special moral considerations which when gauged against one another balance, curtail or defeat arguments against a market in human organs or provide grounds for its prohibition. Against the background of such considerations, one should be better able to appreciate the ways in which the normative assessment of the costs and benefits of an organ market will vary among moral visions and theoretical frameworks. These essays provide resources for more able assessing of current policies regarding organ sales.

Department of Philosophy
Saint Edward's University
Austin, Texas

NOTES

[1]   For a discussion of such proposals see, for example, Caplan and Welvang, 1989; Sells, 1979; Roles *et al.*, 1990; and Monaco, 1990.

[2]   UNOS allows patients to be listed with more than one transplant center, and thus the number of registrations may be somewhat greater than the actual number of patients.

[3]   About 613 hemodialysis machines were imported into India between 1971 and 1988; assuming that all are in working order all the time, an average of 1500 dialyses may be carried out each day; with individual dialysis frequency of bi-weekly and an average period of hemodialysis of three months, the maximum number of patients that can be treated each year in India would be 18,000 or 22.5% of those who require such care (cited in Reddy *et al.*, 1990, p. 910).

[4]   The following chart summarizes the median waiting time for various organ types (UNOS, 1998).

Type of Organ	Median Waiting Time in Days	
	1988	1996
Kidney	400	824
Liver	33	366
Pancreas	189	312
Kidney-Pancreas	383	361
Heart	117	224
Lung	281	566
Heart-Lung	824	887

As obtained from the UNOS 1998 Annual Report, the kidney data represents 1988 and 1994, the last year in which wait time was computed; heart-lung data are for 1988 and 1995, the last year in which wait time was computed; and the kidney-pancreas data are for 1991, the first year wait time was computed, and 1996.

5   For example, in one study of thirty-five patients there were no operative deaths or serious post-operative complications for patients who underwent hepatic resection for tumors (Singer *et al.*, 1989).

6   These prohibitory laws include: *Algeria*: Law No. 85-05, dated 16 Feb 85 'On Health Protection and Promotion,' *Journal official de la République algérienne démocratique et populaire*, 17 Feb 85, No. 8 pp. 122-140.

Argentina: Law No. 21541, dated 21 Mar 77 'On the Removal and Transplantation of Organs and Anatomical Materials,' *Buenos Aires, Ministry of Social Welfare*, 1977, Section ix, paragraph 27 b, c, d.

Australia (Northern Territory): Law No. 121 of 1979 'An Act to Make Provision for an In Relation to the Removal and Use of Human Tissues, for Post-mortem Examinations, for the Definition of Death and for Related Purposes,' *Prohibition of Trading in Tissue*, Part V, paragraph 24(1)-24(5).

Australia (New South Wales): Law No. 164, dated 31 Dec 83 'An Act Relating to the Donation of Tissues by Living Persons, the Removal of Tissue from Deceased Persons, the Conduct of Post-mortem Examinations of Deceased Persons, and Certain Other Matters,' Chapter VI, 'Prohibition of Trading in Tissue,' section 32.

Australia (Queensland): 'The Transplantation and Anatomy Act of 1979-1984,' Part VII, 'Prohibition of Trading in Tissue,' sections 40-44.

Australia (South Australia): 'The Transplantation and Anatomy Act of 1983,' Part VII, 'Prohibition of Trading in Tissue,' section 35(1)-35(6).

Australia (Tasmania): 'The Human Tissue Act of 1985,' Part IV, section 27(1)-27(5).

Australia (Western Australia): 'The Human Tissue and Transplant Act of 1982,' Part V, sections 29-30.

Austria: Federal Law of 1 June 1982 (Serial No. 273) 'Amending the Hospitals Law,' *Bundesgesetzblatt für die Republik Österreich*, 18 June 82, No. 113, pp. 1161-1162. Chapter F: 'Removal of Organs or Parts of Organs from the Bodies of Deceased Persons for Transplantation Purposes,' section 62a(4).

Belgium: Law of 13 June 1986 'On the Removal and Transplantation of Organs,' *Moniteur belge*,14 Feb 87, No. 32, pp. 2129-2132, chapter 1, section 4(1)-4(2).

Brazil: Constitution of the Federative Republic of Brazil, dated 5 Oct 88, *Diário Oficial*, 5 Oct 88, section 199, item 4.

Canada: 'The Uniform Human Tissue Donation Act of 1990,' section 15(1)-15(3).

Canada (New Brunswick): 'An Act (Chapter 44) to Amend the Human Tissue Act,' date of assent: 18 June 86 *Acts of New Brunswick*, 1986, section 8(3).

Canada (Ontario): 'The Human Tissue Gift Act,' 1982, Part III, section 10.

Columbia: Decree No. 1172 of 6 June 89 'For the Partial Implementation of Title IX of Law No. 9 of 1979, as Regards the Procurement, Preservation, Storage, Transplantation, Destination and Final Utilization of Anatomical Organs or Parts, and the Procedures for their Transplantation in Human Beings, as well as of Law No. 73 of 1988,' *Legislation Económica*,Bogotá, 30 June 89, Volume 74, No. 881, pp. 760-776, chapter II, section 17.

Costa Rica: Law No. 5560, dated 20 Aug 74 'On Human Transplants,' *La Gaceta - Diário Oficial*, 3 Sept 74, No. 165, p. 4362, section 14.

Cyprus: Section 4 of Law No. 97 of 1987 'On the Removal and Transplantation of Biological Materials of Human Origin,' *Episêmi Ephimerida tês Kypriakês Dêmokratias*, 22 May 87, No. 2230, supplement No. 1, pp. 921-926, section 4.

Denmark: Law No. 402, dated 13 June 76 'On the Examination of Cadavers, Autopsies, and Transplantation, etc.,' *Lovtidende for Kongeriget Danmark*, 13 June 90, No. 63, Part A, pp. 1331-1334, chapter 6, section 20, subsection 3.

Dominican Republic: Law No. 60-88, dated 30 Aug 88 'On the Donation of Corneas,' *Gaceta Oficial*, 31 Aug 88, No. 9742, pp. 5-8, sections 11-12.

Ecuador: Law No. 64, dated 26 May 87 'Reforming the Health Code,' *Registro Oficial*, 15 June 87, No. 707, section 1: 'Declaration of Death and the Transplantation of Parts, Tissues, and Organs of the Human Body.'

Finland: Law No. 355, dated 26 April 85 'On the Removal of Human Organs and Tissues for Medical Purposes,' *Finlands Forfattningssamling*, 6 May 85, Nos. 355-358, pp. 741-743, section 11.

France: Law No. 76-1181, dated 22 Dec 76 'On the Removal of Organs,' *Journal Officiel de la Republique française, Editiondes Lois et Decrets*, 23 Dec 76, No. 299, p. 7365, section 39.

Greece: Law No. 1383, dated 2 Aug 83 'On the Removal and Transplantation of Human Tissues and Organs,' *Ephêmeris tês Kybernêseôs tês Hellênikês Dêmokratias*, 5 Aug 83, No. 106, part 1, pp. 1917-1920, chapter 1, section 2.

Hong Kong: 'Human Organ Transplant Bill of 1992,' dated 27 March 92.

Hungary: Ordinance No. 18, dated 4 Nov 72 'Of the Minister of Health for the Implementation of the Provisions of Law No. II of 1972 on Health Relating to the Removal and Transplantation of Organs and Tissues,' *Magyar Közlöny*, 4 Nov 72, No. 87, pp. 862-866, part 2, section 1.

Iraq: Decree No. 698, dated 27 Aug 86 'Of the Revolutionary Command Council Promulgating Law No. 85 of 1986 on the Transplantation of Human Organs,' *Alwawai Aliragiya*, 15 Sept 86, No. 3115, p. 559, section 3.

Italy: Law No. 64, dated 2 Dec 75 'Regulating the Removal of Parts of Cadavers for Purposes of Therapeutic Transplantation and Prescribing Rules Governing the Removal of Pituitary Glands from Cadavers with a View to Producing Extracts for Therapeutic Purposes,' *Gazetta Officiale della Repubblica Italiana*, 19 Dec 75, No. 334, Part I, pp. 8869-8871, sections 19-20.

Kuwait: Decree-Law No. 55, dated 20 Dec 87 'On Organ Transplantation,' *Al-Kuwait Al-Yom*, 27 Dec 87, No. 1751, pp. 3-4, section 7.

Lebanon: Decree No. 109, dated 16 Sept 83 'On the Removal of Human Tissues and Organs for Therapeutic and Scientific Purposes,' Al-Jaridah Al-Rasmiyah, 10 Nov 83, No. 45, pp. 645-645, section 1.

Malawi: The Anatomy Act of 1990, dated 18 May 90 'An Act (No. 14 of 1990) to Make Provision for the Donation, Examination and Use of Bodies, or Parts of Bodies, of the Deceased Persons for Education, Scientific, Research, Therapeutic or Diagnostic Purposes: to Re-enact the Law Dealing with Human Tissue and to Provide for Matters Incidental Thereto or Connected Therewith,' *The Malaŵi Gazette Supplement*, 4 June 90, No. 4C, part IV, section 16.

Mexico: Federal Regulations of 16 Aug 76 'On the Use of Human Organs, Tissues and Cadavers,' *Salud Pública de México*, 1977, No. 19, pp. 59-68, sections 10 and 24.

Panama: Law No. 10, dated 11 July 83 'Regulating the Transplantation of Organs and Anatomical Parts, and Laying Down Other Provisions,' *Gaceta Oficial*, 15 July 83, No. 19855, pp. 1-8, part 4.

Romania: Law of July 1978 'On Safeguarding of the Health of the Population, sections 129-137.

Russian Federation: Law of 22 December 92 'Of the Russian Federation on the Transplantation of Human Organs and/or Tissues,' RF 93.12.

Republic of Sigapore: Law No. 15, dated 10 June 87 'Human Organ Transplant Act of 1987,' *Republic of Singapore Government Gazette, Acts Supplement*, 10 July 87, No. 16, pp. 411-421, part IV, sections 14-15.

Spain: Law No. 30, dated 27 Oct 79 'On the Removal and Transplantation of Organs,' *Boletín Oficial de Estado, Gaceta de Madrid*, 13 March 80, No. 63, Serial No. 5627, pp. 5705-5707, chapter 1, section 5.

Sri Lanka: The Transplantation of Human Tissues Act, No. 48 of 1987, dated 11 Dec 87, section 17(1)-17(3).

Switzerland (Ticino): Law of 18 April 89 'On Health Promotion and Coordination in the Health Sector (The Health Law),' section 15.

Syrian Arab Republic: Law No. 31, dated 23 Aug 72 'On the Removal and Transplantation of Organs from the Human Body,' *Recueil des Lois et de la Législation Financière de la République Arabe Syrienne*, Sept 1972, pp. 2-4, part 2, section 6.

Turkey: Law No. 2238, dated 29 May 79 'On the Removal, Storage, Transfer, and Grafting of Organs and Tissues,' *T.C. Resmî Gazete*, 3 June 79, No. 16655, pp. 1-4, chapter 1, sections 3-4.

United Kingdom: The Human Organ Transplants Act of 1989, dated 27 July 89, Part I, section 1a-1d.

USA: Public Law 98-507, The National Organ Transplant Act.

Zimbabwe: Anatomical Donations and Post-mortem Examinations Act No. 34 of 1976, section 17.

Council of Europe: Resolution (78) 29 'On Harmonisation of Legislations of Member States to Removal, Grafting and Transplantation of Human Substances,' 11 May 78, Article 9.

[7] For an historical perspective on the ownership of human bodies see Gracia (1998) and Szawarski (1998). Others such as Fagot (1998) as well as Welie and ten Have (1998) offer legal accounts of body ownership from the perspectives of France and the Netherlands respectively.

[8] See also Schotsmans (1998), Jensen (1998) and Illhardt (1998). In contrast, Wildes (1998) argues that this assumption is misguided and unjustified.

[9] The American Medical Association admits the moral possibility of purchasing organs from cadaveric sources. They require that adequate safeguards be in place to ensure that the health of donors and recipients is not jeopardized and the quality of the organs procured is not degraded. For example, "By entering into a future contract, an adult would agree while still competent to donate his or her organs after death. In return, the donor's family or estate would receive some financial remuneration after the organs have been retrieved and judged medically suitable for transplantation" (1994/95, pp. 26-27).

[10] See also Caplan, 1987; Cooper, 1987; Evans and Yagi 1987; and Callender, 1987; Brecher 1991; 1994.

[11] For a thought-full analysis of the importance of religion to health care policy and bioethics see McKenny, 1997.

[12] The Universal Declaration of Human Rights adopted by the General Assembly of the United Nations in Paris on December 10, 1948 combined forbearance rights (e.g., "Everyone has the right to life, liberty and security of person" (article 3)) and procedural rights (e.g., "No one shall be subjected to arbitrary arrest, detention, or exile" (article 9)) with claim rights (e.g., "Everyone ... has the right to social security and is entitled to realization ... of the economic, social and cultural rights indispensable for his dignity and the free development of his personality" (article 22)). The only clear statement of human dignity appears in the "Preamble": "Whereas recognition of the inherent dignity and of the equal and inalienable rights of all members of the human family is the foundation of freedom, justice and peace in the world..." Therefore, it is unclear which particular articles of this Declaration the WHO believes are contravened by selling organs.

[13] See, for example, Dekkers and ten Have 1998; Blasszauer, 1998.

[14] For religious accounts of organ donation see May, 1985; McCormick, 1978; Rabinovitch, 1979; and Rosner 1979; 1980.

[15] See also Jackson, 1987; Vlastos, 1962; Wartofsky, 1981; Schneider and Flaherty, 1985; Davis, 1987; Mayers, 1987; Healy, 1987.

[16] For a detailed discussion of justice related issues and organ distribution policies see Kamm, 1993, part III; see also, Daniels, 1988; Menzel, 1982.

[17] For an analysis of the difficulties of redistributive justice addressing historical wrongs see Sher, 1979; 1981. For historical analysis of related issues see McCullough, 1981 and Outka, 1974.

[18] The President's Commission reminded policy makers that a reasoned judgment must be concerned "not only about impact of the condition on the welfare and opportunity of the individual but also about the efficacy and the costs of the care itself in relation to other conditions and the efficacy and cost of the care that is available for them ... and the cost of each proposed option in terms of foregone opportunities to apply the same resources to social goals other than that of ensuring access [to health care]" (1983, pp. 36-37). The effects of particular policy decisions can both directly and indirectly impact health care outcomes.

[19] Veatch discusses a similar possibility with regard to directed donation (1998, p. 461): "Occasionally, a sophisticated utilitarian may oppose directed donation claiming that the overall transplant enterprise could be jeopardized if a dramatic directed donation case such as that of the Ku Klux Klan member or donation limited to a gay recipient turned the public against the organ transplant system."

BIBLIOGRAPHY

Abouna, G.M. et al.: 1990, 'Commercialization in human organs: A Middle Eastern perspective,' Transplantation Proceedings 22, 918-921.

Adson, M.A.: 1981, 'Diagnosis and surgical treatment of primary and secondary solid hepatic tumors in the adult,' Surg Clin North Am 61, 181-196.

American Medical Association, Council on Ethical and Judicial Affairs: 1994/95, Code of Medical Ethics, American Medical Association.

Annas, G.: 1983 Report of the Massachusetts Task Force on Organ Transplantation, Boston University School of Public Health, Boston.

Annas, G.: 1987, 'The prostitute, the playboy and the poet: Rationing schemes for organ transplantation,' in D. Cowan *et al.* (eds.), *Human Organ Transplantation: Societal, Medical-Legal, Regulatory, and Reimbursement Issues*, Health Administration Press, Ann Arbor.

Baily, M.A.: 1988, 'Economic issues in organ substitution technology,' in D. Mathieu (ed.), *Organ Substitution Technology: Ethical, Legal and Public Policy Issues*, Westview Press, Boulder, Co.

Basson, M.: 1979, 'Choosing among candidates for a scarce medical resource,' *Journal of Medicine and Philosophy* 4, 313-334.

Beauchamp T. and Childress, J.: 1994, *Principle of Biomedical Ethics*, 4th edition, Oxford University Press, New York.

Blasszauer, B.: 1998, 'Autopsy,' in H. ten Have and J. Welie (eds.), *Ownership of the Human Body*, Kluwer Academic Publishers, Dordrecht.

Blumstein, J.F. and Sloan, F.A. (eds.): 1989, *Organ Transplantation Policy: Issues and Prospects*, Duke University Press, Durham.

Blumstein, J.F.: 1990, 'Government's role in organ transplantation policy,' in J. Blumstein and F. Sloan (eds.), *Organ Transplantation Policy Issues and Prospects*, Duke University Press, Durham.

Blumstein, J.F.: 1992, 'The case of commerce in organ transplantation,' *Transplantation Proceedings* 24, 2190-2197.

Brecher, B.: 1990, 'The kidney trade; Or, the customer is always wrong,' *Journal of Medical Ethics* 16, 123.

Brecher, B.: 1991, 'Buying human kidneys: Autonomy, commodity and power,' *Journal of Medical Ethics* 19, 99.

Brecher, B.: 1994, 'Organs for transplant: Donation or Payment?' in R. Gillon (ed.), *Principles of Health Care Ethics*, John Wiley and Sons Ltd., New York.

Callender, C.: 1987, 'Legal and ethical issues surrounding transplantation: The transplant team perspective,' in D. Cowan *et al.* (eds.), *Human Organ Transplantation: Societal, Medical-Legal, Regulatory, and Reimbursement Issues*, Health Administration Press, Ann Arbor.

Cantarovich, F.: 1990, 'Values sacrificed and values gained by the commerce of organs: The Argentine experience,' *Transplantation Proceedings* 22, 925-927.

Caplan, A.: 1983, 'Organ transplants: The cost of success,' *Hastings Center Report* 13, 23-32.

Caplan, A.: 1984, 'Ethical and policy issues in the procurement of cadaver organs for transplantation,' *New England Journal of Medicine* 311, 981-984.

Caplan, A.: 1987, 'Obtaining and allocating organs for transplantation,' in D. Cowan *et al.* (eds.), *Human Organ Transplantation: Societal, Medical-Legal, Regulatory, and Reimbursement Issues*, Health Administration Press, Ann Arbor.

Caplan, A.: 1992, *If I Were a Rich Man, Could I Buy a Pancreas?*, Indiana University Press, Bloomington.

Childress, J.: 1970, 'Who shall live when all cannot live?' *Soundings* 53, 339-355.

Childress, J.: 1987, 'Some moral connections between organ procurement and organ distribution,' *Journal of Contemporary Health Law and Policy* 3, 85-110

Childress, J.: 1986, 'The implication of major Western religious traditions for policies regarding human biological materials,' Office of Technology Assessment.

Childress, J.: 1989, 'Ethical criteria for procuring and distributing organs for transplantation,' in J. Blumstein and F. Sloan (eds.), *Organ Transplantation Policy: Issues and Prospects*, Duke University Press Durham.

Cooper, T.: 1987, 'Survey of development, current status, and future prospects for organ transplantation,' in D. Cowan *et al.* (eds.), *Human Organ Transplantation: Societal, Medical-Legal, Regulatory, and Reimbursement Issues*, Health Administration Press, Ann Arbor.

Daar, A.S. *et al.*: 1990, 'Ethics and commerce in live donor renal transplantation: Classification of the issues,' *Transplantation Proceedings* 22, 922-924.

Daniels, N.: 1985, *Just Health Care*, Cambridge University Press, New York.

Daniels, N.: 1988, 'Justice and the dissemination of "big-ticket" technologies,' in D. Mathieu (ed.), *Organ Substitution Technology: Ethical, Legal, and Public Policy Issues*, Westview Press, Boulder.

Davis, C.: 1987, 'Paying for organ transplants under Medicare,' in D. Cowan *et al.* (eds.), *Human Organ Transplantation: Societal, Medical-Legal, Regulatory, and Reimbursement Issues*, Health Administration Press, Ann Arbor.

Dekkers, W. and ten Have, H.: 1998, 'Biomedical research with human body "parts"', in H. ten Have and J. Welie (eds.), *Ownership of the Human Body*, Kluwer Academic Publishers, Dordrecht.

Department of Health and Human Services: 1986, *Report of the Task Force on Organ Transplantation: Issues and Recommendations*, Department of Health and Human Services, Washington, D.C.

Dickens, B.M.: 1990, 'Human rights and commerce in health care,' *Transplantation Proceedings* 22, 904-905.

Dossetor J.B. *et al.*: 1990, 'Discussion,' *Transplantation Proceedings* 22, 933-938.

Dossetor J.B. and Stiller, C.R.: 1990, 'Ethics, justice, and commerce in transplantation,' *Transplantation Proceedings* 22, 892-895.

Edelstein, L.: 1967, 'The genuine works of Hippocrates,' *Ancient Medicine*, The John Hopkins Press, Baltimore.

Engelhardt, H.T., Jr.: 1984, 'Allocating scarce medical resources and the availability of organ transplantation: Some moral presuppositions,' *New England Journal of Medicine* 311, 66-71.

Engelhardt, H.T., Jr.: 1987, 'Shattuck Lecture: Allocating scarce medical resources and the availability of organ transplantation,'in D. Cowan *et al.* (eds.), *Human Organ Transplantation: Societal, Medical-Legal, Regulatory, and Reimbursement Issues*, Health Administration Press, Ann Arbor.

Evans, R.: 1987a, 'Health care technology and the inevitability of resource allocation and rationing decisions, part I,' in D. Cowan *et al.* (eds.), *Human Organ Transplantation: Societal, Medical-Legal, Regulatory, and Reimbursement Issues*, Health Administration Press, Ann Arbor.

Evans, R.: 1987b, 'Health care technology and the inevitability of resource allocation and rationing decisions, part II,' in D. Cowan *et al.* (eds.), *Human Organ Transplantation: Societal, Medical-Legal, Regulatory, and Reimbursement Issues*, Health Administration Press, Ann Arbor.

Evans, R. and Yagi, J.: 1987, 'Social and medical considerations affecting selection of transplant recipients: The case of heart transplants,' in D. Cowan *et al.* (eds.), *Human Organ Transplantation: Societal, Medical-Legal, Regulatory, and Reimbursement Issues*, Health Administration Press, Ann Arbor.

Fagot-Largeault, A.: 1998, 'Ownership of the human body: Judicial and legislative responses in France,' in H. ten Have and J. Welie (eds.), *Ownership of the Human Body*, Kluwer Academic Publishers, Dordrecht.

Fox R. and Swazey, J.: 1992, *Spare Parts: Organ Replacement in American Society*, Oxford University Press, New York.

Frier, D. and Mavrodes, G.: 1980, 'The morality of selling human organs,' in M. Basson (ed.), *Ethics, Humanism, and Medicine*, Alan R. Liss, New York.

Gillon, R.: 1990, 'Transplantation: A framework for analysis of the ethical issues,' *Transplantation Proceedings* 22, 902-903.

Gorovitz, S.: 1984, Against selling body parts,' *Report From the Center for Philosophy and Public Policy* 4, 9-12.

Gorovitz, S.: 1987, 'Ethical implications of reimbursement policies,' in D. Cowan *et al.* (eds.), *Human Organ Transplantation: Societal, Medical-Legal, Regulatory, and Reimbursement Issues*, Health Administration Press, Ann Arbor.

Gracia, D.: 1998, 'Ownership of the human body: Some historical remarks,' in H. ten Have and Welie, J. (eds.), *Ownership of the Human Body*, Kluwer Academic Publishers, Dordrecht.

Grubb, A.: 1995, 'The Nuffield Council Report on human tissue,' *Medical Law Review* 3, 235-236.

Hardy, K.J., Fletcher, D.R., and Jones, R.M.: 1998, 'One hundred liver resections including comparison to non-resected liver-mobilized patients, *Aust N Z J Surg* 68, 716-721.

Ignatieff, M.: 1984, *The Needs of Strangers*, Chatto and Windus, London.

Illhardt, F.: 1998, 'Ownership of the human body: Deontological approaches,' in H. ten Have and J. Welie (eds.), *Ownership of the Human Body*, Kluwer Academic Publishers, Dordrecht.

Iwatsuki, S. *et al.*: 1983, 'Experience with 150 liver resections,' *American Surgery* 197, 247-253.

Jackson, D.L.: 1987, 'The role of state health departments in assuring equitable selection and regulation of resources,' in D. Cowan *et al.* (eds.), *Human Organ Transplantation: Societal, Medical-Legal, Regulatory, and Reimbursement Issues*, Health Administration Press, Ann Arbor.

Jensen, U.: 1998, 'Property, rights, and the body: The Danish context — A democratic ethics or recourse to abstract right?' in H. ten Have and J. Welie (eds.), *Ownership of the Human Body*, Kluwer Academic Publishers, Dordrecht.

Johny, K.V. *et al.*: 1990, 'Values gained and lost in live unrelated renal transplantation,' *Transplantation Proceedings* 22, 915-917.

Jonsen, A.: 1977, 'Do no harm: Axiom of medical ethics,' in S. Spicker and H.T. Engelhardt (eds.), *Philosophical Medical Ethics: Its Nature and Significance*, Kluwer Academic Publishers, Dordrecht.

Jonsen, A.: 1987, 'Organ transplants and the principle of fairness,' in D. Cowan *et al.* (eds.), *Human Organ Transplantation: Societal, Medical-Legal, Regulatory, and Reimbursement Issues*, Health Administration Press, Ann Arbor.

Jonsen, A.: 1997, 'Ethical issues in organ transplantation,' in R. Veatch (ed.), *Medical Ethics*, 2nd edition, Jones and Barlett Publishers, Sudbury, MA.

Kamm, F.: 1993, *Morality, Mortality*, volume 1, Oxford University Press, New York.

Kennedy, I.: 1979, 'The donation and transplantation of kidneys: Should the law be changed?' *Journal of Medical Ethics* 5, 13-21.

Keyserlingk, E.: 1990, 'Human dignity and donor altruism -- Are they compatible with efficiency in cadaveric human organ procurement?' *Transplantation Proceedings* 22, 1005-1006.

Massachusetts Task Force on Organ Transplantation: 1984, Report of the Task Force; reprinted in D. Cowan *et al.* (eds.), *Human Organ Transplantation: Societal, Medical-Legal, Regulatory, and Reimbursement Issues*, Health Administration Press, Ann Arbor.

May, W.F.: 1985, 'Religious justifications for donating body parts,' *Hastings Center Report* 15, 38-42.

Mayers, B.: 1987, 'Blue Cross and Blue Shield coverage for major organ transplants,' in D. Cowan *et al.* (eds.), *Human Organ Transplantation: Societal, Medical-Legal, Regulatory, and Reimbursement Issues*, Health Administration Press, Ann Arbor.

McCormick, R.: 1978, 'Organ transplantation: Ethical principles,' in W.T. Reich (ed.), *Encyclopedia of Bioethics*, Macmillan, New York.

McCullough, L.: 1981, 'Justice and health care: Historical perspectives and precedents,' in E. Shelp (ed.), *Justice and Health Care*, Reidel Publishers, Boston, pp. 51-71.

McDermott, W.V. and Ottinger, L.W.: 1966, 'Elective hepatic resection,' *American Journal of Surgery* 112, 376-381.

McKenny, G.: 1997, *To Relieve the Human Condition*, SUNY University Press, Albany.

Meyer, M.J.: 1995, 'Dignity, death, and modern virtue,' *American Philosophical Quarterly* 32.

Monaco, A.P.: 1989, 'A transplant surgeon's views on social factors in organ transplantation,' *Transplantation Proceedings* 21, 3403-3406.

Monaco, A.P.: 1990, 'Transplantation: The state of the art,' *Transplantation Proceedings* 22, 896-901.

Nagao, T. *et al.*: 1985, 'One hundred hepatic resections: Indications and operative results,' *American Surgery* 202, 42-49.

Nuffield Council on Bioethics: 1995, *Human Tissue: Ethical and Legal Issues*, London.

Office of Technology Assessment: 1987, *New Developments in Biotechnology: Ownership of Human Tissues and Cells*, U.S. Government Printing Office, Washington, D.C.

Outka, G.: 1974, 'Social justice and equal access to health care,' *Journal of Religious Ethics* 2, 11-32.

Pauly, M.: 1987, 'Equity and costs,' in D. Cowan *et al.* (eds.), *Human Organ Transplantation: Societal, Medical-Legal, Regulatory, and Reimbursement Issues*, Health Administration Press, Ann Arbor.

Pellegrino, E.D.: 1991, 'Families' self interest and the cadaver's organs: What price consent?' *Journal of the American Medical Association* 265, 1305-1306.

Peters, T.: 1991, 'Life or death: The issue of payment in cadaveric organ donation,' *Journal of the American Medical Association* 265, 1302-1305.

President's Commission for Study of Ethical Problems in Medicine and Biomedical and Behavioral Research: 1983, *Securing Access to Health Care*, U.S. Government Printing Office, Washington, D.C.

Rabinovitch, N.L.: 1979, 'What is the Halakah for organ transplants?' in F. Rosner and J.D. Bleich (eds.), *Jewish Bioethics*, Sanhedrin Press, New York.

Radcliffe-Richards J. *et al.*: 1998, 'The case for allowing kidney sales,' *The Lancet* 351, 1950-1951.

Ramsey, P.: 1970, *The Patient as Person*, Yale University Press, New Haven.

Reddy, K.C. *et al.*: 1990, 'Unconventional renal transplantation in India,' *Transplantation Proceedings* 22, pp. 910-911.

Roles, L. *et al.*: 1990, 'Effect of a presumed consent law on organ retrieval in Belgium,' *Transplantation Proceedings* 22, 2078-2079.

Rosner, F. and Tendler, M.: 1980, *Practical Medical Halacha*, 2nd edition, Feldheim Publishers, New York.

Sadler, A. M. and Sadler, B.: 1984, 'Organ donation: Is volunteerism still valid?' *Hastings Center Report* 14, 6-9.

Schneider, A. and Flaherty, M.P.: 1985, *The Challenge of a Miracle: Selling the Gift*, Pittsburgh Press, Pittsburgh.

Schotsmans, P.: 1998, 'Ownership of the body: A personalist perspective,' in H. ten Have and J. Welie (eds.), *Ownership of the Human Body*, Kluwer Academic Publishers, Dordrecht.

Scorsone, S.: 1990, 'Christianity and the significance of the human body,' *Transplantation Proceedings* 22, 943-944.

Scott, R.: 1981, *The Body as Property*, Viking Press, New York.

Scott, R.: 1990, 'The human body: Belonging and control,' *Transplantation Proceedings* 22, 1002-1004.

Screiner, G.E.: 1968, 'Problems of ethics in relation to haemodialysis and transplantation,' in G. Wolstenhome and M. O'Conner (eds.), *Law and Ethics of Transplantation*, Churchill, London.

Sells, R.A.: 1979, 'Let's not opt out: Kidney donation and transplantation,' *Journal of Medical Ethics* 5, 165-169.

Sells, R.A.: 1990, 'Organ commerce: Ethics and expediency,' *Transplantation Proceedings* 22, 931-932.

Sells, R.A.: 1992, 'The case against buying organs and a future's market in transplantation,' *Transplantation Proceedings* 24, 2198-2022.

Sells, R.A.: 1994, 'Transplants,' in *Principle of Health Care Ethics*, R. Gillon (ed.), pp. 1003-1025.

Sher, G.: 1979, 'Reverse discrimination, the future and the past,' *Ethics* 90, 81-87.

Sher, G.: 1981, 'Ancient wrongs and modern rights,' *Philosophy and Public Affairs* 10, 3-17.

Siegler, M.: 1992, 'Liver transplantation using living donors,' *Transplantation Proceedings* 24, 2223.

Simmons, R.: 1981, 'Psychological reactions to giving a kidney,' in N. Levy (ed.), *Psychonephrology*, Plenum Publishing Co. New York.

Singer, P. *et al.*: 1989, 'Ethics of live transplantation with liver donors,' *New England Journal of Medicine* 321, 620-622.

Singer, P. *et al.*: 1990, 'The ethical assessment of innovative therapies: Liver transplantation using living donors,' *Theoretical Medicine* 11, 87.

Skelly, L.: 1987, 'Practical issues in obtaining organs for transplantation,' in D. Cowan *et al.* (eds.), *Human Organ Transplantation: Societal, Medical-Legal, Regulatory, and Reimbursement Issues*, Health Administration Press, Ann Arbor.

Starzl, T.E. *et al.*: 1980, 'Right trisegmentectomy for hepatic neoplasms,' *Surgery for Gynecology and Obstetrics* 150, 208-214.

Szawarski, Z.: 1998, 'The stick, the eye, and ownership of the body,' in H. ten Have and J. Welie (eds.), *Ownership of the Human Body*, Kluwer Academic Publishers, Dordrecht.

Ten Have, H. and Welie, J.: 1998, 'Medicine, ownership, and the human body,' in H. ten Have and J. Welie (eds.), *Ownership of the Human Body*, Kluwer Academic Publishers, Dordrecht.

Thiagarajan, C.M. *et al.*: 1990, 'The practice of unconventional renal transplantation (UCRT) at a single centre in India,' *Transplantation Proceedings* 22, 912-914.

Townsend, P. and Davidson, N. (eds.): 1982, *Inequalities in Health: The Black Report*, Penguin, Harmondsworth.

Transplantation Society: 1970, 'Statement of the committee on morals and ethics of the Transplantation Society,' *Annals of Internal Medicine* 75, 631-633; reprinted in World Health Organization: 1994, *Legislative Responses to Organ Transplantation*, Martinus Nijhoff, Dordrecht.

Transplantation Society Council: 1985, ' Commercialization in transplantation: The problem and some guidelines for practice,' *Lancet* 2, 715-716; reprinted in World Health Organization: 1994, *Legislative Responses to Organ Transplantation*, Martinus Nijhoff, Dordrecht.

Veatch, R.: 1998, 'Egalitarian and maximin theories of justice: Directed donation of organs for transplant, *The Journal of Medicine and Philosophy* 23, 456-476.

Vlastos, G.: 1962, 'Justice and equality,' in R. Brandt (ed.), Prentice Hall, Englewood Cliffs.

UNESCO: 1989, *Human Rights Aspects of Traffic in Body Parts and Human Fetuses for Research and/or Therapeutic Purposes*, UNESCO.

United Network for Organ Sharing: 1998, *Annual Report*, UNOS.

United States Task Force on Organ Transplantation: 1986, *Organ Transplantation: Issues and Recommendations*, U.S. Department of Health and Human Services, U. S. Government Printing Office, Washington, D.C.

Walsh, P.: 1995, 'Principles and pragmatism,' *Medical Law Review* 3, 237-250.

Wartofsky, M.W.: 1981, 'On doing it for money,' in T. Mappes and J. Zembaty (eds.), *Biomedical Ethics*, McGraw Hill, New York.

Welie, J. and ten Have, H.: 1998, 'Ownership of the human body: the Dutch context,' in H. ten Have and J. Welie (eds.), *Ownership of the Human Body*, Kluwer Academic Publishers, Dordrecht.

Wildes, K.: 1998, 'Libertarianism and ownership of the human body,' in H. ten Have and J. Welie (eds.), *Ownership of the Human Body*, Kluwer Academic Publishers, Dordrecht.

World Health Organisation: 1991, 'Human organ transplantation. A report on the developments under the auspices of the WHO,' reprinted in *International Digest of Health Legislation* 42, 389-396; reprinted in World Health Organization: 1994, *Legislative Responses to Organ Transplantation*, Martinus Nijhoff, Dordrecht.

World Health Organization: 1994, *Legislative Responses to Organ Transplantation*, Martinus Nijhoff, Dordrecht.

Youngner, S.: 1990, 'Organ retrieval: Can we ignore the dark side,' *Transplantation Proceedings* 22, 1014-1015.

SECTION ONE

BODIES AND PERSONS: ONTOLOGICAL QUESTIONS

R.J. HANKINSON

BODY AND SOUL IN GREEK PHILOSOPHY

I. PLATO AND THE BODY

At the beginning of Plato's *Phaedo*, Socrates seeks to reconcile his cheerful acceptance of his own imminent death with his endorsement of a general proscription on suicide. Even if it would be better for Evenus to follow him as quickly as possible out of mortal life (61b), he should not kill himself since suicide cannot be justified (61c). His interlocutor Cebes reasonably inquires: "how can you say, Socrates, that it is not right for the philosopher to do violence to himself, yet he ought willingly to follow the dying man" (61d)? Socrates seeks in what follows to resolve the apparent paradox: we may readily allow that in certain circumstances it would be better to die than to go on living (62a), and hence acquiesce willingly in our own deaths, but we have no right to take steps actively to end our own physical lives, since they are divine gifts, and we are the gods' property, to be disposed of by them and them alone (62b-c). This justification of the prohibition on suicide (or at least on what one might label, by analogy with the euthanasia debate, 'active' suicide) then leads naturally into the central theme of the dialogue, the immortality of the soul.

Socrates' defense of the traditional ban on talking one's own life has, as he himself acknowledges, roots in earlier Greek philosophy, in particular that of the Pythagoreans (61d-e). But it is important to note at the outset that Socrates' reasons for rejecting suicide have nothing to do with any concept of an obligation to the body as such. It is not that our bodies, as individuals, have moral claims of any kind over us – it is simply that ultimately they belong to someone else. Views of that sort, that we are simply the custodians rather than the owners of what we ordinarily think of as our bodies, crop up from time to time in the Christian tradition.[1] But they are not universal in the classical world, and in any case derive not from any special acknowledgment of the rights of the body as a body, but rather from a sense of alienation from it.[2]

M.J. Cherry (ed.), Persons and their Bodies: Rights, Responsibilities, Relationships, 35–56.
© 1999 *Kluwer Academic Publishers. Printed in Great Britain.*

Not everybody in the ancient world, however, endorsed the Socratic view, and some, apparently, misunderstood it: one Cleombrotus is said, in a celebrated epigram of Callimachus (n. 23) much quoted and discussed in later antiquity, to have thrown himself off a high wall as a result of reading the *Phaedo*.[3] Suicide was widespread, and for the most part, particularly in the Roman world, considered acceptable and sometimes even noble. The Stoic philosophers, although differing among themselves on the issue, tended to consider physical death a matter of indifference (rather than something actively to be welcomed, as the Socratic-Platonic tradition had it), and suicide, at least under certain circumstances, not only to be permissible, but actually the right and proper thing to do. Seneca, the Roman Stoic whose writings contain the most extensive Stoic musings on suicide, was eventually obliged by his erstwhile patron, the emperor Nero, to put his precepts into practice, which, according to Tacitus (*Annals* 15 60-7), he duly did, with great courage and dignity.

Moreover, the Stoics, with their Platonically-influenced view that the whole world was a single, integrated organism of which we as individuals were mere parts, allied to their doctrine that true human freedom was to be found in the freedom of the mind, tended to suppose that our bodies were really quite distinct from ourselves, indeed not even really ours at all (although not specifically for Socratic reasons):

> What does Zeus say? 'Epictetus: if I could have done I would have made this little body and property of yours free and not prone to impediment; but make no mistake: it is not yours, but only a subtle mixture of clay' (Epictetus, *Discourses* 1 1).

This is constant Stoic refrain: what we ordinarily take to be our own property, including our own bodies, is not really ours at all, and hence its loss is no real loss to us (*ibid.* 1 19, 24, 25); and such an understanding is the key to freedom from fear (*ibid.* 4 7).[4]

But Socrates' claim that we are not justified in simply disposing of our bodies as we wish does not derive from any inclination on his part to attach any genuine value to the body as such. His famous contention that philosophy is a preparation for (bodily) death (*Phaedo* 64a, 67e) is predicated on the belief that philosophy is the cultivation of the soul, and that such cultivation consists in divorcing, so far as is possible, the soul from the body and its concerns. Death is the separation of soul from body (*ibid.* 64c); and the philosopher, *qua* philosopher at least, should be indifferent to the particular pleasures associated with the body, those of food, drink, sex, and so on (*ibid.* 64c-e): "so is it not clear first of all from these considerations that the philosopher emancipates his soul as far as possible from its association with the body to a far higher degree than other men" (*ibid.* 65a)?

Moreover, the philosopher's great, quasi-erotic, desire is for knowledge (*ibid.* 82b-83a; *Republic* 5, 474b-75e; *Symposium* 204d-12b); and to that end association with the body is an obstacle, since the senses are delusive and reveal at best a partial and shadowy reality; real knowledge is only to be won by the pure, unfettered exercise of reason (*Phaedo* 65a-66a). These are, of course, Platonic commonplaces, closely linked to his general metaphysics of the transcendent, supra-sensible forms, and his rationalistic doctrine of recollection (e.g., *Republic* 5, 476a-80a, 6, 504a-11e; *Meno* 81a-86c; *Phaedo* 73b-76e).

Socrates sums up:

> so it is necessary on all of these grounds that some such opinion is held by all genuine philosophers, so that they say the following sort of thing to one another: there ought to be some way of escape for us, because while we possess a body and our soul is intertwined with an evil of this sort we shall never sufficiently get hold of what we desire, which we hold to be the truth. For our body distracts us in countless ways because of its need for food. Furthermore, if certain diseases fall upon it, they impede us from the search for reality. It fills us with lusts and desires and fears and every kind of illusion and all sorts of nonsense, so that, as it is said, truly and in reality no thought whatever ever comes to us from it. Rather the body and its desires provide us with nothing other than wars and civil strife and battles; for all wars arise as a result of the acquisition of wealth, and we are compelled to acquire wealth by the body, being held in thrall to its service; and as a result of these distractions we have no time for philosophy. And the absolute limit is that if we ever do find some release from it and turn ourselves towards the investigation of something, it still impinges everywhere upon our investigations, it causes disturbance and confusion and distraction, so that we are unable on its account to see the truth (*ibid* 66b-d).

The only alternative is to withdraw as far as possible from bodily pursuits and wait for the day when physical impediments will no longer obstruct our soul's progress towards the truth (*ibid.* 66d-67b); with this in mind, the philosopher must eschew as far as possible the pursuit of physical pleasures which "rivet the soul to the body" (83d) in order that the soul be as pure and unencumbered as possible (*ibid.* 80c-84b).

These views recur elsewhere in the Platonic corpus. In Book 10 of the *Republic*, in arguing that the soul must be immortal because none of its specific vices (injustice, licentiousness, and so on) can be supposed to destroy it (and if they can't nothing can), Socrates nonetheless describes the soul in its embodied state as "maimed" by its association with the body and likens it to marine statues of the sea-god Glaucus which have, over the years, become encrusted with weed, shells, and other detritus (*Republic* 10, 61 1b-d). And in the *Gorgias* (523c-e, 524d-25a), the body is described as getting in the way of a just assessment of a soul's worth, just as clothes impede a proper appreciation of the body; indeed, both the soul judging and the soul being judged should properly be stripped of any association with the body in order to facilitate such judgment.

Thus, for Plato at least, the soul is only temporarily associated with a body (although it may be reincarnated: *ibid.* 617d-21b; cf *Phaedo* 81d-82d; *Phaedrus* 246a-49d; *Timaeus* 91d-92c; however, compare *Gorgias* 523d-26d); and such an association is entirely to its detriment. It is the body which is responsible for the various appetites and lusts which distract the soul from its true purpose of contemplation and reasoning.[5] Moreover, the body is subject to physical ailments, which, when serious enough, prevent the soul from fulfilling its functions. The body should be cared for, but only in order to avoid such ills, and insofar as, in looking after it, one is guarding someone else's property. Most strikingly of all, in the *Gorgias* (492e-93a), Socrates wonders whether what we think of as (embodied) life might not in fact rather be death: "for I have heard this from one of the wise men that we are now dead, and the body (*sōma*) is a tomb (*sēma*) for us."

These views were probably not popular outside strictly philosophical circles (although such circles were wider in the ancient world than they are today); and in any case, not all philosophical persuasions adhered to them, at least in their Platonic form. Yet the sense of alienation from the body, of the body as something to be borne rather than celebrated, recurs in philosophies as widely divergent as Platonism, Stoicism, and Aristotelianism, and was to become an important strand in Christian asceticism. Even the Cynics, while not committed to detachment from the body as such, sought to minimize its claims upon them: witness the famous story of Diogenes, upbraided for public masturbation, replying that it would be a fine thing if he could satisfy the demands of his stomach simply by rubbing it (Diogenes Laertius 6 69). On the other hand, Diogenes insisted upon the importance of physical training "in accordance with which by continuous training impressions are generated which create freedom for virtuous actions" (*ibid.* 70), whatever precisely that may mean; at all events, he placed the physical goods of health and strength on a par with (indeed as being essential preconditions to) those of the soul, a position not unlike Aristotle's.

Yet the following amusing anecdote told of the female Cynic Hipparchia strongly suggests that, for the Cynics at least, the body was personal property, to be disposed of as its possessor thought fit. She is alleged to have confounded one Theodorus the atheist with the following argument: "something which would not be called wrong when done by Theodorus would not be called wrong when done by Hipparchia; Theodorus does no wrong in striking Theodorus; therefore Hipparchia does no wrong in striking Theodorus" (Diogenes Laertius 6 97). The crucial second premise invokes a general belief that what one does with one's body is one's own business: one cannot treat oneself unjustly (cf. Aristotle's view: §6 below).

Nor were the ancient hedonists, the Cyrenaics and Epicureans, in spite of the *canards* circulated by their detractors, simply naïve sensualists, committed to the unrestricted pursuit of physical pleasures. Although the Epicureans did indeed hold that pleasure was the end of life, they considered that the active pleasures of physical engagement, their so-called "dynamic" pleasures, inevitably brought countervailing pains in their train (Plato, *Protagoras* 351c-e, 353c-54e); the wise individual would rather prefer the quiet "static" pleasures of freedom from pain and trouble (Epicurus, *Letter to Menoeceus* 127-32). These involve the body: but they do not consist of its gross indulgence. Moreover, they insisted that death was of no great import to us and held that in certain circumstances (primarily in order to escape unbearable physical pain) suicide was justified.

II. THE NATURE OF THE SOUL

In order to get a clearer idea of the complexities of the various Greek philosophical views of the body and our relations to it, we need to investigate the Greek concept of the soul; and to do that, we must return to the very beginnings of Greek recorded culture: the Homeric poems. In the *Iliad* and the *Odyssey*, the soul, *psuchê*,[6] is conceptualized somewhat vaguely as an internal life-giving force, perhaps akin to a vapor or air, a *pneuma*. At death, the soul leaves the body, often through the puncture of a wound, and "goes gibbering off into Hades". Thus souls do have some existence after physical death; but it is not, at least as far as the Homeric poems are concerned, a consummation devoutly to be wished.

In the curious 11th book of the *Odyssey*, Odysseus descends into Hades to converse with the dead. In order to do so, he first has to slaughter several sheep and let their blood drip into a trench he has dug: only thus fortified with fresh blood can the dead souls speak. Their disembodied existence is shadowy, and for the most part lacks consciousness; Odysseus's mother Anticleia tells him, after he has fruitlessly tried to embrace her, "you are simply witnessing the law of our mortal nature when we come to die. We no longer have sinews holding bones and flesh together, but once the *psuchê* has departed from our white bones all is consumed by the fierce heat of the blazing fire, and the *psuchê* slips away like a dream and flutters on the air" (*Odyssey* 12, 218-222). Later, when Odysseus calls Achilles fortunate for having led a noble life and having met with a glorious end, and for being a prince in the underworld, Achilles replies that he would rather be a slave in the household of a poor man on earth than the greatest king among the dead (*ibid.*, 488-91).

Embodiment and the physical capacities it confers are thus splendid things; even if there is a life after death, it hardly deserves the name of life. Equally, old age is uniformly to be feared for its physical infirmities and ills. Bodies in this tradition are not only good – they are necessary for any worthwhile life,[7] and the better condition they are in, the better the life. This is a world away from the anti-corporeal asceticism of Plato. What has happened in the meantime? One development that may well be relevant to that story is the rise of Orphism, a shadowy body of mystical doctrine associated with its mythical eponym Orpheus. The spread of Orphism in the Greek world is extremely difficult to document and the subject of much scholarly controversy: but it is probable that by the end of the 6th century BC, "Orphic" doctrines included a belief in reincarnation, possibly deriving ultimately from eastern sources.

More reliable are the ascriptions to Pythagoras and his followers of a belief in reincarnation (and perhaps also to an eternal blessed afterlife of the soul [Herodotus 4 95-6] although this is less certain). Pythagoras is a semi-legendary figure, around whose hallowed name later apocrypha were to accrete; but early sources are unanimous in crediting him with a belief in the transmigration of souls from one body to another, both human and animal (Porphyry, *Life of Pythagoras* 19); he himself, according to legend, could remember many previous existences (Diogenes Laertius 7 4-5). Later philosopher-mystics took the belief even further: Empedocles claimed to have been both male and female, as well as a bird, a fish, and even a bush (Fr. 31 B 1 17).

The Pythagoreans considered the soul to be a harmonious arrangement – to have a soul was to be arranged in a certain sort of way.[9] Plato attacked this belief (without attributing it directly to the Pythagoreans) in the *Phaedo* as inconsistent with the doctrine of metempsychosis: an arrangement cannot survive the destruction of that of which it is an arrangement (*Phaedo* 92a b); moreover, arrangements are emergent properties of the things arranged and hence depend upon them, but do not control them as a soul is supposed to (*ibid* 92e-93a). Plato's criticism (echoed by Aristotle: *On the Soul* 1 4) is not as devastating as it might appear at first sight: my particular arrangement might, after my death, be instantiated in some other creature, although it would be a further question whether or not that creature would thereby be me, rather than a distinct token of the same soul-type.[10]

But at all events, the soul as an arrangement requires some material substrate, the only plausible candidate for which is a physical body. Disembodied existence, at least, seems ruled out. If this is right, then Plato does indeed take a radical new step with his supposition that the soul is both immaterial and substantial: a real thing, but not a body at all.[11] Others who treat the soul as substantial, such as Heraclitus, apparently regard it as something physical (in this case a sort of fire) which ramifies through the body, taking care of it: it is like a spider, rushing to repair that part of its web (the body) which has been damaged (Fr. 22 B 67a).

III. THE CULTURE OF THE BODY

So far we have examined, briefly, the antecedents to the Platonic view of the soul; we now turn to an equally compendious compilation of earlier Greek attitudes to the body. Physical culture was important in classical Greece: their cult of the body beautiful being, along with their rampant litigiousness, the characteristic in which they most closely anticipate contemporary Western attitudes. Much of the Hippocratic corpus, a collection of medical writings dating from the 5th to the 3rd century BC, is concerned with the appropriate regimen to be followed in order to keep healthy; and that regimen standardly involves a proper attention to the appropriate types of food and drink and to physical exercise.

Physical education was considered to be at least as important as intellectual training, if not more so; and the victors in the various contests at the Olympic games were as lionized as their modern descendants, their triumphs celebrated in the great lyric poetry of Pindar. Writing at the beginning of the 5th century BC, the philosopher-poet Xenophanes deplores the vulgarity of a public taste that values physical prowess above wisdom:

> If someone were to earn a victory with swiftness of foot,
> or competing in the pentathlon where Zeus's sanctuary
> lies by the springs of Pisa at Olympia, or in wrestling,
> or in boxing's painful skill,
> or in that terrible contest called the pancration,
> to the citizens he would be more glorious to behold,
> and he would earn a place of importance at the games,
> and he would be provided for from the public purse
> of the city, and a gift would be deposited for him.
> And if he did it with horses, he would enjoy all these things,
> although not as worthy as I: for my wisdom
> is better than the strength of men (Xenophanes, Fr. 21 B 2 DK;
> Athenaeus 10, 413f).

Public opinion is confused on the issue, Xenophanes goes on to complain: boxers and runners make no serious contributions to the health of the state. Whatever the justice of Xenophanes' contentions, clearly such a complaint would only arise in a society in which athletic success was revered (a fragment of Euripides' *Autolycus*, quoted in the same context by Athenaeus, and printed by Diels-Kranz as an "imitation" of Xenophanes, makes a similar point about the inflated reputation of athletes). Sparta was particularly noted for the fetish it made of physical education, exciting Aristotle's scorn:

> Nowadays, of those cities with a particular reputation for taking care of their young people, some of them instill in them an athletic condition and damage both the form and the development of their bodies, while the Spartans, although they do not make this mistake, make them

bestial by excessive work, in the belief that this is particularly conducive to courage. But, as we have often said, we should not concern ourselves with one virtue alone, nor particularly with this one, in education (Aristotle, *Politics* 8 4, 1338b9-16).

Moreover, even from its own limited perspective, Aristotle thinks, a single-minded concern with habituation to physical hardship as a means to promoting courage is self-defeating: Spartan youths are not made more courageous by such a regime – only more savage.

Aristotle is not, however, opposed to gymnastic training as such: on the contrary, he thinks it (along with education in literacy and the arts) to be one of the three central planks to any reasonable educational system (*ibid.* 3, 1337b23-7). But it is not to be pursued as an end in itself, or to the detriment of other branches of learning:

> Those who turn over their children excessively to these things, having left them bereft of education in the essentials, make of them simple mechanics, suited to only one of the functions of the state, and even that worse than the others, as our argument shows (*ibid.* 4, 1339a32-6).

Furthermore, Aristotle notes, those who devote themselves more or less exclusively to training tend to burn themselves out physically – such a regime is clearly not in anybody's long-term interests.

Plato too thought that certain sorts of physical regime were excessive and that, in general, care of the body should be subordinate to care of the soul and thus directed by the latter's needs. Although he prescribes physical education in his ideal states both in the *Republic* (3, 403b ff.) and the *Laws* (7, 788c ff., 794c-96e), it is not designed to produce a magnificently-toned body for its own sake, but rather as a way of avoiding the sort of physical ills to which the pursuit of excessive pleasure and idleness render the body subject (*Republic* 3, 403e-405a). Medicine should not be a panacea for the chronic ills of a dissolute life, but rather a temporary method of restoring a body temporarily reduced to sickness by wounds or endemic disease (*ibid.* 405c 408e). Such dissolute medicine is nurtured by an excessive concern for the body which is the greatest enemy of social order and productivity (*ibid.* 3, 406c, 407b-e). Furthermore, a life lived in thrall to such chronic diseases, like the unexamined life, simply isn't worth living (406e-407a) because it renders one incapable of fulfilling one's proper function (which is what, on Plato's account, gives life its meaning).

IV. OBLIGATIONS TO THE BODY

All of this of course implies that the body requires care and that it is wrong to destroy it through licentiousness and excess. But it is wrong only derivatively, since such destruction threatens the well-being of the soul and, hence, that of the society in which that soul finds itself. Here, then, if anywhere, we might expect to discover ancient anticipations of the sort of concerns which form the subject-matter of this volume: Are people obliged in some way to try and live healthy lives, and if so why? Are people's bodies simply their own property, to be disposed of as they see fit? And do physicians have the right to impose the treatments that they see as indicated upon their patients regardless of the latters' wishes and beliefs? The modalities of the ancient discussions are, however, very different from their modern cousins, for two main reasons.

First of all, the notion of public health care is a distinctively modern one: the idea (however attenuated it may be in the United States) that the overall health of a population is a public good to be safeguarded by public investment and the derivative idea that basic health provisions are a social right are both of recent provenance. Equally modern, and for related reasons, is the notion of the general social costs incurred by having sections of the population less healthy and fit than they might be, as well as the derivative notion that correlative to the right to such benefits comes a general social duty to maintain reasonable standards of health.

Equally importantly, it must be remembered just how rudimentary and generally inefficacious ancient medicine really was. Although the ancients were reasonably proficient at certain sorts of surgery, by and large their therapeutic interventions were of little or no benefit, and in some cases were actually harmful; in general, their physiologies and pathologies were false and fantastical, their pharmacologies and therapeutics misguided and useless (this depressing state of affairs is implicitly acknowledged by the Hippocratics, who see their role simply as aiding nature to effect a cure if it can). Of course, this state of affairs was not confined to the ancient world – it is easy to forget just how modern medicine really is: the modern paradigm of disease as caused by invasive pathogenic micro-organisms is barely a century and a half old, as is the central modern notion of immunization; and almost all the effective drug therapies currently available were discovered in this century. Equally, the last few decades have witnessed astounding advances in surgical knowledge and techniques, most dramatically in the case of transplant surgery, which have made it possible to save (or at least prolong) lives which would previously have been lost; and, equally importantly, all of this comes at a price, sometimes a very high price.

These factors, then, have combined to create a situation in which efficacious medical interventions are now possible in an unprecedented variety of cases – but one also in which such interventions are extremely expensive and may, for other reasons, also require rationing (again, transplant surgery is the key case: there are simply not enough available spare organs to go round). Thus we should not expect to find adumbrated in the ancient world anything closely resembling the contours of the modern debates which these distinctively modern circumstances have engendered. And we do not find them. Nonetheless, we do find antecedents to the notion that people can and indeed should be held to account for failing properly to look after their bodies: and for Plato, at least, this is the case for at least three reasons.

We owe our bodies care either (1) because they are not really ours at all (an implication, although not one which Plato expressly draws out, of the Socratic proscription on suicide in the *Phaedo*); or (2) because failure to do so will result in our being an excessive burden on the rest of society (a position again at least hinted at by Plato in the *Republic* and the *Laws*); or (3) because life with a dissipated and destroyed body is not worth living (compare Plato's views on punishment in the *Laws* 9, 862e-63a: we destroy the irremediably reprobate not only for reasons of deterrence and social hygiene, but also because life in such a condition is intrinsically worthless.[13])

Of (1)-(3), only (1) and (2) are clearly moral in character, and only (2) has any serious modern echoes in contemporary debates. (3) seems at first sight a prudential consideration (although Plato thought that we were under some sort of quasi-moral obligation to seek to live a worthwhile life), and it is, furthermore, at least controversial whether or not it is true. But even on the supposition that it is, it is a further question whether that fact justifies any active intervention into anybody's life in order to alter or prevent their self-destructive behavior. I may believe that what you are doing is detrimental, perhaps fatal, to your own best interests; but I may hold perfectly consistency with this belief, and for any one of a variety of reasons of either a consequentialist or a deontological nature, that I have no right, much less any obligation, to intervene to save you from yourself. I shall not here discuss directly the issue of the warrantedness or otherwise of such paternalist intrusions; but it is clear that the more one is inclined to accept such claims as (1), the more legitimacy paternalism gains, since attempting to prevent someone from harming their bodies can then be represented as intervening in order to stop them from damaging someone else's property, an intervention which of course may be warranted quite independently of paternalist considerations.

It is also clear that weaker analogues to (3) may yet possess some persuasive force: I need not think that your self-destructive behavior literally makes your life worthless in order to believe that at the very least I am justified in drawing your attention to its consequences and counseling a change of life-style, if not in actually enforcing such a change. It is thus possible to weaken the force of (3) (and hence increase its plausibility as an operating principle) in two distinct ways, by diminishing its active paternalist content and by broadening its scope.

I noted above that, on Plato's account in the *Republic*. what makes someone's life good is whether or not they fulfill their appropriate function, where appropriateness is determined by what it is that that person is best capable of doing (*Republic* 2, 369c-70c); this is in turn justified by a genetic account of the growth of the state as having arisen in order to supply mutual benefits. But it is, of course, one thing to hold that a state will function better (in the sense of economically more efficiently) if it involves some sort of specialization of labor (and hence that it is, economically, in everyone's benefit to seek the comparative advantage this way) and quite another to hold that people are under some *obligation* to do any such thing, or that they should be *forced* to act in such a way (again these points are quite independent of the further, necessary question of whether such crude principles of specialization really do conduce to overall economic advantage construed in terms of increased productivity). Plato holds that the state is, in some genuine sense, prior to the individuals that make it up; and its interests are thus more than merely the sum of the individual interests of the citizens.

Such a view, anathema to the broad current of modern liberalism (although perhaps now being revived in a certain form by the communitarians), was not at all unusual in the ancient world. Aristotle, for one, explicitly subscribes to it in his teleological account of the nature, origins, and justification of the state at the beginning of the *Politics* (1 2, 1254a24 ff.): parts exist for the sake of the whole; citizens are parts of the state; so citizens exist for the sake of the state. The anti-libertarian nature of this argument can be over-stressed – Aristotle thinks that it is only by being part of a state that we can fully realize our social human natures –

but it is undeniable at the very least that Aristotle is committed to a strongly objective[14] account of what the good life is for human beings, an account presented and argued for primarily in Books 1 and 10 of the *Nicomachean Ethics*.

V. THE GOOD LIFE

Aristotle also appeals to the notion of a function. There is a distinctively human sort of activity which is, in some relatively strong sense, what humans are for (*Ethics* 1 7). It is not that they have been specially made for this end by some intelligent creator, but rather that their unique position in the *scala naturae*, as a species whose distinctness from other types of creature is determined by various functional abilities they possess in virtue of being human, is sufficient to show that there is such a specifically human type of activity (or rather two of them, the political and the contemplative; but we need not broach here the problems of reconciling their apparently competing claims). Thus, it is only by exercising these capacities to the best of our respective abilities that we become as fully human as we are capable of being. For different, although compatible, reasons, both the hermit and gross sensualist have abdicated their essential humanity and are hence incapable of living the good human life (defined simply as one in which those specific capacities are exercised as fully as possible).

Aristotle does not, in the *Ethics*, have a great deal to say about the body as such, but it is clear from his stigmatization of the purely sensual life as "bestial" (1 5), even while admitting that it is the sort of life most people apparently prefer, that he thinks (like Plato) that a genuinely fulfilled human life cannot be carried out in a condition of enslavement to the body and its demands. When Plato distinguishes his three parts of the soul – the rational, the emotional, and the appetitive (*Republic* 4, 437d-442b) – the former is most purely the soul, while the latter is particularly closely enmeshed with the body. In a properly organized individual, the rational soul, with the motivational assistance provided by the emotions, will keep the unruly desiderative part in check: "they will guard over it and see that it is not filled with the so-called bodily pleasures, and by becoming inflated and powerful thereby no longer fulfills its function but tries to enslave and rule over those which it is unfit to rule, and so upsets everybody's entire life (442a-b)". Later (9,

58c-89b), Plato compares the unruly appetites to a hydra-headed monster, which, in the well-run life, must be brought under control and domesticated by the "man within" (reason) and the spirited part, here likened to a lion. And from a metaphysical point of view, the sorts of replenishment with which the body is concerned (food, drink, etc.) are shadowy and unreal by comparison with the metaphysical sustenance of truth which nourish the rational soul.

Aristotle does not, of course, subscribe to that metaphysics; and he is in general more sympathetic to the role that physical pleasures will play in the good life. But even for him, that role is severely circumscribed and subordinated to the other, more fundamentally human goals; moreover it is a mark of the good man that he will take pleasure in the appropriate things – and these are not, at least for the most part, purely physical pleasures (see especially *Ethics* 3 10-11: the pleasures of the true sensualist, like those of animals, are fundamentally to do with contact, taste and especially touch – i.e., they are directly and intimately bodily).

The main difference between Plato and Aristotle (apart from the metaphysics and the question of whether to allow bodily enjoyment some place in the good life) concerns their respective attitudes to their various prescriptions. For Plato, people have an absolute obligation to lead the (Platonic) good life, an obligation which should be backed up by the appropriate social sanctions in the ideal society if necessary, although Plato equally believes that such a life is in the individual's own best interests, since only it involves full psychological health, and health of any kind is clearly in someone's interest (this is, ultimately, his answer to Glaucon's challenge at the beginning of *Republic* to provide an argument that should persuade someone antecedently committed to the advantages of hedonism and injustice). Aristotle, by contrast, does not offer his *Ethics* as a set of normative prescriptions. He *describes* what he takes to be the best human life, and argues for that description, but leaves it up to his audience whether they should decide to subscribe to it or not (although he does believe that education has a powerful role to play in habituating the young and impressionable to appreciating the sorts of pleasures appropriate to the good human being: *Ethics* 2 1-4; *Politics* 7 17, 8 1-7).

Both Plato's and Aristotle's views of the good human life, then, have implications for how we should treat our bodies. We should not, in Plato's language, become enslaved to them and their importunate physical demands but seek rather to bring those appetites under control by rigorous training and domestication. For Aristotle, the necessary conditions that must obtain in order for us to be able properly to flourish as human beings include our having an appropriate relationship with our bodies, being able to satisfy its necessary demands without cultivating and inflaming them.

VI. HARM TO SELF, HARM TO OTHERS

Such views persist in one form or another throughout Greek culture as the model of the temperate man – and they find their way thereby, somewhat modified by other influences, into the Christian tradition; I do not intend to pursue that part of the story very far. However, another passage of Aristotle is worth examining in closer detail. In *Ethics* 5 9, 1136alO-bl4, Aristotle asks whether it is possible willingly to be unjustly treated, and comes to the conclusion that all of unjust treatments must be unwelcome to their recipients (Aristotle thus adumbrates the principle of *"injuria non fit volenti"*: 1136b3-5; compare Hipparchia's argument: §1 above).

He takes up the issue again a few pages later (11, 1138a4-28) when he considers the case of suicide:[15]

> Someone who cuts his own throat as a result of anger acts willingly, contrary to the right principle, and against the law: therefore he does wrong. But to whom? Surely to the state and not to himself, since he suffers voluntarily and no one is voluntarily wronged. Thus is the reason why the state imposes a penalty, and certain dishonor attaches to someone who has taken his own life, as to one who has wronged the state (1138a9-14).

Suicide is a crime because it harms the state and is to be proscribed for that reason. Here, Aristotle clearly appeals to the social consequences of certain, at first sight, apparently self-regarding actions. Suicide is to be condemned not because it is in some sense *intrinsically* wrong, wrong in and of itself, but because of its harmful effects and consequences. Aristotle's position will not bear much analysis, since he gives no hint of what these harmful consequences are supposed to be – and at the very least one might doubt whether, at least from a utilitarian perspective, all suicides were socially damaging (although, on the other hand, there is no reason to treat the case in purely act-utilitarian terms either). But the passage is interesting in that it is one of the relatively rare texts in extant ancient literature in which it is explicitly stated that we have a certain duty to preserve ourselves independently of our own wishes and desires on the issue; as such it supplies (in the case of suicide) an analogue to justification (2) above for the imposition of certain standards of care for the self.[16]

But I can find no trace in the ancient world of any general sentiment to the effect that people should be held responsible for the care of their bodies on similar social grounds of the type which is increasingly prevalent today. Good health, exercise, moderation in physical pleasure, and so on, are all regularly enjoined – but those injunctions are not backed up by the threat of legal sanctions of any kind; they seem rather to be appeals to individuals' senses of self-interest.

VII. DOCTOR-PATIENT RELATIONSHIPS

Equally, there is no discussion in the medical tradition of any right doctors may have to impose particular treatments or regimens upon their patients against their will, much less any such obligation. The fact that their modern counterparts sometimes do has, once again, as much to do with the very different social contexts of medicine in the modern world. There was in antiquity no "health service" as such, no concept of universal coverage in any form, socialized or insurance-driven, no notion of health-care as anything approaching a right: if you were sick and if you were wealthy enough, you could hire a physician (or several); if you were poor, you would have to rely on the community folk-medical tradition, or if you were lucky and happened to be the slave of a rich and philanthropically-inclined master, you might get your treatment paid for by him.[17]

But precisely because health-care was neither an organized industry nor a part of the general socio-political fabric, there was far less chance for the sorts of pressures to develop and the types of cases to occur in which modern doctors might feel themselves obliged to act against their patients' wishes. The contract between doctor and patient was an individual one: the patient hired the doctor best fitted (in his view at least) to effect a cure or prescribe a healthy regimen. To this end, wealthy patients frequently summoned several doctors, of differing theoretical persuasions, to their bedsides, to listen to and adjudicate the medical debate between them.

Galen, writing in the 2nd century AD, frequently records his own participation in several such multiple consultations,[18] invariably to his own advantage; and they had clearly been a part of medical practice since the Hippocratics. The doctor's role in such a debate was that of advocate for his own preferred therapy: he needed to convince the patient that he alone understood the pathological basis of the patient's disease and, hence, that he alone would be able to cure it: "you have made a most philosophical exposition of my case," the Peripatetic philosopher Eudemus tells Galen approvingly after one such multiple consultation: and on those grounds awards Galen the contract.

But while some doctors did gain the reputation for tyrannizing their patients (Galen excoriates them in characteristically ripe terms in his *On the Therapeutic Method*),[19] others (notably Asclepiades: Pliny, *Natural History* 26 7 12-20) came under fire for prescribing the sort of "cure" that patients would find agreeable. Moreover, in Galen's view at least, the degenerate medical practitioners of his time are merely sycophants, pandering to the whims of their rich charges rather than seeking to cure them of illness and prescribing a healthy regimen for them. It is worth quoting at some length a passage from the opening of *On the Therapeutic Method*. After lamenting the degeneracy of the times, principally the fact that most people are concerned with pleasure and wealth to the exclusion of the search for truth, Galen notes acidly that the key to professional success as a physician is careful social cultivation of potential rich clients:

It is for these reasons ... that people are respected, cultivated, and considered to be skillful practitioners, and not as a result of any particular ability they may possess: for there simply isn't anyone capable of assessing that, spending as they do their whole day in leisure pursuits. They indulge in salutation in the mornings, then go their separate ways, a large portion of the tribe repairing to the law-courts, while even more frequent dance-shows and chariot races and another sizeable section busies itself with dicing, sexual encounters, bathing, drinking, carousing, and other sensual pleasures. Finally in the evening they reunite for drinking-parties... drinking toasts to one another and competing to see who can down the largest draughts. Supreme among them is not the one most gifted musically or in philosophical argument but the one who can down the greatest number of wine bowls. Indeed many of them seem still drunk to me in the morning, and some of them stink as powerfully of wine as if they had been drinking only a short while before. So naturally when they fall

sick they call to their aid not the best doctors (whom they never
showed any tendency to be able to identify when they were healthy),
but rather those who are their greatest favorites and flatterers, who will
prescribe cold drinks for them if they are asked to, bathe them if that is
what is required of them, who will offer them snow or wine, and
generally do anything they are told to do, as if they were slaves, in
contrast with those Asclepiad doctors of old who thought it their job to
rule the sick as generals would their armies or kings their subjects ....
Thus it is not the man of superior skill but the cleverer flatterer who is
more honored among them; for such a person the way is easy and
smooth: the doors of houses are opened for him, and he quickly
becomes rich and powerful (Galen, *On the Therapeutic Method* X 2-4
Kühn).

That jeremiad is highly colored and no doubt exaggerated. But it is also
instructive: Galen clearly holds that it is the duty of the doctor to direct
his patients rather than the other way around. Even so, nothing in this
passage, or for that matter anywhere else in Galen's voluminous *oeuvre*,
suggests that this duty to lead should extend to deliberately ignoring the
patient's own wishes in certain cases. Rather, it is the good doctor's duty
merely to attempt to persuade the patient that a certain course of
treatment is in their own best interests. There is, interestingly, little or no
suggestion as to what ought to be the procedure in cases where, for
whatever reason, the patient is incompetent to decide:[20] Galen simply
behaves as though there will be somebody, a family member or the
master in the case of slaves, whose responsibility it will be to authorize
the appropriate decisions.

Equally, while we may find passages in which Galen expresses his
disapproval of bodily licentiousness, he stops short of couching that
disapproval in terms of a general and enforceable moral injunction
against such activity. The following passage, from his "Hymn to Nature"
and its providence in *On the Function of the Parts*[21] is typical, and
provides a fitting epilogue to this paper:

It is now time for you, my reader, to decide which chorus you will join,
the one that gather round Plato, Hippocrates, and the others who
admire the works of nature, or the one consisting of those who blame
her because she has not arranged to have waste products discharged
through the feet.[22] Anyone who has the temerity to say such things to
me has been corrupted by luxury to such an extent that he considers it a

hardship to rise from his bed when he voids, thinking that man would be better constructed if he could simply extend his foot and discharge the excrement through it. How do you imagine such a man feels and acts in private? How wantonly he uses all the openings of his body! How he maltreats and ruins the noblest qualities of the soul, crippling and blinding that divine faculty by which alone nature allows a man to behold the truth, and allowing his worst and most bestial faculty to grow huge, strong, and insatiable of lawless pleasures and to hold him in wicked servitude! (Galen, *On the Function of the Parts* I 173-4 Helmreich, trans. after May)

If we behave in such a bestial fashion, we are indeed abusing the gifts of nature, and destroying our own humanity. We can be condemned for it, not least for our foolishness in failing to appreciate what is as a matter of fact the good life for a human being: all of this, of course, has long roots in the tradition we have been examining, reaching back at least to Plato and Aristotle. We are to be exhorted and castigated, but we are not, here as elsewhere, to be forced to alter our behavior in the name of any supposed general social benefits that will accrue as a result of such a conversion. That, again, remains a distinctively modern notion, generated by a distinctively modern set of circumstances and their attendant concerns.

Department of Philosophy
University of Texas
Austin, Texas

NOTES

[1]   Cf. Aquinas, quoted below, n. 16.
[2]   Compare the insistence of various early Christian sects on the mortification of the flesh, a trend perhaps carried to its furthest extremes by the desert fathers, in particular the stylites: see B. Ward (trans.) *The Wisdom of the Desert Fathers* (Oxford, 1985); and B. Ward, *Harlots of the Desert: A Study of Repentance in Early Monastic Sources* (Kalamazoo, 1987). But although a distaste for and fear of the flesh permeates Christianity from St. Paul onwards (cf e.g. *Epistle to the Galatians* 5, 19-21), it is by no means uniform in its vehemence or its metaphysical underpinnings. Augustine, for instance, whose struggles with the flesh were the stuff of heroic epic, refuses to see all created matter as intrinsically evil, as the Gnostics and various Neoplatonists had it: rather, everything created by God is good (*City of God* 12, 4), although it can be misused and corrupted through human sin, which is, however, to be ascribed to the soul rather than the body as such (*ibid.* 14, 1-5).

[3]   For an illuminating discussion of the Cleombrotus story and its progress in the ancient world, see Stephen A. White, 'Callimachus on Plato and Cleombrotus', (1994, pp. 135-61).

[4]   This doctrine is in some tension with the Stoics' notion that every animal is born with an innate sense of "appropriation", the instinctive drive towards self-preservation which obviously includes bodily preservation (e.g. Diogenes Laertius 7 85-6, Seneca *Letters* 121 6-15; Hierocles 2.1-9). That tension is not unresolvable, however: crudely, the Stoics thought that ideally we progressed from the irrational and instinctive animal drive to physical self preservation to the rational realization that we are all parts of the same great functioning whole: hence our sense of "appropriation" ought, ultimately, be directed towards the cosmos as a whole, with an obvious concomitant devaluation of our ordinary "irrational" special concern for ourselves: see Cicero *On Ends* 3 62-8; Anonymous *On the 'Theaetetus'* 5.18-6.31.

[5]   Elsewhere, as is well known, Plato ascribes emotions and desires to other parts of the soul: *Republic* 4, 436a-42d; *Timaeus* 69b-72d.

[6]   I shall continue, for want of any better acceptable alternative, to translate *psuchê* as "soul". "Soul", with its Judaeo-Christian connotations of immateriality and immortality, is thoroughly misleading: Greek *psuchai* were neither necessarily immaterial, nor necessarily (or even generally: Plato had to argue strenuously for it) thought to be immortal. At its most general, a *psuchê* is simply whatever it is in virtue of which whatever is alive is alive – it's a placeholder, in fact.

[7]   It is worth comparing this perspective with that of certain recent philosophers on the undesirability of disembodied existence: B. Williams, 'Persons, character and morality'; S. Shoemaker, 'Embodiment and behavior', in A. Rorty (ed.) *The Identities of Persons* (1976).

[8]   Herodotus says that it arrived in the Greek world from Egypt (*Histories* 2 123), although such evidence is suspect.

[9]   This view was by no means restricted to them, however, it is to be found in the intriguing Hippocratic treatise *On Regimen* (1 7-9, 24); Empedocles subscribed to a version of it (Aristotle, *On the Soul* 1 4, 408a18-29); and, in spite of his protestations to the contrary, Aristotle's own theory is perhaps best described as a harmony-theory.

[10] The resolution of this issue is difficult, and leads straight to the heart of much contemporary debate on personal identity. We need not attempt to resolve it here, but much depends upon just how much is built into the individual harmony in question. If it simply involves such things as character-types and dispositions, the temptation to say that successive instantiations of it in different bodies are distinct individuals will be all the stronger. If, on the other hand, it includes certain semi-permanent dispositional states of consciousness (in particular, but not exclusively, experiential memories), then the case becomes less clear, and it is not known how the Pythagoreans would have stood on this question (although Pythagoras himself is alleged to have understood the criterial role played by memory in determining personal identity).

[11] This step may, however, have been anticipated by Alcmaeon of Croton, a Pythagorean-influenced philosopher and physiologist from the early 5th century BC; Alcmaeon is the author of the earliest surviving argument for the soul's immortality, which concludes that the soul is immortal because it is in perpetual motion, like the heavenly bodies (Aristotle, *On the Soul* 1 2, 40Sa29-bl; Diogenes Laertius 8 83). However that argument is to be filled out, (Plato himself offers an interesting expansion of it: *Phaedrus* 245c-46a), Alcmaeon appears to have conceived of the soul as a motor, literally imparting its motion to the body, and as such, for whatever reasons (causal or logical, see R.J. Hankinson, 'Greek Medical Models of

the Mind', in S. Everson (ed.) *Psychology Companions to Ancient Thought,* pp. 194-6), incapable of ceasing to move and hence in perpetual motion. Nothing whatever is known about the nature of Alcmaeon's psychic motions (although it is plausible that he thought them to be revolutions, such as those of the heavenly bodies); but he was of course well aware of human physical mortality ("human beings die because they cannot join the beginning to the end", Aristotle, *Problems* 17 3, 916a33-7), and hence it is hard to see how he could have thought that these psychic orbits were themselves physical; moreover, since their motion is perpetual, they cannot submit to discontinuous existence, such as the Pythagorean souls might; they are therefore most likely to be immaterial substances.

¹² *Politics* 2 9, 1271a41-b6; 7 14-15.

¹³ A position adopted and expressed with characteristic vehemence by the great later physician Galen, *That the Character of the Soul is Consequent upon the Structure of the Body* IV 814~16 Kühn, see further R.J. Hankinson, 'Actions and passions: Affection, emotion, and moral self-management in Galen's philosophical psychology', in J. Brunschwig and M.C. Nussbaum (eds.) *Passions & Perceptions* (1993).

¹⁴ A view of the good life is "strongly objective" in this sense if its espouser believes such a life (however it may in fact be delineated) to be good for any individual, regardless of their particular tastes or preferences: see Hankinson, art.cit., n. 13 above.

¹⁵ The text is disputed, but the controversy does not affect the overall sense.

¹⁶ This position was also to become, by way of Aquinas's endorsement of it, Catholic orthodoxy: see *Summa Theologiae* Qu 64, Sect. 5: "suicide is totally wrong for three reasons: First, everything naturally loves itself, and therefore it naturally seeks to preserve itself and to resist what would injure it as much as it can. Therefore suicide is against natural inclination, and is also opposed to charity, in accordance with which one should love oneself... Secondly every part that exists is a part of a whole; man is a part of the community, and the fact that he exists affects the community; hence if he kills himself he injures the community, as the Philosopher [i.e. Aristotle] makes clear. Thirdly, life is a gift given by God to man, and is subject to his power as the one who takes away and gives life. Therefore a man who takes his own life sins against God, just as someone who kills a slave injures his master.... God alone has authority of life and death". Thus, Aquinas marries the Platonic and the Aristotelian rationales for the proscription of suicide into a single doctrine.

¹⁷ Galen in his medical autobiography *On Prognosis* mentions cases in which he treated such slaves.

¹⁸ Again the text *On Prognosis* is typical: note in particular his lengthy record of the case of the Peripatetic philosopher Eudemus: *Prognosis* XIV 609-19 Kühn.

¹⁹ The first two methodological books of which are translated with a commentary in R.J. Hankinson, *Galen on the Therapeutic Method* (1991).

²⁰ Galen certainly allows that there are such cases: e.g., the discussion of various forms of mental illness in *On the Affected Parts* (VIII 160 ff., 217 Kühn).

²¹ *On the Function of the Parts* is edited by G. Helmreich: *Galenus: de Usu Partium* (1907-9), and translated into English by M.T. May: *Galen: On the Usefulness of the Parts of the Body* (1968).

²² Galen does not say who it was who made this claim, but the context is the ramified ancient debate between teleological and mechanistic styles of natural explanation: the opponents are clearly mechanists, seeking to impugn the claim of teleologists like Galen that Nature really has arranged everything for the best. On Galen's teleology, see R.J. Hankinson, 'Galen and the best of all possible worlds', pp. 206-27. This view of the fundamental goodness of created

nature derives ultimately from Plato's *Timaeus* and is echoed by the Stoics (cf. Cicero, *On the Nature of the Gods* 2, pp. 154-67) and in the Augustinian Christian tradition (*City of God*, 12, 4).

BIBLIOGRAPHY

Aquinas, St. T.: 1942, *Summa Theologiae*, Institute of Medieval Studies of Ottawa, Ottawa.

Galen: 1907-9, *Galenus: de Usu Partium*, (2 volumes), G. Helmreich (trans.), Leipzig.

Galen: 1968, *Galen: On the Usefulness of the Parts of the Body*, (2 volumes), M.T. May (trans.), Cornell University Press, Ithaca.

Hankinson, R.J.: 1989, 'Galen and the best of all possible worlds,' *Classical Quarterly* 39, 206-227.

Hankinson, R.J.: 1991, 'Greek medical models of the mind,' in S. Everson (ed.), *Psychology, Companions to Ancient Thought* 2, Cambridge University Press, New York, pp. 194-196.

Hankinson, R.J.: *Galen on the Therapeutic Method*, Oxford University Press, New York.

Hankinson, R.J.: 1993, 'Actions and passions: Affection, emotion, and moral self-management in Galen's philosophical psychology,' in Brunschwig and M.C. Nussbaum (eds.), *Passions & Percentions*, Cambridge University Press, Cambridge.

Shoemaker, S.: 1976, 'Embodiment and behavior,' in R. Rorty (ed.), *The Identities of Persons*, University of California Press, Berkeley.

Ward, B. (trans.): 1985, *The Wisdom of the Desert Fathers*, Oxford University Press, New York.

Ward, B. (trans.): 1987, *Harlots of the Desert: A Study of Repentance in Early Monastic Sources*, Mowbray, London.

White, S.A.: 1994, 'Callimachus on Plato and Cleombrotus,' *Transactions and Proceedings of the American Philological Society* 124, 135-161.

Williams, B.: 1976, 'Persons, character and morality,' in R. Rorty (ed.), *The Identities of Persons,* University of California Press, Berkeley.

WENDY DONNER

A MILLIAN PERSPECTIVE ON THE RELATIONSHIP
BETWEEN PERSONS AND THEIR BODIES

I. MILL'S UTILITARIANISM[1]

John Stuart Mill's distinctive variant of utilitarianism is an attractive
framework for deliberations in medical ethics. Unlike many other
utilitarians, notably Mill's predecessor Jeremy Bentham and, currently,
Peter Singer, Mill carves out a central place for rights in his theory. Yet
rights are not foundational elements in Mill's system. These are utilitarian
rights, and as such are grounded in utility and well-being. While Mill's
utilitarian foundation for rights does not offer as secure a basis as some
wish, because they are not "trumps", his defense is robust and his rights,
although not absolute, are weighty and not easily overturned.

The foundation of Mill's utilitarian system is the principle of utility.
The principle of utility is, according to Mill, the supreme principle of
morality and thus justifies, directly or indirectly, all obligations and
secondary moral principles. The classic statement of the principle occurs
early in *Utilitarianism*, the essay Mill devotes to a defense of his system.
He says,

> The creed which accepts as the foundation of morals, Utility, or the
> Greatest Happiness Principle, holds that actions are right in proportion
> as they tend to promote happiness, wrong as they tend to produce the
> reverse of happiness (*Utilitarianism,* 1974-91, vol. 10, p. 210).[2]

In this theory the good is prior to the right, and so moral rules which set
out rights and obligations are constructed and defended on the basis of the
good. It has sometimes been overlooked that the principle of utility is, in
the first instance, a principle of the good rather than simply a foundational
moral principle. As the principle of utility is most directly a principle of
the good, it provides the grounding for all practical reasoning,
prominently including but by no means limited to moral reasoning.
Another formulation illuminates the status of this principle more
transparently:

M.J. Cherry (ed.), Persons and their Bodies: Rights, Responsibilities, Relationships, 57–72.
© 1999 *Kluwer Academic Publishers. Printed in Great Britain.*

> The utilitarian doctrine is, that happiness is desirable, and the only thing desirable, as an end; all other things being only desirable as means to that end (*Utilitarianism*, 1974-91, vol. 10, p. 234).

This structure is explained more fully in the *Logic*, and is alluded to in other writings. In the *Logic* Mill says,

> There are not only first principles of Knowledge, but first principles of Conduct. There must be some standard by which to determine the goodness or badness, absolute and comparative, of ends, or objects of desire.
>
> ... For the ... practice of life some general principle, or standard, must still be sought; and if that principle be rightly chosen, it will be found, I apprehend, to serve quite as well for the ultimate principle of Morality, as for that of Prudence, Policy, or Taste.
>
> ... The general principle to which all rules of practice ought to conform, and the test by which they should be tried, is that of conduciveness to the happiness of mankind, or rather, of all sentient beings: in other words, that the promotion of happiness is the ultimate principle of Teleology (*A System of Logic*, 1974-91, vol. 8, p. 951).

The principle of utility serves to ground all three areas of what Mill calls the Art of Life – morality, prudence or self-interest, and beauty or nobility – as well as more particular moral arts such as the "hygienic and medical arts" which have as their ends, respectively, "the preservation of health" and "the cure of disease"(*A System of Logic*, 1974-91, vol. 8, p. 949).

The principle of utility is thus the foundation. But Mill's moral philosophy has a complex structure. For example, debates continue about whether his theory is properly interpreted as a form of act utilitarianism, rule utilitarianism, or neither of these alternatives. While I side step this issue here, I do take up the crucial exploration of the place of rights within the framework of his theory, which is elaborated in the last chapter of *Utilitarianism*, entitled "On the Connexion Between Justice and Utility" (1974-91, vol. 10, pp. 240-259). In the course of setting out the architecture of his moral philosophy, Mill presents both an analysis of the concept of a right and a utilitarian defense of rights. My previous discussion of the status of the principle of utility provides the background for the presentation of the structure of his moral theory in *Utilitarianism*.

II. MORAL RULES, JUSTICE AND RIGHTS

Mill begins exploring the structure of his theory by distinguishing the category of morality from the broader category of expediency or the general promotion of utility of which it is a sub-class. The principle of utility most generally governs expediency. He says,

> We do not call anything wrong, unless we mean to imply that a person ought to be punished in some way or other for doing it; if not by law, by the opinion of his fellow creatures; if not by opinion, by the reproaches of his own conscience. This seems the real turning point of the distinction between morality and simple expediency (*Utilitarianism*, 1974-91, vol. 10, p. 246).

Mill's emphasis on the coercive element of moral sanctions is instructive for questions of medical ethics.

> It is part of the notion of Duty in every one of its forms, that a person may rightfully be compelled to fulfill it ... I think there is no doubt that this distinction lies at the bottom of the notions of right and wrong; that we call any conduct wrong, or employ, instead, some other term of dislike or disparagement, according as we think that the person ought, or ought not, to be punished for it; and we say that it would be right to do so and so, or merely that it would be desirable or laudable, according as we would wish to see the person whom it concerns, compelled, or only persuaded and exhorted, to act in that manner (*Utilitarianism*, 1974-91, vol. 10, p. 246).

David Lyons has drawn out the implications and brought to center stage the analysis of moral rules in play here. According to Lyons's interpretation of Mill, not every action which fails to maximize utility is wrong. To be wrong, an act must further be liable to punishment, and this has led Lyons to interpret Mill as propounding "a model based on coercive social rules" (1976, p. 108).[3] In this interpretation, moral obligation and coercive sanctions are conceptually linked. These sanctions can be legal, social, or internal feelings of guilt. In sum, Lyons claims that

> These considerations suggest that Mill had a view something like this. To call an act wrong is to imply that guilt feelings, and perhaps other sanctions, would be warranted against it. But sanctions assume coercive rules. To show an act wrong, therefore, is to show that a

coercive rule against it would be justified. The justification of a coercive social rule establishes a moral obligation, breach of which is wrong (Lyons, 1976, p. 109).

Thus the utilitarian costs of setting up, maintaining and enforcing a coercive social rule, namely, restrictions on freedom and the sanctions for violation, are all calculated in deciding whether a particular moral rule is justified. Wayne Sumner, in scrutinizing Lyons' account, explains that "if we employ the notion of a positive balance of utility to mean an excess of benefits over costs then the existence of a coercive rule against doing some kind of act would be justified if and only if it would yield a positive balance of utility" (Sumner, 1979, pp. 104-105).

So in this first distinction Mill's moral theory marks off moral rules of obligation from the broader class of rules of general promotion of the good. But Mill also demarcates a further sub-class within this class of moral rules, consisting of rules of justice which "involve the idea of a personal right – a claim on the part of one or more individuals" (*Utilitarianism*, 1974-91, vol. 10, p. 247). Injustice "implies two things – a wrong done, and some assignable person who is wronged" (*Utilitarianism*, 1974-91, vol. 10, p. 247). Mill's definition of a right follows:

> When we call anything a person's right, we mean that he has a valid claim on society to protect him in the possession of it, either by the force of law, or by that of education and opinion. If he has what we consider a sufficient claim, on whatever account, to have something guaranteed to him by society, we say that he has a right to it (*Utilitarianism*, 1974-91, vol. 10, p. 250).

He reiterates this central point:

> To have a right, then, is, I conceive, to have something which society ought to defend me in the possession of (*Utilitarianism*, 1974-91, vol. 10, p. 250).

Lyons points out that in the first part of the passage Mill presents an analysis of a right in general. This analysis is distinct from his utilitarian defense of rights and as such can be accepted as a compelling analysis of rights by utilitarians and non-utilitarians alike (Lyons, 1976, pp. 10-11). But Mill also presents his utilitarian defense of rights in the same paragraph as this analysis, which can and has led to confusion. He says that "if the objector goes on to ask why it ought, I can give him no other reason than general utility." The justification is based on "the extraordinarily important and impressive kind of utility which is conceived" (*Utilitarianism*, 1974-91, vol. 10, pp. 250-251). Mill is concerned here to defend his claim that justice and utility are not in conflict. In his view, rules of justice must be founded on well-being.

> While I dispute the pretensions of any theory which sets up an imaginary standard of justice not grounded on utility, I account the justice which is grounded on utility to be the chief part, and incomparably the most sacred and binding part, of all morality. Justice is a name for certain classes of moral rules, which concern the essentials of human well-being more nearly, and are therefore of more absolute obligation, than any other rules for the guidance of life (*Utilitarianism*, 1974-91, vol. 10, p. 255).

> Justice is a name for certain moral requirements, which, regarded collectively, stand higher in the scale of social utility, and are therefore of more paramount obligation, than any others; although particular cases may occur in which some other social duty is so important, as to overrule any one of the general maxims of justice (*Utilitarianism*, 1974-91, vol. 10, p. 259).

In addition to his analysis of the concept of a right and his utilitarian justification for rights, Mill also has a substantive theory of justice and rights.[4] There are many rights in his system, but in *Utilitarianism* Mill names the right to security and the right to liberty (including the right to liberty of self-development) as the two most basic rights. Mill's analysis of rights illuminates that in his view they involve claims which are socially guaranteed by social and political institutions collectively set up and maintained effectively to carry out these guarantees. Rights, as the above quotes demonstrate, protect those interests which are most vital to well-being.

Mill's conception of value also is distinctive, for his conception essentially involves a notion of self-development. As a classical utilitarian, Mill locates value in states of happiness or satisfaction. However, he is a qualitative hedonist and maintains that the kind of satisfaction must be taken into account in assessing its value. The most valuable kinds of happiness are those in which humans engage themselves in developing and exercising certain capacities. The most valuable forms of happiness engage our intellectual, affective and moral or caring capacities. According to Mill, society has an obligation to nurture these human capacities in its members, who have a right to development and self-development. The development of our human capacities in childhood socialization is the first part of this process. In the usual course of events, when we reach adulthood we ourselves take control of this process which then continues as one of self-development. In this continuation, higher-order capacities of individuality, autonomy and sociality are developed. Individualism is the process in which we discover our endowment or balance of talents based on the generic capacities. This mix of talents and endowments is not firmly set, but rather consists of a range of possibilities from within which we make choices. Autonomy is the capacity critically to reflect upon and endorse our commitments, our pursuits, our character, and our way of life and to reject or revise them if necessary. Mill believes that the most worthwhile lives result from the use of individuality and autonomy to discover and shape our possibilities and choose our commitments and way of life on their foundation. This core value of self-development, with the essential component of autonomy, explains the great weight which Mill attaches to individual liberty and the central place of the principle of liberty within his moral philosophy.

III. LIBERTY

Mill's principle of liberty is frequently invoked in a variety of discussions in which freedom comes into play. But quite often this principle is misused because its place within the complex structure of his theory is misunderstood. One of the more flagrant examples of this occurs when the liberty principle is misperceived as a principle competing with the principle of utility and of equal status with it. Mill is not the libertarian that this confusion presupposes; although liberty is a strong value within his system, it must be balanced by other values which are, in his view, equally important. This misunderstanding can be avoided if we integrate Mill's discussion of principles of justice and rights in *Utilitarianism* and his examination of the liberty principle in *On Liberty*. These discussions are properly seen as showing two sides of the same coin, rather than two conflicting views.

I have argued in my previous discussion that Mill grounds rights within a utilitarian system on vital interests. In *On Liberty* he links interests, especially vital interests, and harms. If we bring together the arguments of the two essays, the principle of liberty and the right it defends can correctly be seen as embedded in Mill's theory of justice.

Mill also provides a statement of the principle of liberty early in the essay devoted to its defense. He says,

> The object of this essay is to assert one very simple principle, as entitled to govern absolutely the dealings of society with the individual in the way of compulsion and control...That principle is, that the sole end for which mankind are warranted, individually or collectively, in interfering with the liberty of action of any of their number, is self-protection. That the only purpose for which power can be rightfully exercised over any member of a civilized community, against his will, is to prevent harm to others (*On Liberty*, 1974-91, vol. 18, p. 223).

The also classic anti-paternalist statement follows:

> His own good, either physical or moral, is not a sufficient warrant. He cannot rightfully be compelled to do or forbear because it will be better for him to do so, because it will make him happier, because, in the opinion of others, to do so would be wise, or even right. These are good reasons for remonstrating with him, or reasoning with him, or persuading him, or entreating him, but not for compelling him, or visiting him with any evil in case he do otherwise. To justify that, the conduct from which it is desired to deter him, must be calculated to produce evil to some one else. The only part of the conduct of any one, for which he is amenable to society, is that which concerns others. In the part which merely concerns himself, his independence is, of right, absolute. Over himself, over his own body and mind, the individual is sovereign (*On Liberty*, 1974-91, vol. 18, pp. 223-224).

But Mill's rhetorical flourishes, as well as the distressing tendency of readers to examine only small selections of his essays, rather than articles in their entirety, have perhaps served to obscure some important aspects. In the last chapter of this essay, he restates the principle, but as two maxims which need to be balanced. Here he says that he wants to "bring into greater clearness the meaning and limits of the two maxims which together form the entire doctrine of this Essay, and to assist the judgment in holding the balance between them, in the cases where it appears doubtful which of them is applicable to the case" (*On Liberty*, 1974-91, vol. 18, p. 292). He says,

The maxims are, first, that the individual is not accountable to society for his actions, in so far as these concern the interests of no person but himself...Secondly, that for such actions as are prejudicial to the interests of others, the individual is accountable, and may be subjected either to social or legal punishment, if society is of the opinion that the one or the other is requisite for its protection (*On Liberty*, 1974-91, vol. 18, p. 292).

This is not one principle which is to be flatly applied. Mill has brought to the fore a most contemporary point, namely, that principles need to be examined and applied in their contexts, and that we need considerable detail and skill at balancing to apply these maxims (or this principle) with accuracy.

The principle of liberty is most plausibly viewed as defending the crucial right to liberty within Mill's theory. The principle maintains that liberty can be restricted only to prevent harm to others. This raises the central interpretive question of what is Mill's conception of harm. The temptations of viewing the liberty principle in terms of self-regarding and other-regarding action, that is, of protecting freedom in self-regarding but not other-regarding actions, should be avoided, as this removes most of the strength and life of the principle.[5] It is more plausible to view this principle as regarding interests, especially vital interests. Since Mill usually discusses harms in terms of interests, it is most helpful to interpret or analyze harm as injury to interests. The interests Mill discusses in both *Utilitarianism* and *On Liberty* are the same, and we have seen how in his analysis and justification of rights in *Utilitarianism* he makes interests the foundation of his theory of justice.

Fred Berger argues convincingly that Mill has a theory of justice and that the central theme of *On Liberty* is the positive defense of the right to individual self-development or autonomy. Berger says that "[*On Liberty*] is aimed at providing a rule of conduct for society that is designed to protect what Mill regards as a vital interest of persons – autonomous development and activity...people have a right to individuality. Mill's theory of liberty, then, is an application of his theory of justice" (1984, p. 229). Mill's positive defense of this right to liberty of self-development is brought into clear focus if it is examined in the light of his explanation of harm in terms of injury to interests and his view that interests, especially vital interests, ground rights. Mill says,

The fact of living in society renders it indispensable that each should be bound to observe a certain line of conduct towards the rest. This conduct consists, first, in not injuring the interests of one another, or rather certain interests, which...ought to be considered as rights (*On Liberty*, 1974-91, vol. 18, p. 276).

Thus the principle of liberty can be seen as defending a right to liberty. This right is one of the weightiest in Mill's system. But it is a secondary principle in his theory, not of equal status with the principle of utility, and so it is not absolute. When this right conflicts with other principles of justice or other secondary moral principles, the general apparatus of his complex utilitarian theory is relied upon to resolve the conflict or dilemma.

IV. APPLICATIONS

How does this Millian framework apply to certain questions about the relationship between persons and their bodies which pertain to medical ethics? Should persons have the freedom to sell their body parts? What about the freedom to sell reproductive material, such as eggs or sperm, or reproductive services?

It appears that Mill provides a clear affirmative answer to these questions in his initial statements of the principle of liberty. He says that "in the part [of conduct] which merely concerns himself, his independence is, of right, absolute. Over himself, over his own body and mind, the individual is sovereign" (*On Liberty*, 1974-91, vol. 18, p. 224). If the selling of body parts or reproductive material concerns only him or herself, then this sale is not prohibited. But we should also take note of Mill's qualifications of his own scrutiny of examples of application of the liberty principle. Although many commentators have taken these examples to be part of the definitive statement of Mill's theory, he is much more guarded. He says that "the few observations I propose to make on questions of detail, are designed to illustrate the principles, rather than to follow them out to their consequences"(*On Liberty*, 1974-91, vol. 18, p. 292).

Mill states plainly that individuals have control over their own bodies and thus what will be done to them. As the foregoing account of the structure of his moral theory makes clear, Mill emphatically rejects the notion of duties to oneself. We have an inviolable sphere of freedom in those matters which concern our own interests and we have no duties to ourselves which could, in principle, prohibit us from selling body parts or reproductive material. To have a duty not to sell body parts means that it would be wrong to do so. If it is wrong to do so then we are appropriately subject to coercion and sanctions if we do so. This is firmly ruled out by the principle of liberty. It is a contradiction to maintain both that we have liberty in regard to those matters which affect only our own interests, and that we are appropriately subject to sanctions if we make certain decisions about matters within this sphere. Mill says,

> What are called duties to ourselves are not socially obligatory, unless circumstances render them at the same time duties to others. The term duty to oneself, when it means anything more than prudence, means self-respect or self-development; and for none of these is any one accountable to his fellow creatures, because for none of them is it for the good of mankind that he be held accountable to them (*On Liberty*, 1974-91, vol. 18, p. 279).

But this does not entirely end the matter. Mill definitively rules out any place for self-regarding duties and punishment for acting in certain ways in the self-regarding realm. But *On Liberty*, in addition to being an eloquent tract in defense of liberty, also contains a strong exhortation for much more active engagement with "each other's conduct in life" (*On Liberty*, 1974-91, vol. 18, p. 277). While Mill is properly viewed as a strong anti-paternalist, this does not mean that he believes that people should be disengaged from each other. He rules out self-regarding duties, but he argues for a greater place for what he calls the "self-regarding virtues"; these correspond to the sphere of beauty and nobility in the Art of Life (*On Liberty*, 1974-91, vol. 18, p. 277). He says,

> Instead of any diminution, there is need of a great increase of disinterested exertion to promote the good of others...Human beings owe to each others help to distinguish the better from the worse, and encouragement to choose the former and avoid the latter. They should be for ever stimulating each other to increased exercise of their higher faculties, and increased direction of their feelings and aims towards wise instead of foolish, elevating instead of degrading, objects and contemplations (*On Liberty*, 1974-91, vol. 18, p. 277).

> If he is grossly deficient in those [admirable] qualities a sentiment the opposite of admiration will follow. There is a degree of folly, and a

degree of what may be called (though the phrase is not unobjectionable) lowness or depravation of taste, which, though it cannot justify doing harm to the person who manifests it, renders him necessarily and properly a subject of distaste, or, in extreme cases, even of contempt ... It would be well, indeed, if this good office were much more freely rendered than the common notions of politeness at present permit, and if one person could honestly point out to another that he thinks him in fault, without being considered unmannerly or presuming (*On Liberty*, 1974-91, vol. 18, p. 278).

But these efforts are limited by the respect for the liberty of others.

Moral vices...may be proofs of any amount of folly, or want of personal dignity and self-respect; but they are only a subject of moral reprobation when they involve a breach of duty to others, for whose sake the individual is bound to have care for himself (*On Liberty*, 1974-91, vol. 18, p. 279).

Considerations to aid his judgment, exhortations to strengthen his will, may be offered to him, even obtruded on him, by others; but he himself is the final judge (*On Liberty*, 1974-91, vol. 18, p. 277).

While freedom over one's own affairs is the prevailing value, Mill's encouragement of other forms of engagement with the actions and characters of others will strike some libertarians (and even some others) as being excessive. But his concern with the general promotion of good, coupled with his view of the good for humans as intimately connected with the development and exercise of their higher faculties, precludes the lack of engagement which the strong libertarian view upholds.

It is somewhat speculative to consider whether Mill would think the selling of body parts or reproductive material to be the sort of action which, though not prohibited, would leave the agent open to judgments of being foolish, degraded, or even depraved. But Mill's theory has some salient differences with other philosophical traditions on the question of the commodification of body parts and these differences may lead him to view their sale as degrading.

One political philosophy lineage, tracing its origins to Locke, uses the concept of self-ownership to explain and justify persons' control over themselves and their bodies. This concept of self-ownership is appealed to by Robert Nozick's libertarian theory, and it is invoked by G.A. Cohen in a Marxist critique of Nozick (Nozick, 1974, pp. 149-82; Cohen, 1986a;

1986b). Will Kymlicka interprets Nozick's self-ownership argument "as an appeal to the idea of treating people as equals" (1990, p.98). Kymlicka says that,

> For Nozick ... the most important rights are rights over oneself – the rights which constitute "self-ownership" ... The basic idea of self-ownership can be understood by comparison with slavery – to have self-ownership is to have the rights over one's person that a slaveholder has over a chattel slave (1990, p. 105).

Kymlicka continues that self-ownership claims "are intended to model the claim that no one is the possession of any other" (1990, p. 105). The denial of self-ownership "makes some people mere resources for the lives of others" (1990, p. 107). G.A. Cohen brings out a Kantian point against the moral acceptability of claims of self-ownership. Cohen says that Kant argues that "it is morally unacceptable for human beings to sell parts of themselves, engage in prostitution, etc. Kant's idea is that ... acting as though one owned one's parts and powers is immoral ... human beings are ends-in-themselves, and parts of ends-in themselves ought not to be sold for gain" (Cohen, 1995, p. 212).

This appeal to Kant points us in the right direction. Mill would agree that it is a serious misstep to appeal to such a concept for this purpose, for it inevitably undermines the very end which it is supposed to promote. In contrast, Mill's liberty principle and his core value of self-development are also designed to justify persons' control over themselves, including their minds and bodies. But this justification is carried out by appeal to such concepts as well-being and vital interests. Millian concepts of self-development, autonomy, and freedom contain no references to concepts of property and commodities. This allows him to mount a robust defense of the right to liberty without taking the step onto Nozick's and Cohen's slippery slope. For, once liberty and control are grounded on references to the person as property, albeit self-owned, it is hard to avoid all sorts of questions about what parts can be sold, under what circumstances, for what ends, and so on. But it may appropriately be asked, and is quite compatible with the Millian framework to ask, whether it is not demeaning and degrading to view oneself as property and commodity which can be bought and sold. Kymlicka's interpretation misses the point that it is the very viewing of slaves as commodities that makes slavery such a depraved institution. If respect for persons and control and liberty over their lives can be powerfully defended without viewing persons as property, even if owned by themselves, then there seems to be little reason to invoke the notion of self-ownership which has so much dangerous potential.

These admittedly speculative lines of thought emphasize the central point that it is troubling to view persons and their minds and bodies as commodities. While entering into contracts to sell body parts may not be prohibited, there are reasons to believe that Mill would have regarded these activities as properly subject to judgments of being degrading.

In addition, these ideas resonate with current discussions within feminist bioethics focussing on problems with the commodification of reproduction. This is an ongoing and wide ranging debate, but one consistent thread is an objection to applying property concepts to certain things. For example, Sara Ann Ketchum's argument against the commodification of babies and women's bodies in paid surrogate (or contract) motherhood shows much greater appreciation of the spirit of the Kantian framework than Kymlicka's comments. According to Ketchum, on the Kantian argument, "selling people is objectionable because it is treating them as means rather than as ends, as objects rather than as persons ... allowing women's bodies to be bought and sold (or 'rented' if you prefer) adds to the inequality between men and women" (1992, p. 286).[6] She adds,

Suppose we do regard mother contracts as contracts for the sale or rental of reproductive capacities. Is there good reason for including reproductive capacities among those things or activities that ought not to be bought and sold? ... A Kantian might argue that there are some activities that are close to our personhood and that a commercial traffic in these activities constitutes treating the person as less than an end (or less than a person) (1992, pp. 287-288).

The underlying claim is that there are certain sorts of things that cannot appropriately be commodified, and that such commodification is degrading and disrespectful even if it is not legally prohibited.

I turn now to the question of whether persons have duties actively to care for their bodies according to the Millian framework. Once again, an apparently clear answer is given by appeal to the liberty principle as defending an important right. Since duties to oneself are clearly ruled out by the liberty principle, this also seems to rule out the specific duty actively to care for one's body. But once again, there are some complicating factors which need to be more fully explored.

Recalling Mill's claim that his examples are designed to be illustrative of, rather than definitive of, the liberty principle, some of these examples might serve to draw out the implications of this principle. He says,

> The making himself drunk, in a person whom drunkenness excites to do harm to others, is a crime against others. So, again, idleness ... cannot without tyranny be made a subject of legal punishment; but if, either from idleness or from any other avoidable cause, a man fails to perform his legal duties to others, as for instance to support his children, it is no tyranny to force him to fulfill that obligation (*On Liberty*, 1974-91, vol. 18, p. 295).

Mill also remarks that "every one who receives the protection of society owes a return for the benefit;" in this case the benefit is access to institutions of medical care set up and maintained effectively to guarantee the person's right to adequate health care (*On Liberty*, 1974-91, vol. 18, p. 276). Two related issues are here intertwined. (1) Can the failure of persons to care for their bodies reasonably be construed as a case of harming, or injuring the vital interests of others, if, for example, this failure has the consequence that the person becomes ill and cannot earn the income that his or her dependents require, or, further, the person needs expensive medical care? (2) Can this failure reasonably be construed as a failure to provide a return for the benefits of living in society if the person is unable, because of illness, to contribute a fair share to the cooperative enterprises needed to maintain its institutions?

Although these complications must be considered, I believe that on reflection the Millian conclusion must be that persons cannot be required actively to care for their bodies, and that the degree of coercion that would be needed to enforce this would tip the balance perilously away from liberty. Mill would likely consider this an example of undue encroachment of the "moral police" (*On Liberty*, 1974-91, vol. 18, p. 284) upon the legitimate sphere of individual liberty, and an ineffective way to ensure the cooperation needed to maintain the adequate health care institutions of a decent and humane society.

The Millian framework of utilitarian rights and the principle of liberty is illuminating and helpful for contemporary problems in health care and medical ethics. It provides a strong defense of individual liberty, but it does so by looking at applied questions within a social context and with essential reference to human well-being.

Department of Philosophy
Carleton University
Ottawa, Canada

NOTES

[1] For further development of some of the points of Mill's moral and political philosophy which I raise here, see Wendy Donner, *The Liberal Self: John Stuart Mill's Moral and Political Philosophy* (1991); Wendy Donner, 'Utilitarianism,' in *The Cambridge Companion to Mill*, John Skorupski (ed.), (1997).

[2] John Stuart Mill, *Utilitarianism*, in *The Collected Works of John Stuart Mill*, 33 vols., ed. John M. Robson (1974-91). Mill's *Collected Works* hereafter cited by volume and page only.

[3] David Lyons, 'Mill's theory of morality,' (1976, p. 108). For articles on related themes also see: David Lyons, 'Human rights and the general welfare,' (1977, pp. 113-129); David Lyons, 'Liberty and harm to others,' (1979, pp. 1-19); David Lyons, 'Mill's theory of justice,' (1978, pp. 1-20).

[4] For a fuller treatment of Mill's theory of justice, see Fred Berger, *Happiness, Justice, and Freedom: The Moral and Political Philosophy of John Stuart Mill* (1984), especially pages 123-225.

[5] For further discussion of the problems with this interpretation, see C.L. Ten, *Mill on Liberty*, (1980, pp. 10-14); J.C.Rees, 'A re-reading of Mill on liberty' (1960, pp. 113-29); John Gray, *Mill on Liberty: A Defense*, (1983, pp. 48-57). For a more general discussion, also see John Skorupski, *John Stuart Mill*, (1989, pp. 337-88).

[6] Sara Ann Ketchum, 'Selling babies and selling bodies,' (1992,p. 286). See also Christine Overall, *Ethics and Human Reproduction*, (1987).

BIBLIOGRAPHY

Berger, F.: 1984, *Happiness, Justice, and Freedom: The Moral and Political Philosophy of John Stuart Mill*, University of California Press, Berkeley.

Cohen, G.A.: 1986a, 'Self-ownership, world-ownership, and equality,' in *Justice and Equality Here and Now*, F. Lucash (ed.), Cornell University Press, Ithaca, pp. 108-135.

Cohen, G.A.: 1986b, 'Self-ownership, world-ownership, and equality: Part II,' in E. Paul *et al.* (eds.), *Marxism and Liberalism*, Oxford University Press, Oxford.

Cohen, G.A.: 1995, *Self-Ownership, Freedom and Equality*, Cambridge University Press, Cambridge.

Donner, W.: 1991, *The Liberal Self: John Stuart Mill's Moral and Political Philosophy*, Cornell University Press, Ithaca.

Donner, W.: 1997, 'Utilitarianism,' in *The Cambridge Companion to Mill*, J. Skorupski (ed.), Cambridge University Press, Cambridge, pp. 255-292.

Gray, J.: 1983, *Mill on Liberty: A Defense*, Routledge & Kegan Paul, London.

Ketchum, S.A.: 1992, 'Selling babies and selling bodies,' in *Feminist Perspectives in Medical Ethics*, H.B. Homes and L. M. Purdy (eds.), Indiana University Press, Bloomington, pp. 284-294.

Kymlicka, W.: 1990, *Contemporary Political Philosophy*, Clarendon Press, Oxford.

Lyons, D.: 1976, 'Mill's theory of morality,' *Nous* 10, 101-120.

Lyons, D.: 1977, 'Human rights and the general welfare,' *Philosophy and Public Affairs*, 6, 113-129.

Lyons, D.: 1978, 'Mill's theory of justice,' in *Values and Morals*, A.I. Goldman and J. Kim (eds.), D. Reidel, Dordrecht, pp. 1-20.

Lyons, D.: 1979, 'Liberty and harm to others,' in *New Essays on John Stuart Mill and Utilitarianism*, W.E. Cooper, K. Nielsen and S. C. Patten (eds.), *Canadian Journal of Philosophy Supplementary Volume 5*, 1-19.

Mill, J.S.: 1974-91, *The Collected Works of John Stuart Mill*, 33 volumes, J. M. Robson (ed.), University of Toronto Press, Toronto.

Overall, C.: 1987, *Ethics and Human Reproduction*, Allen & Unwin, Boston.

Nozick, R.: 1974, *Anarchy, State, and Utopia*, Basic Books, New York.

Rees, J.C.: 1960, 'A re-reading of Mill on liberty,' *Political Studies* 8, 113-129.

Skorupski, J.: 1989, *John Stuart Mill*, Routledge & Kegan Paul, London.

Sumner, W.: 1979, 'The good and the right,' in *New Essays on John Stuart Mill*, W.E. Cooper, K. Nielsen and S. C. Patten (eds.), *Canadian Journal of Philosophy Supplementary Volume 5*, 99-114.

Ten, C.L.: 1980, *Mill on Liberty*, Clarendon Press, Oxford.

S. KAY TOOMBS

WHAT DOES IT MEAN TO BE SOME*BODY*? PHENOMENOLOGICAL REFLECTIONS AND ETHICAL QUANDARIES

The practical realm of health care almost inevitably draws our attention to the distinctive relationship between bodies and persons. When we are sick, or when we watch a loved one struggle with pain and debility, we recognize immediately that bodies and persons are inextricably interconnected. In clinical practice disease is not simply something that affects a biological body in the abstract; rather illness is uniquely experienced by the particular person whose body it happens to be. What, then, is the distinctive relation between persons and their bodies?

In this essay I shall consider the phenomenological analysis of body provided in the works of Edmund Husserl, Maurice Merleau-Ponty and Jean-Paul Sartre[1]. I shall suggest that this analysis provides valuable insights with respect to health care issues such as organ transplantation, individual and societal responsibilities for maintaining health, and the debate concerning the desirability and necessity of surgical procedures that result in changes in physical appearance.

I. OBJECTIVE RELATION WITH BODY

In reflecting upon bodily experience phenomenologists distinguish the *lived body*, the body of an experiencing subject, from other physical or animate bodies. According to Edmund Husserl (1982, p. 97) phenomenological reflection discloses the lived body as a privileged kind of object (the "primordial" object) which is distinctively different from all other physical objects in the world. The lived body is distinct in a variety of ways . Most importantly, I am able to do things with it so that it "holds sway" over the surrounding world and I am related to all other realities through the medium of my body.

I am able to do things with my body because – unlike other "material things" – the lived body is experienced as freely and spontaneously moveable, "an organ of the will," the one and only object which one governs (Husserl, 1982, p. 97; Husserl, 1989, p. 159). For example, under

M.J. Cherry (ed.), Persons and their Bodies: Rights, Responsibilities, Relationships, 73–94.
© 1999 *Kluwer Academic Publishers. Printed in Great Britain.*

normal circumstances I can "will" my body to move from my study to the kitchen, or cause my head to turn so that I can look at the tree outside the window. Furthermore, if I wish to examine the back side of the tree which is presently hidden from my view, I can move outside and walk around the trunk in order to make it clearly visible.

As the freely moveable organ of sensation, the lived body is the necessary means by which the world and the objects within it are apprehensible to me. Every "thing appearance" is correlated with certain kinesthetic and tactual sensations:

> In seeing, the eyes are directed upon the seen and run over its edges, surfaces, etc. When it touches objects, the hand slides over them. Moving myself, I bring my ear closer in order to hear. Perceptual apprehension presupposes sensation-contents, which play their necessary role for the constitution of the schemata and, so, for the constitution of the appearances of the real things themselves (Husserl, 1989, p. 61).

The necessary "involvement" of the body in this regard discloses an "if-then" relation with respect to perceived things. "*If* I turn my head and direct my vision to a point outside the window, I can see the tree; *if* I move into the garden, I can hear the birds singing." Thus, every perception is both a sensory experience of the thing perceived *and* an experiential consciousness of the body as one's own.

The lived body is also distinct from other physical objects in that I have limited access to my body from *within*. Kinesthetic sensation gives me a sense of where my body is in space, as well as an immediate sense of connectedness to the body. For instance, when I move my leg, I am conscious of the changing position of the limb (I "know" where the limb is in space) and I am also aware that it is "my" leg (and not another's) that moves. So certain is this experience, that I do not have to look to see where my leg is when I step up the curb, nor do I have any doubt when I reach for a coffee cup that it is "my" arm that reaches.

This consciousness of spatial location, as well as the sense of "ownedness" of the lived body is explicitly recognized when kinesthetic sensation is disturbed. One has only to think of the mundane experience of waking in the night and discovering that an arm has "gone to sleep." In those instances, one experiences the arm as profoundly other, as "deadened" – an object that seems no longer a part of one's body, no longer one's own. This experience of alienation from one's own body is particularly profound in the case of prolonged or permanent sensory loss due to physical trauma or disease (Toombs, 1992b). Individuals who have lost sensation in a limb report that the limb seems "wooden", object-like, detached – so much so that one is apt to refer to "the" leg rather than "my" leg when referring to a "de-sensitized" limb (Murphy, 1987; Toombs, 1995b; Sacks, 1984).

In addition to kinesthetic sensation, I also have some awareness of the interior of the body through visceral sensations such as stomach-aches, pangs of hunger, and so forth. Leder (1990, pp. 38 ff) distinguishes three categories of sensory awareness – "interoception" referring to all sensations of the viscera or internal organs, "exteroception" referring to our five senses that are open to the external world, and "proprioception" referring to our sense of balance, position and muscular tension. Obviously, however, many bodily processes are not available to consciousness (Zaner, 1981, pp. 92-114; Leder, 1992).

The body is also distinguished from other objects in that it has the capacity to be reflexively related to itself. For instance, it is possible to perceive one's hand by means of one's eyes, and to feel one's right arm by touching it with the fingers of one's left hand. Tactual perception reveals a double-sensing. If I stroke my cheek with the fingers of my left hand, the sense of touch is both "on" my cheek and "in" my fingers (Husserl, 1982, p. 96).

With respect to the surrounding world, phenomenological reflection reveals the body as orientational locus, the spatial and temporal center around which the rest of the world is grouped. My body has the central mode of givenness of "Here", whereas all other things (including my fellow human beings) are given as located "There" in relation to my body (Husserl, 1982, pp. 116-117; Husserl, 1989, pp. 61, 165-166). When sitting at my desk, for instance, I note that my coffee cup is "on the right" (i.e., located to the right side of my body within reach of my right hand), yet if I stand up and turn around to search for a book on the shelf, the cup is located "on the left" so that I will need to reach out my left hand to pick it up. As Husserl (1989, p. 166) notes, expressions of orientation, such as "near," "far," "high," "low," refer back to this primary bodily orientation. To say that the coffee cup is "near" is to assert that it is in close proximity to my body and, thus, reachable. The distinctions of "near", "far", "high", "low", always refer not just to one's bodily placement but also to one's peculiar bodily capacities or physique. This reference to bodily capacity becomes explicit in the event that one's body undergoes physical change.

For example, when one uses a wheelchair for mobility, one experiences the third shelf in the grocery store as "high," yet for an upright person it is well within reach and, therefore, unremarkable in this way (Toombs, 1995b). Similarly, for a person six foot seven inches tall, the height of a regular doorframe is "low" requiring a bodily adjustment (bending one's head) in order to go through it, whereas a young child experiences the top of the door as impossibly "high" to be reached only by jumping and stretching the arm above the head.

As the "freely moveable organ of sensation" (Husserl, 1989, p. 61) the body allows one to change position and view physical objects *in an originary fashion* from different spatial perspectives. However, under normal circumstances, one does not have this same freedom with respect to one's own body. Unlike the tree that I can walk around in order to see it from all sides, areas of my body (such as the back of my head) will always be beyond my view (Husserl, 1989, p. 167). I may, perhaps, use a mirror or some other device to "look at" such "hidden" areas of my body but I cannot see them directly. This fundamental limitation extends to the body's interior. Though I may "see" my arteries on the T.V. screen during a medical procedure such as an arteriogram, this "seeing" is unlike the originary perception I have of the palm of my hand as I bring my fingers up towards my nose in order to scratch it. There is a sense in which I have to "take it on trust" that the arteries on the screen are "mine," whereas I have no doubt that "my" fingers are doing the scratching.[2]

With respect to spatial orientation, the impossibility of distancing oneself from one's body is felt in other ways. Although, under normal circumstances, one has the freedom to move away from all other objects in the world, one does not have the power to withdraw from one's body. Thus, there is a symbiotic relation with the lived body – a relation that can be problematic or even threatening in the event that one's body malfunctions (Toombs, 1992a). In this regard Engelhardt (1973, p. 41) argues that the organs of the body are differentiated in terms of the capability that one has to distance oneself from them. It is possible to recognize a distance between oneself and all *replaceable* organs, for instance, yet one cannot separate oneself from one's central nervous system and remain intact.[3] Nor, obviously, can one separate oneself from one's body *in toto*.

The body is distinct from other physical objects in that bodily capacities reveal the experiencing subject's involvement in the world, the domain of practical possibility: the realm of the "I can."

In the physical sphere, all my abilities are mediated by the "operation of my body," by my Bodily abilities and faculties. I know through experience that the parts of my Body move in *that* special way which distinguishes them from all other things and motions of things (physical, mechanical motions); i.e., they have the character of *subjective* movement, of the "I move." And from the very outset this can be apprehended as something practically possible ..... Originally, it is only here that the "I will" emerges (Husserl, 1989, p. 271) [*emphasis mine*].

In distinguishing itself from all other things in this manner, the body announces itself as *belonging to* the experiencing subject – "*I* move," "*I* can." Consequently, just as the sense of "ownedness" of the lived body is disrupted with the loss of kinesthetic sensation, so it is also the case that one becomes subjectively alienated from the body if one loses the ability to move. As a multiple sclerosis patient I can no longer move my legs – although I retain sensation in them. If I sit on the bed and try to lift my legs, I note to myself that "the" legs, rather than "my" legs remain still.[4]

In Husserl's schema, in instances such as this, the lived body as "organ of the will" encounters resistance. With resistance, one concretely experiences the frustration of practical possibility. This frustration can take the form of setting the body in opposition to oneself as experiencing subject:

In the physical sphere, all my abilities are mediated by the "operation of my Body" ... the parts of my Body ... have the character of the "I move".... Originally it is only here that the "I will" emerges .... But here I can also come up against resistance. My hand is "asleep"; I cannot move it, it is temporarily paralyzed, etc. I can also experience resistance in the area of external "consequences" of my Bodily movements. The hand pushes aside what stands in its way, it "gets there." At times it gets there "with difficulty," "with not much difficulty," or "with no resistance." At times it does not succeed at all, the resistance being insurmountable despite all the pushing force (Husserl, 1989, p. 271).

In the experience of "resistance," the experiencing subject comes "up against" the body, or "up against" the world.

As the freely moveable organ of perception, the lived body (the "I can") is, thus, not only the medium through which I apprehend the world but also the means by which I purposefully interact with it. As such, my body is that which enables me to actualize the existential projects that constitute, and express, my personhood and my personal life (Husserl, 1989, pp. 195-96). This facet of the lived body is explicitly recognized when "practical possibility" is frustrated through bodily incapacity. In this instance one experiences not only resistance to bodily involvements but the concrete loss of existential possibility. In thwarting the "I can," the incapacitated body severely disrupts those goals and projects that reflect one's personhood.

The lived body is also experienced as "my" body in the sense that bodily comportment – gestures, ways of moving, facial expressions, tones of voice, and so on – identify the body as "mine" so that I recognize this particular body as my own. Such ways of bodily being also represent my self expression to others (Behnke, 1989). In this regard, Husserl (1989, p. 282) notes the extent to which the social and cultural milieu influences and shapes individual development including modes of bodily comportment:

> The development of a person is determined by the influence of others, by the influence of their thoughts, their feelings ... their commandments. This influence determines personal development, whether or not the person himself subsequently realizes it, remembers it, or is capable of determining the degree of the influence and its character....Besides the tendencies that proceed from other individual persons, there are ... demands of morality, of custom, of tradition, of the spiritual milieu: "one" judges in this way, "one" *has to hold his fork like this*, and so on – i.e., demands of the social group, of the class, etc. [*emphasis mine*].

Thus, bodily comportment is always expressive not merely of the individual concerned but also of the cultural/social/familial milieu that has shaped this particular person (Behnke, 1989).

Husserl's analysis gives us a starting point for thinking about the relation between persons and their bodies: (1) although the lived body may be apprehended as an object (the "primordial object"), it is distinct from other "merely physical" objects in the world, being an "organ of the will" and the medium through which, and by means of which, I apprehend the world and interact with it; (2) the lived body is "owned" in the sense that in a variety of ways it is apprehended as belonging to the experiencing subject, i.e., I am aware that it is "my" body. However, this relation of "ownership" that I have with respect to my body is not a relation of possession in the same way that I "own" other objects such as my automobile or my house; (3) an important feature that differentiates the lived body is its "proximity." I am not able physically to distance myself from my body in the manner that I can distance myself from other objects of the world (although I may separate myself from certain parts of it under certain circumstances); (4) in addition, the body's orientational locus makes it impossible for me to apprehend all aspects of my own body directly, and this limits the experiential awareness I have of my body in a variety of ways; (5) bodily capacities not only reveal the body as "belonging to" the experiencing subject but, in disclosing the realm of practical possibility (the "I can" and "I will"), demonstrate the bodily involvement that makes possible the existential projects that are expressive of personhood (i.e., the "I choose" is to a large extent dependent upon the "I can"); and (6) bodily comportment not only identifies the lived body as "mine" (to me, as well as to other people), but also reflects the body as a social and cultural entity.

II. EXISTENTIAL RELATION WITH BODY

In their phenomenological analysis of the body, Jean-Paul Sartre (1956) and Maurice Merleau-Ponty (1962) each argue that the primary relation with lived body is an existential – rather than an objective – relation. Rather than being an object *of* the world, the lived body represents my particular point of view *on* the world – a point of view on which, as the experiencing subject, I can take no point of view.

In the normal course of events one does not apprehend one's body as a peculiar kind of material object among other objects in the world (Sartre, 1956, pp. 429-30, 436). Indeed, for the most part, one is not conscious of one's body at all. The body *as lived* is that which is perpetually overlooked or "surpassed" in carrying out one's projects. In the act of typing this manuscript, for example, I am not explicitly conscious of my fingers striking the computer keyboard, nor am I aware of the manner in which my eyes function in the act of reading the words as they appear on the screen. In all my bodily involvements the lived body is given only *implicitly* as the "total center of reference" for my world, a center which is indicated but never grasped as such (Sartre, 1956, p. 422).

> ... the body is present in every action although invisible, for the act reveals the hammer and the nails, the brake and the change of speed,

not the foot which brakes or the hammer which hammers. The body is
lived and not *known* (Sartre, 1956, p. 427).

Although the lived body is not the explicit focus of attention, however,
one does have a pre-reflective consciousness of the body as one's own.
My body is not "in front of" me in the same way that objects in the world
are "in front of me," yet it is always "with" me (Merleau-Ponty, 1962).
As I sit here in my study preoccupied with the project at hand, there is a
sense in which I am also attuned to the fact that it is "my" body sitting
here "doing the work."

Not only is the lived body the center of reference which things
indicate, it is the instrument of one's actions, the means by which one
apprehends and interacts with the world. Nevertheless, the instrumental
relation is not an objective one. Under normal circumstances I do not
perceive my body to be an instrument similar to other instruments that I
utilize (the telephone, the pen on the desk, the cup containing my coffee):

> I am not in relation to my hand in the same utilizing attitude as I am in
> relation to the pen; I am my hand ... I can apprehend it – at least in so
> far as it is acting – only as the perpetual, evanescent reference of the
> whole series ... my hand has vanished; it is lost in the complex system
> of instrumentality in order that the system may exist. It is simply the
> meaning and orientation of the system ... . "We do not use this
> instrument for we *are* it" (Sartre, 1956, pp. 426-27).

As orientational locus, the body orients one to the world around by
means of the senses and positions the world in accord with one's bodily
placement. Moreover, the lived body is the locus of one's intentions.
Rather than being an exclusively physical thing devoid of intentionality,
the lived body is an embodied consciousness which engages and is
engaged in the surrounding world (Merleau-Ponty, 1962, p. 79). Not only
do I constantly find myself within the world but I continually move
towards it and organize it in terms of my projects. As orientational and
intentional locus, the lived body represents not simply my bodily being
but my *being-in-the-world*. My embodying organism is always
experienced as "in the midst of environing things, in this or that situation
of action, positioned and positioning relative to some task at hand"
(Zaner, 1981, p. 97).

Merleau-Ponty (1962, p. 138) notes that, in its directedness towards the
world, the lived body exhibits a bodily intentionality. As a practical field
of significance, the world arouses in the lived body certain habitual

intentions (for example, manipulatory movements such as grasping, and so forth). The parts of the body may be understood as "intentional threads" linking it to objects which surround it. Surrounding objects are apprehended as manipulatable or utilizable by the body (Merleau-Ponty, 1962, pp. 81-82). In the action of the hand as it reaches for the book is contained a reference to the object, not as object represented, but as that highly specific thing at which I aim "*in order to*" effect some action. I reach for the book "in order to" look up a particular reference or "in order to" replace it on the shelf in front of me. My body appears to me as "an attitude directed towards a certain existing or possible task" (Merleau-Ponty, 1962, p. 100). The parts of the body, thus, form an *intentional unity* in the worldly engagement of the experiencing subject (Rawlinson, 1986, pp. 42-43).

The parts of the body also form an intentional unity in the sense that the lived body is experienced as an integrated system of coordinated body movements which are distributed among the various body segments. "I do not bring together one by one the parts of my body; this translation and this unification are performed once and for all within me; they *are* my body itself" (Merleau-Ponty, 1962, p. 149.) For example, when I reach for a book on the shelf in front of me, my body coordinates not only the physical movements of the arm but also links tactile and visual sensations – without any conscious effort on my part to effect such coordination. Under normal circumstances this aspect of bodily unity is unnoticed and taken-for-granted. However, when one experiences physical impairment or dysfunction, the disruption of coordination and concurrent experience of disunity is profoundly disturbing (Toombs, 1995b; Toombs, 1994).

As the locus of intentionality, the body has a fundamental value sense (Straus, 1966; Zaner, 1981, pp. 60-63) as that which enables one to carry out one's projects in the world. With respect to bodily comportment, upright posture is of particular significance since it is "at once distinctive of man's corporeal endowment and makes possible the specifically human modes of deportment and the specifically human world" (Zaner, 1981, p. 60; Straus, 1966, pp. 247-248).

In getting up, standing is achieved; and with it, not only does walking become available, but the distinctive forms of seeing, hearing, reaching, grasping, and holding, as well as the primary forms of spatiality, gesturing, and social deportments (Zaner, 1981, p. 61).

The loss of upright posture is, therefore, of particular significance in illness (Toombs, 1992a; Toombs, 1992b; Toombs, 1995b; Straus, 1966, p. 139; Zaner, 1981, p. 264).

The existential relation with body is also evidenced in the distinct bodily patterns that identify my body as peculiarly *me*. That is, certain unique ways of moving, gesturing, speaking, laughing, and so forth, express *my* bodily being (Merleau-Ponty, 1962, p. 150). When I catch sight of my reflection as I move past a store window, or check my appearance in a mirror, I immediately recognize not only "a" physical body, or even "my" physical body, but rather I see an embodied being who is *ME* and nobody else. In this respect, Merleau-Ponty (1962, pp. 194-95) notes that there is a certain ambiguity with respect to the lived body: it is at once subject and object being both the expression and the expressed of my existence. While the lived body is that which is most intimately *me* and *mine* (the body which *I am*), it is an object for others.

Sartre (1956) also notes that, while the primary pre-reflective relation with one's body is an existential one, there are ways in which the lived body is apprehended as an object.[5] While I do not intend to discuss Sartre's analysis in detail, for the purposes of this paper it is important to note Sartre's insight that the lived body is apprehended as an object whenever it is conceived to be a neurophysiological organism. In the normal course of events, I do not experience my body as a physiological organism (a skeleton, brain, nerve endings and so forth). Indeed, there is a sense in which it is quite impossible for me to do so, given that there are many bodily processes and functions of which I can have no direct awareness (Leder, 1990; Sartre, 1956, p. 466). Consequently, in those circumstances when I do conceive it to be a physiological organism, I have shifted from the level of immediate experiencing to a level of objectification and abstraction that is dependent upon acquired knowledge such as the principles of physiology and pathology.

As a multiple sclerosis patient, for example, I know (because a doctor has told me and I have researched the matter) that my inability to move my legs is the result of a demyelinating process taking place in my central nervous system. However, I do not experience the demyelinating process itself, nor do I experience the lesions in my central nervous system – indeed, there is no possibility that I could ever do so. Even if the lesions are pointed out to me on a CT scan, they remain ineffable. I have to take it on trust that the picture on the screen is, in fact, a depiction of *my* central nervous system, as opposed to someone else's, and that the lesions are occurring in *my* body.

Sartre contends that whenever one apprehends the lived body as an object, one experiences a sense of alienation from one's own body.[6] I would add that this sense of alienation is, perhaps, particularly manifest when one explicitly recognizes the "hidden" physiological processes occurring within one's body. As Straus (1970, p. 139) notes:

> My hands are parts of my body. Yet other parts – the heart, the adrenal glands, the reticular activating system – are not mine in the same sense, for two reasons: (1) Not immediately acquainted with them I only know them through anatomical studies and instruction; (2) *I do not master them; as a creature they possess me [emphasis mine].*

The awareness of alienation from body is particularly profound in illness when the sick person apprehends the lived body not just as an object – a neurophysiological organism – but, more importantly, as a *malfunctioning* neurophysiological organism that is beyond personal control. There is a chilling sense in which one is "at the mercy of" one's body. I do not simply "have" a serious illness, it also "has" me. Consequently, although I find myself alienated from my body, I am paradoxically aware that there is a symbiotic relation between body and self. If I discover that there is a malignant lump in my breast, for example, I do not immediately think with reference to myself, "My *body* has cancer. My *body* is going to die."

In summary, the existential relation with body is disclosed in the following manner: (1) in the normal course of events I do not apprehend my body as an object for me-as-subject. Indeed, I am not explicitly conscious of my body as I interact with the world. My attention is directed to the task at hand, rather than to my body's involvement in carrying out the task. Rather than being an object *of* the world, my body is my particular point of view *on* the world – the means by which I apprehend and interact with the surrounding environment; (2) although my body is an "instrument" I do not "use" it in the same way that I "use" other tools; (3) in all its worldly involvements, the lived body exhibits a bodily intentionality that reveals a dynamic relation between body/world; (4) the parts of the body form an intentional unity that *is* my body and makes possible my worldly involvements; (5) as intentional unity the body has a fundamental value sense especially with respect to upright posture; (6) as an embodied being I exhibit a certain corporeal style –

ways of walking, gesturing, and so forth, that identify the lived body as
ME; (7) the "ownedness" of body is evident but not explicitly thematized
since my body is not *in front of* me yet always *with* me. This
"ownedness" is not a relation of possession but rather one of existing; (8)
the primary existential relation with body is ambiguous in the sense that
the body can be subject for me and object for others; and (9) when one's
body is conceived as a neurophysiological organism (a "hidden presence"
that is necessarily beyond one's control) the existential relation is
disrupted in that the lived body is apprehended as an object by the
experiencing subject. The apprehension of body-as-object can be deeply
alienating, especially if it is the result of bodily breakdown and
malfunction.

III. IMPLICATIONS FOR HEALTH CARE AND MEDICAL ETHICS

The foregoing phenomenological analysis of embodiment is relevant to
controversies in medical ethics and health care that raise issues about the
relationship between persons and their bodies. In the remainder of this
essay, I should like to reflect upon some of these controversies.

A. Organ Transplantation

Developments in medicine have made it possible to donate body parts
(such as a spare kidney), and to receive body parts from another person
(e.g., by undergoing a heart transplant). As Renee Fox and Judith Swazey
(1992, p. 13) pointed out in their essay, 'Leaving the field,' these
developments have resulted in an aggressive "movement toward the
'commodification' and 'marketification' of the organs." There are
ongoing discussions about the merits of markets in human body parts or
"HBPs" as they are often referred to by medical professionals (Fox and
Swazey, 1992, p. 10). Given the scarcity of available organs for
transplantation, in March 1995 the Council on Ethical and Judicial Affairs
of the AMA (Glasson, 1995) proposed that some sort of future contracts
program for cadaveric donors be instituted that offers financial incentives
for organ donation.

Such developments inevitably raise the issue of the relation between
human beings and their bodies. Are bodies and body parts akin to other
items of personal "property" that can be sold or traded by the "owner"

(presuming he or she does so as an act of free will)? Is the human body merely a biological entity, so that organs are "just organs" as opposed to living parts of a person that "resonate with the symbolic meaning of our relationship to our bodies, ourselves, and to each other" (Fox and Swazey, p. 13)?

The phenomenological analysis of the lived body makes it clear that: (1) by its very nature the lived body is distinct from all other kinds of material objects or instruments in the world and it cannot simply be conceived as akin to other items of personal property; (2) to claim that the lived body is "mine" is not to claim that I "own" it as a physical possession, in the same way that I "own" my dog or my house. For Husserl, the recognition that the lived body is "mine" is a felt apprehension or experiential awareness of the lived body.[7] For Sartre and Merleau-Ponty the "ownedness" of lived body is a fundamental feature of embodiment that characterizes being-in-the-world and that represents a relation of existing rather than having; (3) phenomenological analysis discloses that, as lived, the body is not simply a machine-like entity comprised of separate organ systems and parts. Rather the lived body is an *intentional unity* that makes possible the worldly involvement of a unique experiencing human being. Consequently, if one conceives of transplanted organs as simply replaceable "HBPs" or "just organs," one has moved from the level of immediate experience to a level of abstraction that ignores what it means to be an embodied human being. This reduction is problematic in the context of clinical care which, by its very nature, is devoted to the care of sick *persons* and not simply biological entities.

B. In-corporation of Transplanted Organs

An important question regarding the relation between persons and their bodies concerns the in-corporation of transplanted organs. Is it the case that a donor's organs are (or can be) fully in-corporated into the recipient's body in such a way that the transplanted organ is apprehended by the transplant recipient as fully and wholly "mine"? This question regarding the recipient's bodily experience is important in terms of the meaning that such procedures have for individual patients – a meaning that will, of course, affect a patient's response to the transplant procedure.

With respect to the in-corporation of organs, in writing about the human experience of heart transplant, Clouse (1989, p. 18) notes:

There have been cases of men getting worried and depressed after they learned they were carrying a heart transplanted from a woman. Bigots have worried they might receive a heart from a person of another race. And doctors have reported some patients wondering if the donor "had Jesus in his heart."

Five years after his own successful transplant, Clouse uses the terminology "*the* 'old' heart," and "*the* 'new' heart," and "my transplanted organ" to refer to his (own) heart.

The phenomenological account of lived body reveals that, under normal circumstances, one experiences a sense of distance from those parts of the body that are experientially "absent." Not having immediate experience of the heart, the liver, the kidneys, I do not have the same sense of "ownedness," as I do with, say, "my" arm as I move it in reaching for a book. Moreover, *malfunctioning* organs are experienced as other-than-me in the sense that they "have me in their grips," and are essentially beyond "my" control.

What appears to be distinct about transplanted organs is that, in addition to being other-than-me in the sense that they represent the "hidden presence" of the body, they retain a peculiar kind of "otherness" by virtue of the fact that they are perceived as having "belonged to" another human being. The difficulty of in-corporation of such organs is, therefore, profound since it is rooted in the human experience of embodiment.

In contrast, one might note the manner in which certain prosthetic devices are in-corporated into the body. Merleau-Ponty (1962, p. 143) notes that, in the normal course of events through the performance of habitual tasks, we in-corporate objects into bodily space. For example, a woman who routinely wears a hat with a long feather on it intuitively allows for the extension of the feather when she goes through a doorway – just as a 6' 7" basketball player allows for his height by routinely bending his head forwards to avoid bumping into the door frame. Similarly, with habitual use, prosthetic devices are in-corporated; for instance, a blind person's cane is experienced as an extension of the body increasing bodily range. The point of the stick is "an area of sensitivity, extending the scope and active radius of touch, and providing a parallel to sight" (Merleau-Ponty, 1962, p. 143).

As a person who routinely uses a wheelchair, I have in-corporated the device into my body in the sense that I routinely allow for the extra width of the wheels when going through a doorway, I no longer have to "think" what body movements to make in order to move backwards or forwards. My wheelchair has become an integral part of my body – so much so that, if someone comes up to me and starts pushing it without my permission (as they do quite often in airports, shopping malls, etc.), I feel that my bodily space has been violated.

I have argued elsewhere that health care professionals should pay close attention to the manner in which prosthetic devices are in-corporated since this has important implications for therapy (Toombs, 1994). I would suggest that it is equally important that therapists develop a thorough understanding of the ways in which the human experience of in-corporation differs with respect to various types of transplanted organs, "internally" situated prosthetic devices (such as pacemakers), and "externally" located prostheses (such as artificial limbs, canes, wheelchairs, etc.).

C. Obligations with Respect to Maintaining "Health" and a "Healthy Body"

An enduring controversy with respect to health care is whether or not persons have an obligation to "take care of" their own bodies and to live more "healthful" lives. While I do not intend to focus directly on the question of whether or not individuals are obliged to care for their bodies in certain ways, I will show that different conceptions of body have implications in the debate over "health" and personal responsibility.

The phenomenological account of lived body provides the insight that the relation between person and body is a relation of *existing* (and not "having"). This insight can motivate a sense of personal responsibility for the choices that characterize the unique ways in which one "lives" one's body. The recognition that *I AM* my body focuses one's attention on the role of personal participation in the prevention of illness and treatment of disease. In addition, in rendering explicit the dynamic relation between body/self/world, the phenomenological account underscores the necessity of paying explicit regard to the role of the environment in enhancing (or endangering) the "health" of individual members of society – thus, motivating a societal, as well as individual, sense of responsibility for personal well-being. Moreover, phenomenology reminds us that the notion of "environment" includes more than simply physical environs. Rather, if we are to take seriously the role of the individual's environment in the promotion of "health," we must pay explicit attention to the social, psychological, spiritual (as well as physical) dimensions of the person's life.

It should be noted, however, that the prevailing Western biomedical model of disease, along with the concurrent emphasis on technology dominated medicine, does not encourage individuals to take personal responsibility for their own "health." Rather, in construing the body in largely mechanistic terms, Western "scientific" medicine encourages individuals to *give up* personal control and to "hand over" their bodies to medical experts to be "fixed" and returned to them in "good working order." In addition, the biomedical model de-emphasizes the crucial interconnectedness between body/self/world (being-in-the-world) with the consequence that little attention is paid to the multifaceted ways in which the lifeworld of a particular individual has a direct impact on his or her well-being (Leder, 1992b).

Scientific and cultural definitions of what constitutes "health" are, of course, intimately related to scientific and cultural conceptions of body. In our society "health" is equated with an ideal of complete physical and mental well-being (Toombs, 1995a, p. 13). This societal view of what constitutes "wellness" is, for the most part, predicated on the biomedical model of illness and the corresponding conception of the body as a purely biophysiological entity. This particular view of "health" significantly influences the allocation and utilization of financial resources devoted to health care and medical research, in addition to determining the focus of clinical medicine and the priorities within a particular health care system (Sheets-Johnstone, 1992; Leder, 1992a). Thus, the predominant focus of medical care, research, and medical education in our society is directed towards acute care, with a corresponding emphasis on "rescue," "waging war" on disease, and prolonging biological life at all costs. This emphasis on acute illness and the aggressive intervention against disease shifts the primary focus of medical care away from prevention of illness and, thus, does little to motivate individuals to take responsibility for their own well-being.

D. Bodily Alteration and Personal Identity

There is considerable debate about the desirability and necessity of some surgical procedures and medical treatments that alter physical appearance (such as breast augmentation, radical mastectomy, liposuction, growth hormone therapy, and so forth). In what follows I will suggest that a phenomenological account of embodiment provides valuable insights that must be taken into account when considering these types of interventions.

As Merleau-Ponty has noted (1962, p.150), the lived body manifests one's being-in-the-world in the sense that distinct bodily patterns (ways of moving, gesturing, etc.) and a certain physical appearance express a unique corporeal style that identifies the lived body as peculiarly "me." Thus, not only do I experience my body as that which is most intimately *me* and *mine* (the body which *I AM*), but my body is that which identifies "me" to others. Moreover, personal bodily comportment reflects the social and cultural milieu since ways of being-in-the-world are, to some degree, necessarily molded by the demands and values of the social group (Husserl, 1989; Merleau-Ponty, 1962). In the context of medical decision making, two aspects of Merleau-Ponty's analysis should be considered: (1) the personal dimension of embodiment (i.e., the strong sense that one has of personal/bodily identity; (2) the interpersonal dimension of embodiment.

With reference to the personal dimension of embodiment, I have argued elsewhere that diseases or medical treatments that alter corporeal style represent a threat to one's sense of personal identity (Toombs, 1992a). Altered patterns of movement or changes in physical appearance cause one to experience the body as unfamiliar and unrecognizable – as no longer one's own (Toombs, 1995b, p. 16). As a person with progressive neurological disease, for example, I have difficulty identifying with my changed (and continually changing) ways of moving. When I see myself on a home video, and observe my disordered body style, I experience a sense of puzzlement. I catch myself wondering not so much whether the *body* projected on the screen is my body but, rather, if that person in the video is really *ME*?

It should be noted that it is the *change* in corporeal style that precipitates loss of self identity. For instance, a woman with a congenital disability explains why she *resisted* her mother's attempts to make her appear more typical:

> She (her mother) made numerous attempts over the years of my childhood to have me go for physical therapy and to practice walking "more normally" at home. I vehemently refused all her efforts. She could not understand why I would not walk straight. ... My disability, with my different walk and talk and my involuntary movements, having been with me all my life, was part of me, part of my identity. With these disability features, I felt complete and whole. My mother's

attempt to change my walk, strange as it may seem, felt like an assault on myself, an incomplete acceptance of all of me, an attempt to make me over (Rousso, 1984).

It is essential that medical professionals be fully aware of the profound identity issues that are connected with bodily alteration. Treatments that result in physical change often directly threaten personal integrity (Asch and Fine, 1988, pp. 23-26).

As Iris Young (1992, p. 227) notes in the case of mastectomy, "for many, if not most, women, breasts are an important aspect of identity." Yet, the "integration of breasts with a woman's self is seriously denied."

> Phenomenologically, the chest is a center of a person's being in the world and the way she presents herself in the world, so breasts cannot fail to be an aspect of her bodily habitus. For many women, breasts are a source of sexual pleasure or bodily pride. Many women emotionally locate important episodes in their life history – such as coming to adulthood or having children – in their breasts. ... (but) ... In conformity with Western medicine's tendency to objectify the body and to treat the body as a conglomerate of fixable or replaceable parts, a woman's breast is considered to be detachable, dispensable (Young, 1992, p. 227).

Young notes that it is vital to recognize that breast loss is intimately related to issues of personal identity and bodily self image since studies show that many women who lose a breast suffer serious emotional distress that is undetected and untreated.

With respect to the interpersonal dimension of corporeal style, it is perhaps obvious that cultural norms and ideals with respect to physical appearance can (and do) threaten self integrity. While in the cultural imagination persons and their bodies are in some ways considered to be distinct – so that, if asked, the "man in the street" will avow wholeheartedly that he is "more than his body" – yet, at the same time people act towards strangers as if *persons* and *bodies* were one and the same. Language reflects this perceived symbiosis in the common utterance: "I want to be '*somebody*' one day."

This identification of body with person is so natural that the defense in the now notorious O. J. Simpson trial objected to an unflattering photograph of Simpson's face on the grounds that seeing his physical appearance would, in and of itself, cause the jury to make a negative judgment about his character. Similarly, trial commentators observed that if the prosecution were to get a conviction from the jury they would first have to "overcome" Simpson's winning smile and his appealing bodily presence in the courtroom.

The distinction between body and person is routinely overlooked in normative judgments with respect to bodily appearance. Negative judgments about someone's physical appearance are, more often than not, accompanied by negative judgments concerning personhood (Leder, 1992a).

Such negative judgments are overtly displayed in the social response to persons with disabilities. Ours is a culture that places great emphasis on physical fitness, sexuality, productivity, and youth – an ideal that is associated with a particular bodily physique. Consequently, there is little tolerance for physical difference or deviation from the ideal (Asch and Fine, 1988). As I have noted elsewhere regarding the response of others to persons with disabilities (Toombs,1995a):

> When strangers look at me what they see is the wheelchair. They make the immediate judgment that my quality of life is diminished and that my situation is an essentially negative and unhappy one. In the first place, because my mobility is limited, strangers invariably conclude that my intellect must be likewise affected. For instance, in my presence, strangers usually address questions to my husband and refer to me in the third person – "What would *she* like to drink? Where would *she* like to sit?" Many people who do not know me believe that, because I cannot walk, I am largely dependent on others and unable to engage in professional activities.

Obviously, such common responses to one's physical appearance directly cause one to experience feelings of diminished self-worth (Goffman, 1963; Toombs, 1995c).

Negative judgments about body/self are, of course, not confined to persons with disabilities. Feminist thought illuminates the myriad of ways in which women's bodies are objectified and judged in a culture that uses bodily appearance as a criterion of success and self worth,[8] and essays on aging draw attention to the loss of self esteem experienced with the physical changes engendered by age (Cole and Gadow, 1986).

This negative response from others is, of course, a major determinant in the personal decision to undergo some kind of surgical procedure to change one's physical appearance. A television news broadcast in August 1995 noted a major increase in the number of U.S. males who undergo elective cosmetic surgery. The major reason given by those interviewed was: the desire to look "younger" in order to be competitive in the job market.

As Young (1992, p. 226) notes with respect to breast augmentation surgery for women, while the decision to choose elective surgery may be a "rational response" given prevailing social pressures, there is some question as to the extent to which those undergoing surgical procedures of this nature can be said to be exercising choice. With respect to augmentation surgery:

> Phallocentric norms do not value a variety of breast forms, but rather elevate a standard; women are presented culturally with no choice but to regard our given breasts as inferior, puny, deflated, floppy.

Similarly this culture – with its emphasis on youthfulness and productivity – does not value individuals of either sex who have passed a certain age limit.

Given the interpersonal dimension of embodiment, it is obvious that societal judgments with respect to body/self have a major influence in patients' decisions to undergo certain kinds of procedures (such as breast augmentation, liposuction, cosmetic surgery). Societal judgments also deeply affect patients' responses to disease or medical treatments that result in bodily change. At the same time, given the personal dimension of embodiment, it is very important to note that one necessarily experiences *any* significant alteration in the body as a change in corporeal style – a change that raises profound identity issues. Both these dimensions of embodiment must be taken into account when considering medical interventions that alter "physical" appearance.

Department of Philosophy
Baylor University
Waco, Texas

NOTES

[1]   In the interests of limiting the project, I have chosen to focus the major part of my analysis only on the work of these particular philosophers.

[2]   For a fascinating discussion of this difference see Robillard's discussion of monitoring his body's functioning on the T.V. screens in the I.C.U. unit (Robillard, 1994).

3   As is the case with all organs that are irreplaceable, the sensed "inseparability" of the central nervous system is important with respect to diseases involving such organs and systems (Toombs, 1992b).

4   The "I cannot" is also evidenced in the inability to control inner structures of the body – the viscera. For an extended discussion with respect to the "inner body," see Leder, 1990, pp. 36-68.

5   For a full discussion of the various ways in which the body is experienced as an object, see Toombs, 1992a, pp. 51-88.

6   There are experiences of bodily objectification that are not necessarily alienating – such as the reflective awareness of body that occurs in such circumstances as participating in dance, sporting activity, sexual experience, and so forth. However, in illness, bodily objectification is particularly alienating (Toombs, 1992a, pp. 70-81.)

7   For an excellent discussion with respect to "ownership" of the lived body, see Gallagher, 1986.

8   While there are any number of essays on this topic, for a good example with respect to medical thought see Bordo, 1992.

BIBLIOGRAPHY

Asch, A. and Fine, M.: 1988, 'Introduction: Beyond pedestals,' in M. Fine and A. Asch (eds.), *Women With Disabilities: Essays in Psychology, Culture and Politics*, Temple University Press, Philadelphia, pp. 1-37

Behnke, E.A.: 1989, 'Edmund Husserl's contribution to phenomenology of the body in Ideas II,' *SPPB Newsletter* 2, 5-18.

Bordo, S.: 1992, 'Eating disorders: The feminist challenge to the concept of pathology,' in D. Leder (ed.), *The Body in Medical Thought and Practice*, Kluwer Academic Publishers, Dordrecht, The Netherlands, pp. 197-213.

Cole, T.R. and Gadow S. (eds.).: 1986, *What Does It Mean to Grow Old: Reflections from the Humanities*, Duke University Press, Durham.

Clouse, R.G.: 1989, 'A new heart in the face of old ethical problems,' *Second Opinion* 12, 11-26.

Engelhardt, H.T., Jr.: 1973, *Mind-Body: A Categorial Relation*, Martinus Nijhoff, The Hague, The Netherlands.

Fox, R.C. and Swazey, J.P.: 1992, 'Leaving the field,' *Hastings Center Report* 22, 9-15.

Gallagher, S.: 1986, 'Lived body and environment,' *Research in Phenomenology* 16, 139-170.

Glasson, J., *et al.*: 'Financial incentives for organ procurement,' *Archives of Internal Medicine* 155, 581-589.

Goffman, E.: 1963, *Stigma: Notes on the Management of Spoiled Identity*, Prentice Hall, Inc., New Jersey.

Husserl, E.: 1989, *Ideas Pertaining to a Pure Phenomenology and to a Phenomenological Philosophy: Second Book: Studies in the Phenomenology of Constitution*, R. Rojcewicz and A. Schuwer (trans.), Kluwer Academic Publishers, Dordrecht, The Netherlands.

Husserl, E.: 1982, *Cartesian Meditations: An Introduction to Phenomenology*, D. Cairns (trans.), 7th Impression, Martinus Nijhoff, The Hague, The Netherlands.

Leder, D.: 1992a, 'A tale of two bodies: The Cartesian corpse and the lived body,' in D. Leder (ed.), *The Body in Medical Thought and Practice*, Kluwer Academic Publishers, Dordrecht, The Netherlands, pp. 17-35.

Leder, D.: 1992b, 'Introduction,' in D. Leder (ed.), *The Body in Medical Thought and Practice*, Kluwer Academic Publishers, Dordrecht, Holland, pp. 1-12.

Leder, D.: 1990, *The Absent Body*, The University of Chicago Press, Chicago.

Merleau-Ponty, M.: 1962, *Phenomenology of Perception*, C. Smith (trans.), Routledge and Kegan Paul, London.

Murphy, R.F.: 1987, *The Body Silent*, Henry Holt, New York.

Rawlinson, M.C.: 1986, 'The sense of suffering,' *The Journal of Medicine and Philosophy* 11, 39-62.

Robillard, A.B.: 1994, 'Communication problems in the intensive care unit,' *Qualitative Sociology* 17, 383-395.

Rousso, H., 1984: 'Fostering healthy self-esteem,' *The Exceptional Parent*, pp. 9-14.

Sacks, O.: 1984, *A Leg to Stand On*, Summit Books, New York.

Sartre, J.P.: 1956, *Being and Nothingness: A Phenomenological Essay on Ontology*, H.E. Barnes (trans.), Pocket Books, New York.

Sheets-Johnstone, M.: 1992. 'The materialization of the body: A history of Western medicine, a history in process,' in M. Sheets-Johnstone (ed.), *Giving the Body Its Due*, SUNY Press, New York, pp. 132-158.

Straus, E. W.: 1970, 'The phantom limb.' in E. Straus and R. Griffith (eds.), *Aisthesis and Aesthetics*, Duquesne University Press: Pittsburgh.

Straus, E. W.: 1966, *Phenomenological Psychology. Selected Papers*, E. Eng (trans.), Basic Books, New York.

Toombs, S.K.: 1995a, 'Chronic illness and the goals of medicine,' *Second Opinion* 21, 11-19.

Toombs, S.K.: 1995b, 'The lived experience of disability', *Human Studies* 18, 9-23.

Toombs, S.K.: 1995c, 'Sufficient unto the day: A life with multiple sclerosis,' in S.K. Toombs, D. Barnard and R.A. Carson (eds.), *Chronic Illness: From Experience to Policy*, Indiana University Press, Indiana, pp. 3-24.

Toombs, S.K.: 1994, 'Disability and the self,' in T.M. Brinthaupt and R.P. Lipka (eds.), *Changing the Self: Philosophies, Techniques, and Experiences*, SUNY Press, New York, pp. 337-357.

Toombs, S.K.: 1992a, *The Meaning of Illness: A Phenomenological Account of the Different Perspectives of Physician and Patient*, Kluwer Academic Publishers, Dordrecht, The Netherlands.

Toombs, S.K.: 1992b, 'The body in multiple sclerosis: A patient's perspective,' in D. Leder (ed.), *The Body in Medical Thought and Practice*, Kluwer Academic Publishers, Dordrecht, The Netherlands, pp. 127-137.

Young, I.M.: 1992, 'Breasted experience: The look and the feeling,' in D. Leder (ed.), *The Body in Medical Thought and Practice*, Kluwer Academic Publishers, Dordrecht, Holland, 215-230.

Zaner, R.M.: 1981, *The Context of Self: A Phenomenological Inquiry Using Medicine as a Clue*, Ohio University Press, Ohio.

ALLYNE L. SMITH, JR.

AN ORTHODOX CHRISTIAN VIEW
OF PERSONS AND BODIES

What is the Christian view of the relationship between persons and their bodies? The question as formulated contains two problematic notions. First, there is arguably no one Christian view because there is not one Christian denomination but many, each with its own distinctive theological tradition and sometimes even different scriptures.[1] Thus there is no one Christian approach to answering this question. Just as there is no generic Christianity, there can be no generic Christian theological anthropology or ethic of embodiment. The only way to proceed in this essay is for the author to speak from his own tradition, that of Orthodox Catholicism. This is the faith associated with the seven ecumenical councils of the first eight centuries of the Christian era and the patriarchates of the East. One advantage of speaking from this tradition is that Orthodox theology is heavily dependent upon patristic texts which are shared and valued (though to lesser degrees) by Roman Catholics, Anglicans, and some Protestants.

The second problem with the question as posed is that it skews the answer, for it will be the task of this essay to show that the Orthodox viewpoint insists that one cannot legitimately speak of persons as distinct from their bodies; rather, the Orthodox Christian views the person as an irreducible whole of body, soul, and spirit (or, more commonly, body and soul).

I. THE SCRIPTURAL ACCOUNT

The people of the first Covenant, the Jews, have their story recorded in the Hebrew Testament and in the Greek translation of it known as the Septuagint. Jewish conceptions of embodiment underlie Christian notions as they are developed in the New Testament and later, and they therefore serve as the originating hermeneutic field for subsequent Christian understandings.

Although an exegesis of the relevant Old Testament passages cannot be accomplished in the present essay, it is nonetheless possible to

M.J. Cherry (ed.), Persons and their Bodies: Rights, Responsibilities, Relationships, 95–108.
© 1999 *Kluwer Academic Publishers. Printed in Great Britain.*

summarize relevant aspects of the first Testament's theological anthropology.[2] The Jewish authors use a variety of terms to translate different aspects of the human person in different contexts, including terms that are usually rendered in English as "flesh", "soul", and "spirit". The multiplication of terms can be misleading, and the (mis-) translation of *nephesh* as "soul" implicitly insinuates an alien view (that of pagan Greek dualism) into many accounts of the Hebrew texts. Whatever else more one might wish to say about the view of the first Testament, it is at least clear that it holds the human person to be a psychosomatic unity. This Hebrew insight into the nature of the human person must be kept in mind as we turn to the New Testament.

Again, exegesis is not possible in this essay, but it is possible to summarize the New Testament account. The Apostle Paul, writing in 1 Thessalonians, speaks of "spirit and soul and body" (5.23), translating the Greek words *pneuma*, *psyche*, and *soma*. More commonly in the tradition, one finds references to simply "body and soul", the latter term collapsing the meanings of spirit (the spiritual aspect of the person) and soul (the mind). Nothing in this text, however, licenses the simplistic reading which would have Paul describing the human person as composed of three distinct and separable parts, much less the current notion that the person is something other than an indivisible unity of body and soul.[3] Rather these are three different aspects of the person, three different ways of looking at this mystery made in the image and likeness of God (Genesis 1.26).[4] Indeed, in his classic study of Paul's anthropology, the late Anglican New Testament scholar John A. T. Robinson concluded that for Paul, the human person is "flesh-animated-by-soul, the whole conceived as a psycho-physical unity" (Robinson, p. 14). The body (*soma*) is a psychosomatic unity, an inseparable unity of soul (*psyche*) and flesh (*sarx*).

Much more could be said about the relevant terms, but "a biblical anthropology cannot rest on philology alone, for the usage is finally elastic and eclectic.[5] A biblical view of persons and of their embodiment is not given with vocabulary and concepts alone; it is given with a narrative" (Verhey, p.10). In his essay on the biblical view of the body, Allen Verhey summarizes this narrative thusly: "To be formed by this literature is not to yearn for an escape from the world or for rescue from our bodies; it is to hope for the redemption of the world and of embodied selves" (Verhey, p. 14).

II. PATRISTIC WITNESS

That Paul should be taken as affirming this unity is confirmed not only by its congruence with the rabbinical tradition in which he was trained but also by an examination of the patristic tradition. One Greek Orthodox theologian, after surveying the patristic testimony, summarized and affirmed that tradition as holding it as a "central Christian truth that the body and the soul together 'constitute' the 'natural' man" (Nellas, p. 46). That he has judged correctly may be ascertained by several examples from the tradition. An example from the early apologists is Justin Martyr (2nd century) who writes:

> For what is man but the rational animal composed of body and soul? Is the soul by itself man? No, but the soul of man. Would the body be called man? No, but it is called the body of man. If, then, neither of these is by itself man, but that which is made up of the two together is called man, and God has called man to life and resurrection, he has called not a part but the whole, which is the soul and body.[6]

One may go further. The Church fathers held differing opinions as to the nature of the *imago Dei*, a term

> ... enriched with the most varied meanings, corresponding each time with the problems which have to be faced. Sometimes, for example, the expression "in the image" refers to man's free will, or to his rational faculty, or to his characteristic of self-determination, sometimes to the soul along with the body, sometimes to the mind, sometimes to the distinction between nature and person, etc., and sometimes comprehensively to the whole man (Nellas, p.22).

It is to this last understanding of image that I wish to point as the fullest Orthodox conception of the meaning of the person (and therefore of the understanding of the body).[7] One finds this in various fathers, including the first great theologian of the West, St. Irenaeus of Lyons (2nd century):

> By the hands of the Father, that is, by the Son and the Spirit, the human person was created in the likeness of God. The person was so created, not just a part of the person. Now soul and spirit are certainly a part of the person, but they are not the person as such. For the complete person consists in the commingling and union of the soul that receives

the spirit (or breath) of the Father, together with the flesh (or physical nature) that is fashioned according to God's image.[8]

This view is echoed later in the Orthodox tradition in such writers as Niketas Choniates (d.127) who wrote that "The term human being is applied not to the soul alone or to the body alone, but to both of them together; and so it is with reference to both together that God is said to have created the human person in His image."[9] And Kallistos Ware, summarizing the teaching of St. Gregory Palamas, says that for Palamas "it is the total person – body, soul, and spirit together – that participates in the vision of the Divine Light; the divine energies transfigure not only our inner self but also our physicality (Ware 1996a, p. 9).

Moreover, one finds in the Orthodox tradition an acknowledgment that since the human *psyche* is in the divine image, this includes not only the conscious mind but the unconscious as well. Ware notes that

> ... as persons in the divine image we relate to God not only through the feelings and emotions of which we are fully aware but also through the deeper levels of our inner self that elude the scrutiny of the conscious, thinking mind. God speaks to us more particularly through our dreams, as Scripture frequently indicates; there is in the patristic tradition, especially in the writings of Evagrios of Pontos (d. 399), an elaborate discussion of dream interpretation that contemporary Orthodox psychologists might profitably explore (Ware 1996a, p. 8).

While acknowledging that some Church fathers "such as Clement of Alexandria and Evagrios of Pontos, adopt a negative, Stoic view of the passions, condemning them as a sinful distortion of true personhood," the Orthodox tradition also includes a much more positive assessment. For others

> the passions are impulses implanted in our nature by God that are open to misuse but are also capable of being employed to God's glory. A monastic author such as Abba Isaias (d. ca. 491) holds that anger, for example, can be employed in a positive way against the demons; jealousy can be transformed into zeal for righteousness; even pride can be put to good use if it leads us to affirm our meaning and value in God's eyes when assailed by self-hatred and despair (p. 26). For St. John Klimakos (7th century), physical *eros* is a true "paradigm" of our love for God (p. 27). St. Maximos refers to "the blessed passion of holy love," (p. 28) and St. Gregory Palamas speaks of "divine and

blessed passions" and maintains that our aim should be not the mortification" of the passions but their redirection or "transposition (Ware, 1996a, pp. 9-10).

III. CONTEMPORARY ORTHODOX

A. Christos Yannaras

We will answer the question, "What is the body and what is the soul of man?" with the criteria of the ecclesial tradition: Both the body and the soul are energies of human nature, that is the modes by which the event of the hypostasis (or personality, the ego, the identity of the subject) is given effect. What each specific man *is*, his real existence or his hypostasis, this inmost *I* which constitutes him as an existential event, is identified neither with the body nor with the soul. The soul and the body only reveal and disclose what man *is*; they form energies, manifestation, expressions, functions to reveal the hypostasis of man (Yannaras, p. 63).

B. Kallistos Ware

Kallistos Ware summarizes the Orthodox tradition concerning the body thusly:

Whatever our specific theology of the divine image, one thing is surely evident. The body is integral to our personhood and central to our life in Christ. It is a "temple of the Holy Spirit" (1 Cor. 6: 19) through which we "glorify God" (1 Cor. 6:20) and which we offer to Him as a "living sacrifice" (Rom. 12:1). All the sacraments of the Church involve the body's participation. Furthermore, unity of soul and body continues into eternity, for at the resurrection of the dead on the last day, we shall be reunited to our bodies. In this present life, then, we need to listen to the body – to its rhythms, its dreams, its modes of understanding – for the body does not lie. Our human physicality bears God' s imprint and His seal, and can be used as a means of communion with Him. Is it not wiser, then, to give to the divine image a maximalist rather than a minimalist sense (Ware, 1996a, p. 10)?

This understanding of the person as an irreducible psychosomatic unity is part and parcel of the salvific process. For Orthodox, salvation is something, that deals with every day of our lives as well as personal existence after death. Just as the body is included in the Orthodox understanding of the *imago Dei*, so too is it included in the likeness of God into which we are to grow. Salvation is the process of "growing up in every way into Him Who is the head, into Christ" (Ephesians 4.15). This process is termed *theosis* or deification, and it is summed up in the patristic maxim found in Irenaeus in the West and Athanasius in the East: "God became man in order that men might become God." Of course, Orthodox understand this as meaning "participation not in the nature or substance of God, but in His personal existence" (Zizioulas, p. 50). Salvation must be understood as holistic, as evidenced in the Christian belief in the resurrection of the body of the human person. As St Cyril of Jerusalem writes, "Therefore, since it is the body that has served to every work, the body will have its share in what comes to pass in the hereafter" (quoted in Harakas 1990, p. 32). Maximos Confessor, explicitly holds that the body is to be deified.[10]

One contemporary Protestant theologian who is deeply influenced by patristic theology is Colin Gunton. Commenting upon the *imago Dei*, he writes:

> It is important also to realize that this being in the image of God will embrace both what we have been used to call spiritual and our bodiliness. The merit of the approach to anthropology by means of the concept of the person is that it relativizes so many inherited dualisms. Relations are of the whole person, not of minds or bodies alone, so that from all those created in the image of God there is something to be received, and to them something to be given. When the image is located in reason, or for that matter in any internal qualification like consciousness, problems like those of "other minds" are unavoidable. The person as a being in relation is one whose materiality is in no way *ontologically* problematic, whatever problems derive from the way in which we relate in actual fact to others (Gunton, pp. 117-118).

Although it has not been a feature of our account of the body, it should be noted that the notion of body is polyvalent in Orthodox theology. From the beginning, that is, in the Pauline epistles themselves, Christian reflection on the body has spoken of the relatedness of both the human body and the Body of Christ which is his Church. Just as one cannot separate the body from the person, one cannot separate the Christian's human body which has been incorporated into the Body of Christ by baptism. This is not simile but metaphor.

Paul assumes the church is Christ's body in such a way that immorality is not *like* the body becoming ill or polluted; it *is* the body becoming ill or polluted. So questions of a man having sexual relations with his stepmother (1 Cor. 5:1), of Christian men visiting prostitutes (6:12-20), of eating meat sacrificed to idols (chapters 8-10), and of the proper eating of the Lord's Supper (11:17-34) are all connected. For Paul, each of these are questions of the purity of the body and consequently of the avoidance of pollution. A Christian man visiting a prostitute is the exact equivalent of the body being invaded by a disease that threatens all its members, since in fact every member is the body (Hauerwas and Shuman, pp. 61-62).

IV. THE BODY ABSTRACTED

Having shown that the Orthodox tradition views the human person as an indivisible whole, it is nonetheless the case that some patristic authors do address the physical aspect of human being. The *locus classicus* for Orthodox teaching on the body may be found in the catechetical teachings (i.e., instruction given to candidates for baptism) of St Cyril, bishop of Jerusalem, around the middle of the fourth century. Here we find perspicuous teaching on the Christian view of the body. Stanley Harakas has summarized Cyril's teaching under four themes: the goodness of the body, the relationship of the body to sin, the body as the tool of the soul, and care for the body. In this section we follow his analysis closely.

First, it was important for Cyril to affirm the goodness of the body because the contrary view arose quickly in the primitive Church and, even though it was condemned in the New Testament (1 Timothy 4.1-5), it has always remained a temptation for Christians. Against those who would denigrate the body, Cyril warns against such dualist thinking and exhorts the catechumens: "Do not bear with anyone if he says that the body is alien to God" (quoted by Harakas, 1990, p. 29). Cyril goes on to point to the contrary affirmation of the tradition.

But why have they depreciated this marvelous body? Wherein does it fall short in dignity? What is there about its construction that is not a work of art? Ought not the alienators of the body from God to have taken knowledge of the brilliant ordaining of eyes? or how the ears are placed right and left, and so receive hearing with nothing in the way? Or how the sense of smell can distinguish one vapour from another, and receives exhalations? Or how the tongue has a double ministry in maintaining the faculty of taste and the activity of speech? How the lungs, hidden away out of sight, keep up the breathing of air with never a pause? Who established the ceaseless beating of the heart? Who distributed blood into so many veins and arteries? Who was so wise as to hot bones to muscles? Who was it that assigned us part of our food for sustenance, and separated off part to firm a decent secretion (perspiration is meant by this phrase). while hiding away the indecent members in the more fitting positions? Who, when man was constituted that he will die, ensured the continuance of his kind by making intercourse so ready (Cyril of Jerusalem, *Catecheses*, Lecture 4, in Telfer, p. 111)?

In discussing the relationship of the body to sin, Cyril sets about to refute the ubiquitous notion that it is human physical nature which corrupts human spiritual nature captured in the proverbial "The spirit is willing but the flesh is weak." On the contrary, Cyril writes,

Do not tell me that the body is the cause of sin. For were the body the cause of sin, why does no corpse sin? Put a sword in the right hand of a man who has just died, and no murder takes place. Let beauty in every guise pass before a youth just dead, and he will not he moved to fornication. Wherefore so? Because the body does not sin of itself, but it is the soul that sins, using the body (Cyril of Jerusalem, Catecheses, Lecture 4, in Telfer, p. 111).

Acknowledging that there are New Testament passages which suggest rather nasty things about the "flesh" (especially in Romans 7-8), Harakas makes an important distinction.

But what is being spoken of here, in both the Scriptures and the patristic teaching, is not the body. In Greek, the word *soma* is used most often to refer to the body in a positive or ethically neutral way. The word *sarx*, frequently translated "flesh," is used in two very different ways. Often it refers to the body or human life in general, carrying with it no negative connotation, but on the contrary even a positive one. Thus "the two shall he one flesh" (Matthew 19:5); "and all flesh shall see the salvation of God" (Luke 3:6); "the Word became flesh and dwelt among us" (John 1:14); "the bread which I shall give for the life of the world is my flesh" (John 6:51); "no man hates his own flesh, but nourishes and cherishes it" (Ephesians 5:29) (Harakas 1990, pp. 31).

This serves as a transition to the third feature of Cyril's teaching, that of the soul's use of the body. One issue which is heuristic of Cyril's view is the matter of marriage and sex.

He is opposed to those rigorists who condemn sex, even marriage, and give honor only to those who have controlled themselves to God in sexual abstinence. Speaking to the latter, after praising them for their commitment to the angelic life of chastity, he says. 'do not make the opposite mistake of guarding chastity successfully and being puffed up with disdain of those who have come down to marrying... and are using (wedlock) aright." Marriage is the right place for sex. "But every other kind of sex-relation must be put right away, fornication, adultery, and every form of debauchery. The body is to be kept pure for the Lord, that the Lord may look favourably upon the body" (17, *ibid.*, p. 113). A similar approach is taken to food and clothing (Harakas 1990, pp. 31-32).

Finally, St Cyril addresses the proper care of the body.

So brethren, let us take good care of our bodies and not misuse them as if they were no part of ourselves. Do not let us say what the heretics say, that the body is a garment, and not part of us, but let us take care of it as being our own. For we shall have to give account to the Lord for everything we have done with the body as instrument" (quoted in Harakas 1990, p. 33).

Of course, one may find examples of ascetics both East and West whose rigors were intended to injure the body rather than to care for it. But as Harakas notes, "this is not a main line of Orthodox teaching regarding ascetic discipline and its relationship to the body. Rather, all ascetic disciplines as they affect the body are perceived as means to an end – the submission of life to God's purposes and the transformation of human life into the fullness of the image and likeness of God" (Harakas 1990, p. 33).

Two early monastic writers, Evagrius Pontius (346-399) and John Cassian (c.360-435) have been examined by a contemporary Western theologian of the ascetic tradition, Margaret Miles. She writes that while they valued ascetic practices regarding the body, they were nonetheless opposed to "any abuse of the body." Moreover, "the ascetic who understands the 'tools of the trade' "will benefit both body and soul by the ascetic practices employed" (Miles pp. 139-140). One may find similar teachings in the writings of St. John Climacus, Macarius the Great, and St. Isaac the Syrian. The ascetic literature is dotted throughout with aphorisms such as "The body too is not yours but a work of God," and "the soul has its being in the body" (Tsirpanlis, p. 187).

Harakas writes for the Orthodox tradition when he sums up:

> Thus, over a millennium and a half ago, Cyril, bishop of Jerusalem, taught his new converts a series of truths about the body which have remained "core teaching" for Eastern Orthodoxy: the body is not alien to God; the body is not in itself a cause of sin; the soul uses the body for good or for evil and may thus be not only a mere tool but a temple as well; the body is a permanent and essential aspect of the whole human nature; as a result it shares in the God-likeness to which human beings are called; and consequently, one has a responsibility to care for its well-being (Harakas 1990, pp. 33-34).

Even the later Western medieval tradition which is often thought of as denigrating the body in favor of the soul, has been misunderstood. One writer has argued that " ... theorists in the High Middle Ages did not see body primarily as the enemy of the soul, the container of the soul or the servant of the soul; rather, they saw the person as a psychosomatic unity, as body and soul together" (Bynum, p. 188).

Kallistos Ware summarizes:

> The aim of *asceses* is to secure our freedom. Ascetic rules are often expressed in a negative form – don't smoke, don't drink, don't eat meat on this day – but behind the negative rules there is the supremely positive aim. The Russian theologian and priest Father Serge Bulgakov used to say, "Kill the flesh in order to acquire a body." That is exactly what *asceses* is doing. Using the Pauline distinction between *sarx* and *soma*, the ascetic kills the flesh in order to acquire a body (Ware 1996b, p. 101).

V. ARE THERE DUTIES TO THE BODY?

For the Orthodox Christian, there are, strictly speaking, no duties to the body insofar as one's duties are to God who is the owner of the body. As St. Paul writes to the church in Corinth, "Do you not know that your body is a temple of the Holy Spirit within you, which you have from God? You are not your own; you were bought with a price. So glorify God in your body." (1 Cor 6.19-20) For the Christian, the rubric for understanding the relation is not ownership but stewardship, i.e., the Christian is responsible for the proper use of himself (including his body) because his owner (God) will hold him to account for his stewardship.

Having said that, how would an Orthodox Christian approach two of the issues which essays in this volume have addressed, namely, organ transplantation and the sale of organs? First, it must be acknowledged that the Orthodox Church has not yet spoken authoritatively on these matters in that they have not been acted upon by any synod of bishops or made the subject of new canon law. Nonetheless, a number of Orthodox have addressed these two concerns.

On the matter of organ transplantation, a consensus has yet to be formed. There are Orthodox who object to organ transplantation as immoral in that it is "a violation of the integrity of the body" (Harakas 1995, p. 646). Orthodox ethicists, however, have generally been supportive of organ transplantation in general. Harakas notes, e.g., that the objection based upon the integrity of the body is significant but "does not outweigh the value of concern for the neighbor, especially since organs for transplants are generally donated by persons who are philanthropically motivated for the protection of life" (p. 646). He also notes that in 1989 the Orthodox bishop of Athens, Greece, issued a statement to the press "that he had made provision for the donation of his eyes after his death" (p. 646).

On the issue of the sale of organs, Orthodox reflection has been uniformly negative. As Harakas writes, "The sale of organs is seen as commercializing human body parts and therefore unworthy, and is prohibited by a concern for the protection of life and its dignity" (p. 646).

On other issues related to the body such as the revived debate on human cloning, the Orthodox view on the psychosomatic unity of the human person will no doubt be a part of any Orthodox response.

Ultimately, theology's contribution to the cloning debate revolves around what it means to be human. From the Jewish and Christian perspective, our vision of a man or woman is that of a unity of body, mind, and spirit made in God's image, created to live in relationship to God and others. There is a reciprocal relationship between the three parts of our nature. Christian theology teaches that the immaterial soul during a person's life unites with the material body in an indissoluble, substantial way. It has an analogical relation to the union of the human and divine nature in Jesus Christ ... We err gravely in defining a human when we take one of these dimensions and make it the whole or when we separate the dimensions from one another (Evans, p. 28).

As biotechnology continues to advance, the Orthodox Church will no doubt be called upon increasingly to speak with authority for its members on matters pertaining to the human person (including the body). Its reflection upon these matters and ultimately its moral teaching will, however, be guided by its fundamental belief in the indissoluble unity of the human person, body and soul.

University of Osteopathic Medicine and Health Sciences
Des Moines, Iowa

NOTES

[1] Although Christians agree on the canon of the New Testament, there are at least three different canons of the Old Testament. Protestants in general have adopted the Jewish canon in Hebrew. Orthodox and Roman Catholics both follow the Septuagint, a Greek Old Testament that contains a number of books for which no Hebrew originals exist. These deuterocanonical ("second canon") books are sometimes referred to as "Apocrypha" by Protestants. Orthodox Catholics have a slightly larger canon than Roman Catholics do.

[2] For an interesting examination of embodiment by contemporary Jewish scholars, see the collection by Eilberg-Schwartz (Eilberg-Schwartz 1992).

[3] For a rich and textured account of the Pauline use of "body", see especially Martin 1995.

[4] Here Orthodox understand the parallelism of the Genesis verse as distinguishing between that in which we are created (image) and that into which, by grace, we are to grow (likeness).

[5] The reader is referred to the articles on *soma* and *sarx*, among others, in the *Theological Dictionary of the New Testament*.

[6] Justin Martyr, *On the Resurrection*, 8. Text in Migne's *Patrologia Graeca* 6.1585 and translation in *Ante-Nicene Fathers* 1.297.

[7] As to why many of the fathers identified the image as connected with the soul but not the body, Kallistos Ware suggests that there are several reasons. "In some cases the writers in question have been heavily influenced by a Platonic separatist view of the human person, which fails to allow sufficiently for the interdependence of body and soul. Also, the fathers needed to guard against a crudely anthropomorphic view of the deity; if it were said in an

unqualified way that the human body is in the divine image, simpler believers might have taken this to mean that God has a physical body like our own and is literally an old man up in the sky" (Ware, p. 9).

8   Irenaeus, *Against the Heresies* V, vi, 1.
9   *Prosopopeiai, PG* 150:1361C, a passage sometimes attributed to St Gregory Palamas. Translation in Ware.
10 See, e.g., his *Ambigua* 1336AB and 1364CDf.

BIBLIOGRAPHY

Brown, P.: 1988, *The Body and Society: Men, Women and Sexual Renunciation in Early Christianity*, Columbia University Press, New York.

Bynum, C. W.: 1989, 'The female body and religious practice in the later Middle Ages,' in M. Feher (ed.), *Fragments for a History of the Human Body*, Urzone, New York.

Cahill, L. S. and Farley, M. A.: 1995, *Embodiment, Morality, and Medicine*, Kluwer Academic Publishers, Dordrecht.

Cahill, L. S.: 1995a, '"Embodiment" and moral critique: A Christian social perspective,' in L. S. Cahill and M. A. Farley, *Embodiment, Morality, and Medicine*, Kluwer Academic Publishers, Dordrecht

Chirban, J. (ed.): 1996, *Personhood: Orthodox Christianity and the Connection Between Body, Mind, and Soul*, Bergin & Garvey, Westport, CT.

Cole-Turner, R.: 1997, *Human Cloning: Religious Responses,* Westminster John Know Press, Louisville, KY.

Culianu, I. P.: 1995, 'The body reexamined,' in J. Law (ed.), *Religious Reflections on the Human Body*, Indiana University Press, Bloomington.

Cyril of Jerusalem, St.: 1969, volume 1, *The Fathers of the Church* (vol. 61), L. P. McCauley SJ, and A. A. Stephenson (trans.), Catholic University of America Press, Washington, D.C., pp. 110-111, 128, 130-134, 181, 193, 204, 243, 248.

Cyril of Jerusalem, St.: 1970, volume 2, *The Fathers of the Church* (vol. 64), L. P. McCauley SJ, and A. A. Stephenson (trans.), Catholic University of America Press, Washington, D.C., pp. 131-132.

Eilberg-Schwartz, H.: 1992, *People of the Body: Jews and Judaism from an Embodied Perspective*, State University of New York Press, Albany, NY.

Evans, A. R.: 1997, 'Saying no to human cloning,' in R. Cole-Turner (ed.), *Human Cloning: Religious Responses,* Westminster John Know Press, Louisville.

Feher, M. (ed.): 1989, *Fragments for a History of the Human Body*: Part One, Urzone, New York.

Foucault, M.: 1986, *The Care of the Self*, Vintage Books, New York.

Gunton, C. E.: 1991, *The Promise of Trinitarian Theology*, T&T Clark, Edinburgh.

Harakas, S. S.: 1995, 'Eastern Orthodox Christianity,' in W. T. Reich, *Encyclopedia of Bioethics*, 2nd edition, Simon & Schuster MacMillan, New York.

Harakas, S. S.: 1990, *Health and Medicine in the Eastern Orthodox Tradition: Faith, Liturgy, and Wholeness*, Crossroad, New York.

Hauerwas, S., and Shuman, J.: 1997, 'Cloning the human body,' in R. Cole-Turner (ed.), *Human Cloning: Religious Responses,* Westminster John Know Press, Louisville.

Law, J. M.: 1995, *Religious Reflections on the Human Body*, Indiana University Press, Bloomington.

Levin, E.: 1989, *Sex and Society in the World of the Orthodox Slaves, 900-1700*, Cornell University Press, Ithaca, NY.

Martin, D.: 1995, *The Corinthian Body*, Yale University Press, New Haven, CT.

McPartlan, P.: 1993, *The Eucharist Makes the Church: Henri de Lubac and John Zizioulas in Dialogue*, T&T Clark, Edinburgh.

Miles, M. R.: 1981, *Historical Foundations for a New Asceticism*, Westminster Press, Philadelphia.

Nellas, P.: 1987, *Deification in Christ: The Nature of the Human Person*, St. Vladimir's Seminary Press, Crestwood, NY.

Robinson, J. A. T.: 1977, *The Body: A Study in Pauline Theology*, Westminster Press, Philadelphia.

Tsirpanlis, C.. N.: 1991, *Introduction to Eastern Patristic Thought and Orthodox Theology*, Michael Glazer/Liturgical Press, Collegeville, MN.

Verhey, A.: 1995, 'The body and the Bible: Life in the flesh according to the spirit,' in L. Cahill and M. Farley (eds.), *Embodiment, Morality and Medicine*, Kluwer Academic Publishers, Dordrecht.

Ware, K.: 1996a, '"In the image and likeness": The uniqueness of the human person,' in J. Chirban (ed.), *Personhood: Orthodox Christianity and the Connection Between Body, Mind, and Soul,* Bergin and Garvey, Westport, pp. 1-13.

Ware, K.: 1996b, '"Panel on personhood: Medicine, psychology, and religion,' in J. Chirban (ed.), *Personhood: Orthodox Christianity and the Connection Between Body, Mind, and Soul,* Bergin and Garvey, Westport, pp. 97-104.

Yannaras, C.: 1991, *Elements of Faith: An Introduction to Orthodox Theology*, T&T Clark, Edinburgh.

Zion, W. B.: 1992, *Eros and Transformation: Sexuality and Marriage: An Eastern Orthodox Perspective*, University Press of America, Lanham, MD.

Zizioulas, J. D.: 1985, *Being as Communion: Studies in Personhood and the Church*, St Vladimir's Seminary Press, Crestwood, NY.

SECTION TWO

NATURAL LAW AND NATURAL RIGHTS

JOSEPH BOYLE

PERSONAL RESPONSIBILITY AND FREEDOM
IN HEALTH CARE:
A CONTEMPORARY NATURAL LAW PERSPECTIVE

I. INTRODUCTION

Natural law theorizing about moral and social problems is rooted in the ethical and political writings of Thomas Aquinas. Aquinas' theorizing, in turn, is indebted to Greek philosophy, especially Aristotle, and to Christian morality as expressed in the Scriptures and the earlier tradition, especially as articulated by Augustine and the other Fathers of the Church (Boyle, 1987, pp. 703-708). Thomistic natural law theory provides one of the most sophisticated and historically important interpretations of what Alan Donagan calls "common morality," that part of Hebrew-Christian moral tradition which is non-religious in two ways: (1) it is comprised of precepts guiding actions which are not religious but secular, that is, its subject matter is actions affecting oneself and other human beings, not God, and (2) these precepts are held to be knowable by human reason without religious revelation or theistic conviction (Donagan, 1977, pp. 6-8, 26-31).

Philosophers and theologians working within the natural law tradition have not articulated a developed account of the moral issues surrounding personal freedom and responsibility in health care. But the tradition does contain a general account of personal freedom and responsibility which has implications for the exercise of personal discretion in health care. In this paper, I will try to spell out some of these implications.

Part of the interest of this exercise is that natural law provides a distinctive, and nowadays unfashionable, picture of individual and social existence: all of individual and social life is tightly governed by moral norms, and little room is left for persons to do simply as they please. But moral norms are not, on that picture, external constraints on action but requirements of practical reason which become embodied in the lives of virtuous people in such a way that what they want and what is morally required come to be the same. Virtuous living requires not only the

M.J. Cherry (ed.), Persons and their Bodies: Rights, Responsibilities, Relationships, 111–141.
© 1999 *Kluwer Academic Publishers. Printed in Great Britain.*

acceptance of responsibility but considerable personal discretion about how to integrate and carry out responsibilities.

Developing the implications of this picture of human life which bear upon the problems of personal responsibility and freedom in health care matters is not likely to generate a list of moral judgments and a rationale for them which will command wide acceptance. Nevertheless, the exercise of drawing these implications should be of some normative interest. For natural law theorizing is continuous with a developed body of casuistical thinking which has sought to bring the norms of traditional morality to bear on moral and social problems. These norms are bound to remain part of bioethical discussion, and the effort to bring them to bear on questions of personal freedom and responsibility in health care is likely to reveal morally relevant aspects of the issues which other moral approaches overlook or downplay.

In this paper, I will not focus on two issues which usually receive considerable attention in discussion of personal freedom and responsibility in health care matters: suicide and abortion. These issues have been thoroughly discussed, and the natural law thinking about them is much better known and developed than that on other relevant matters. Furthermore, the discussion of abortion from the natural law perspective is complicated by the fact that it is regarded not simply as a matter of the health, bodily integrity or life of the pregnant woman, but as a conflict of rights or interests between two persons.

Suicide also differs from most other matters involving a person's care for his or her life, health and bodily integrity. For in the case of suicide the outcome of the successful act does not require further actions and the endurance of further miseries; the suicide no longer exists, whereas in the case of many other acts and omissions plausibly governed by a person's responsibilities toward his or her health and bodily integrity, the outcomes of these actions often do require further actions, and, in the cases where these responsibilities are not fulfilled, the endurance of significant miseries which often affect others and call for their response.

Another, closely related, consideration further differentiates the moral issues raised by suicide from those raised by other decisions about one's life, health and bodily integrity. In the clearest cases of suicide the agent intends his or her death; usually the suicide chooses to bring about his or her death so that, being dead, he or she will no longer undergo the evils being alive involves. But people do not have a similar interest in being unhealthy. The harms to health which people often bring upon themselves are not ordinarily (and perhaps not ever) thought of as benefits, even instrumental benefits. There are possible cases of mutilation which approximate the intentional structure of suicide, but these are rare. For example, a person who chooses mutilation to make money from his or her mutilated condition as a beggar, circus attraction, or eunuch.

More generally, one's position on the moral or social acceptability of suicide does not appear sufficient to settle the character of one's personal and social obligations towards one's body and health, nor do these obligations appear sufficient to fix a determinate position on suicide. Donagan, for example, holds a rather permissive position on suicide and a rather strict and traditional position on a person's responsibilities towards his or her own health (1977, pp. 76-81). Inferences from either set of obligations to the other are chancy at best.[1]

My discussion of personal freedom and responsibility in health care will be carried out in four steps. First, I will discuss some aspects of what is often called patient autonomy. I maintain that there is a judgment about the allocation of decision making authority in health care matters which is so widely accepted that it can be treated as a considered judgment of reasonable people. Second, I will propose a natural law account of this considered judgment. This account includes two components: (1) a view about the special status of the human body and the moral implications of this status, and (2) a view about how decision making authority and discretion are to be handled socially. Third, I will consider the limitations on patient autonomy which this account justifies in virtue of considerations about (1) the direct impact of a person's health care decisions on others, (2) more paternalistic and moralistic considerations, and (3) the commercialization of the use of the human body. Fourth, I will conclude by briefly considering how the natural law approach is to be evaluated in comparison with other approaches to the issues of personal responsibility and freedom in health care, and how it can be expected to make a contribution to current discussions.

II. A CONSIDERED JUDGMENT ABOUT PATIENT AUTONOMY

The literature of bioethics contains a great deal of analysis of the right of competent, adult patients to refuse medical treatment and of the closely related obligation on the part of health care personnel to get informed consent from those they assist. Much of this literature focuses on the limits of this right and this obligation and on their application to decisions in which the refusal of treatment will predictably lead to death. But discussions of the limits, applications, and possible extensions of this right and this duty presuppose that there is considerable agreement about some central aspects of patient autonomy. I think such agreement exists, and that it can provide a useful starting point for reflection upon personal freedom and responsibility in health care.

Most of us think that it is the responsibility of a competent adult to seek out health care at his or her discretion and to accept or reject the advice of physicians and others asked to provide diagnosis and treatment. Health care professionals are not, nor do we think they should be, anything like health care police who have authority to command people to seek out appropriate health care or to accept the health care which the professionals think best, or to prosecute those who fail in these regards. Indeed, not only do we think that health care professionals should not act in these ways, we also think that nobody else should either, including political authorities.

We believe, I think, that a competent adult's decisions about his or her own health care are not generally a matter for public decision and that among the private parties involved in such decisions, the person whose health is at stake should have the final say. The health care professionals are seen as *agents* or helpers of the person seeking health care assistance: we seek their help when we think we need it and they provide it at our request. They are, in effect, our employees, to whom we are related as to anyone else with whom we would contract for a job. We perhaps feel compelled to give the health care professionals more discretion than we would to others with whom we contract, and to put more trust in their professional judgment, but the relationship is contractual and not like that between one who has authority and one who is bound by that authority.

This picture plainly includes a right of competent adults to refuse medical treatment, and an obligation on the part of health care professionals to seek informed consent of those they treat. If those seeking health care had no right to refuse treatment, the professionals would, in effect, be authorities in this matter, and if the professionals were not obliged to get informed consent, the relationship could hardly be contractual.

Some of these ideas about patient autonomy can be gathered in the following proposition: there is generally no authority governing the decisions of a competent adult concerning his or her own health care.

I believe this proposition to be a considered judgment of reasonable people and so a useful starting point for the analysis of personal freedom and responsibility in health care. My proposition contains the notion of authority as understood within the natural law tradition. But this does not disqualify it from expressing a considered judgment of reasonable people generally.

On the natural law account, one person has authority over another whenever the other has a moral obligation to act in accord with the authority's decision in a matter concerning which the person under authority would otherwise (that is, in the absence of the decision taken to be authoritative) be free to do as he or she sees fit (Finnis, 1980, pp. 233-234; Simon, 1951, pp. 7-35). Thus my proposition means that other persons, including family members, health care professionals and legal or political authorities, do not ordinarily have authority in these decisions; the person whose health is at stake has the final authority, which in natural law terms means there is no authority, except insofar as others are obliged to respect the person's decision in their actions related to his or her treatment.

Putting things in this way does not disqualify my proposition from being a considered judgment, because the idea it expresses is close enough to other ways of stating core aspects of what most people would accept in the idea of patient autonomy that the needed implications between such alternative expressions could be developed. Thus, there is no objection to formulating the matter in terms of rights, just as long as this formulation is not theory laden in a way that would either beg theoretical questions or overstate what is agreed to by most people. The evidence for the claim that my proposition qualifies as a considered judgment cannot in the nature of the case be decisive. If someone has reason to judge the proposition false or knows others who do, then the claim potentially is in need of revision.

I have three reasons for this claim. First, this proposition does not appear to be significantly contested in the bioethical literature, or by the views or attitudes of most who have dealt with these decisions either theoretically or practically. This seems to be a matter about which there is surprising agreement, though not, of course, concerning the precise formulation I have given it. Such paternalistic attitudes as exist among physicians and other health care professionals either do not reach as far as this core of patient autonomy or are quite implausible if they seek to. Second, those who seek to restrict the range of patient autonomy do not deny this proposition, but others. Third, this proposition expresses ideas which have acceptance not simply in the context of modern health care and bioethics but in a rather long legal tradition as well as within traditional morality.

Each of these considerations needs a bit of development. Concerning the first, some health care professionals surely are paternalistic, but it is not clear that even those with paternalistic attitudes would extend them so far as to deny my proposition if starkly presented to them. Many may not want to be bothered about more than the legalistic minimum concerning informed consent, but that does not involve the rather extravagant claim to authority which denying my proposition would involve.

Further, some health care professionals may oppose patient autonomy because they do not want to acquiesce in a view of their role which makes them mere technicians, totally subservient to the will of the patient. But my proposition does not have this implication. Patients have the obligation to take seriously the expert advice of health care professionals, and the latter are not ordinarily required to cooperate in actions they feel to be immoral or contrary to good medical practice. Medical personnel and others are not prohibited by my proposition, nor by any other obvious moral principle, from seeking to persuade competent adults to choose the health care options they think best.

Concerning the second consideration, the well known article of Kenny Hegland is instructive. Hegland argues for restricting the application of the right to refuse treatment in cases in which the refusal of treatment will lead to death. But he does not deny the core ideas of patient autonomy, or anything like my proposition. Indeed, he accepts them as the framework for his argument. He says:

> Common law recognizes the right to refuse medical treatment at least in the non-emergency situation. Tort liability is imposed on the physician who renders treatment without his patient's authorization, or having once obtained it, goes beyond it by rendering treatment different from or more extensive than that authorized (1966, p. 355).

This quotation is also relevant to the third consideration: something like my proposition is well established in the legal tradition. One classic statement is the 1914 decision of Benjamin Cardozo while still on the New York bench.

> Every human being of adult years and sound mind has a right to determine what shall be done with his own body; and a surgeon who

performs an operation without his patient's consent commits an assault, for which he is liable in damages (1914, p. 93).

A similar position is also to be found in the moral tradition as represented by statements of Pope Pius XII during the 1950's. Pius said:

First of all, one must suppose that the doctor, as a private person, cannot take any measure or try any intervention without the consent of the patient. The doctor has that power over the patient which the latter gives him, be it explicitly, or implicitly and tacitly (1960, p. 201).

Similarly, some years later, Puis said:

The rights and duties of the doctor are strictly correlative to those of the patient. The doctor, in fact, has no separate or independent right where the patient is concerned. In general he can take action only if the patient explicitly or implicitly, directly or indirectly, gives him permission.... The rights and duties of the family depend in general upon the presumed will of the patient if he is of age and *sui juris* (1958, p. 397).

Pius goes beyond what is central to patient autonomy in holding that the patient has duties, that is, moral responsibilities, concerning his or her health. But this further claim is consistent with the aspect of patient autonomy expressed in my proposition. His statements can reasonably be presumed to assume that normally the acceptance or rejection of health care by a competent adult are not pubic matters in which legal authorities can rightly interfere. Thus, the view of the moral tradition he represents is, not surprisingly, quite similar to the common law view expressed by Cardozo.

III. A NATURAL LAW ACCOUNT OF PATIENT AUTONOMY

The account of this core of patient autonomy which emerges from natural law considerations can be developed by considering the idea of bodily intangibility suggested by Cardozo's statement and prominent in the common law tradition. The idea seems to be that, except in rare circumstances like public emergencies and police activity against people who are acting improperly, human beings should not be touched without their own permission or that of someone qualified to give permission. An account sufficient to explain and justify this idea of bodily intangibility

could be developed by conjoining a claim about the special status of the
human body with a claim about the appropriate locus of decision making
authority for actions concerned with human bodies. I believe that the
natural law tradition contains, at least implicitly, an account of this kind.

A. The Unique Status of the Human Body

According to natural law tradition, the human body has a unique status
among the things in the physical universe. The reason for this special
status is that the person's body is taken to be a part of the person.
Aquinas, although holding for the existence of a nonmaterial soul, held a
version of Aristotle's view that the human soul, analogously to the souls
of other living things, is the substantial form of the bodily human
individual. He takes it for granted that humans are animals – rational
animals, but animals nonetheless. Not surprisingly, he explicitly denies
that a person's self can be identified with his or her soul:

> ... a human being naturally desires the salvation of his or her very self,
> but the soul, since it is part of the human body, is not the entire human
> being, and my soul is not myself; whence although the soul achieves
> salvation in another life, neither I nor any human thereby does so
> (*Super Primam Epistolam ad Corinthios Lectura*, XV, lect. ii, author's
> translation).

There are surely contrary strains of thought in Christianity, especially in
the theology and spirituality influenced by neoplatonism. But Aquinas
represents the tradition accurately; the doctrine of the resurrection of the
body, which St. Paul was discussing in the text on which Aquinas was
commenting in the above quotation, indicates a regard for the human
body which many Christians influenced by neoplatonism strained to
accommodate and often failed to explain satisfactorily. But they accepted
it. For this doctrine is surely one of the ways in which the early Church
separated its views from those of the more "spiritual" philosophers and
sects. Christian regard for the body shows up in other contexts as well;
for example, in the idea of the importance of bodily contact with Christ
which underlies the Eucharist, in the idea that the human body is the
temple of the Holy Spirit, which is used as a premise in St. Paul's
rejection of fornication, and for his judgment that sexual sin has a special
malice (1 *Corinthians* 6: 15-19), and in the doctrine that marriage is an
image of the relation between Christ and the Church (*Ephesians* 5: 32).

So, the human body has special status in natural law thinking and this reflects very basic Christian beliefs. This special status and importance has surely been part of the justification within the natural law tradition of moral obligations towards human bodies and toward human bodily life, including an individual's obligations towards his or her own life and health. The rejection of most kinds of killing, including, at a fairly early period, the rejection of suicide, the strict sexual morality of Christianity and the contention of Pius that individuals have duties towards their body relevant to decisions about health care are all examples of this.

It would be a mistake to suppose that this evaluation of the human body is strictly religious. Unlike Locke's duty of self-preservation which is a duty to God (Mack, 1999, pp. 143-176), natural law theory, and common morality more generally, regard one's responsibilities to the human body, including one's own, as responsibilities grounded in the respect for human beings. Respect for others' bodies is a part of the respect owed them as rational creatures, and respect for one's own body, though not a matter of justice, is similarly grounded. It is among the duties to oneself (Donagan, 1977, pp. 79-80). Thus, according to common morality and natural law, failing in one's duties towards one's health is an offense against God, but is sinful because it is immoral, because it is in some way a failure to be reasonable.

This idea that there are responsibilities to oneself may seem odd and indefensible to many: the very ideas of duty and obligation suggest a social context. Thus, Mill appears to hold that in the private arena in which personal health care decisions are made moral sanctions are inappropriate. But common morality, including natural law, appears to have a somewhat different view of the purpose of moral sanctions than does Mill. For, according to common morality, moral judgments are the implications of being thoroughly practically reasonable, and can come into play whenever there are choices to be guided. So morality is not limited to constraints on individual behavior for the sake of social benefit and cooperation. Furthermore, in the natural law tradition in particular, the goal to which practical reason directs the actions of individuals is their full flourishing as human beings insofar as they can achieve that by voluntary action. A central part of that flourishing is living a life in accord with the virtues, and these are not simply other regarding.

Donagan states succinctly the central moral norm in common morality's view of a person's responsibility for his or her bodily welfare:

Since man is a rational creature who is a rational animal, respect for man as the rational creature he is implies respect for the integrity and health of the body. Hence *it is impermissible,* according to the first principle [of respect for rational creatures], *for anyone to mutilate himself at will, or to do at will anything that will impair his health* (1977, p. 79).

Thus, human beings are to be respected because of their rational nature, but Donagan does not think that this means that only rationality has intrinsic value. Human beings have it. Donagan, and the philosophical tradition he represents, are making no simple mistakes in thinking that although humans are to be respected in virtue of their rationality, it is not simply the rationality of humans but they themselves who are to be respected.

B. Decision Making Authority and Discretion

If one's own bodily welfare is taken as a ground for moral obligations toward oneself, the question arises as to why these obligations should not be subject to authority in ways incompatible with the idea of bodily intangibility and my proposition about patient autonomy. This question can arise in the following way: surely, some of the norms which flow from the special status of bodily life are properly enforced by authority, for example, some of those governing actions in which one person harms the person of another. And, some of these norms surely imply limitations on bodily intangibility, for example, when a person is treated against his or her will because public authorities judge the treatment necessary for the public welfare. Many people, including natural law theorists, would regard such authoritative limitations on personal discretion justified. But natural law theorists have a special problem to deal with here. Why should authority in health care matters not be extended farther, beyond decisions having a clear public impact? For natural law has not provided a clean, clear line between the areas of public authority and private discretion (Aquinas, *ST, Prima Secundae*, q. 96, aa. 2-3), and political leaders are not the only legitimate authorities on the natural law account. Even more importantly, natural law as already noted, presents a thoroughly moralized view of human life and relationships in which the room for acting simply as one pleases, without regard to some moral responsibilities, is quite limited. What grounds compatible with such a view could justify demarcating a sphere of personal discretion which is virtually immune from the authority of others?

To answer this question it is useful to consider the natural law account of property. Aquinas and the tradition following him simply assume that only nonhuman things can be property. The things over which humans have natural dominion and thus can rightly possess either in common or individually are *res exteriores*, not human beings. The indefensible statements about slavery in much of the older tradition did not assume that slaves were the property of the owners, but generally assumed the paternalism of Aristotle's views about natural slaves. Similarly, the natural law tradition contains no concept of self-ownership. The idea that the relationship of a person to one of his or her parts could be like the relationship of use which defines ownership relationships is incompatible with the special status the tradition gives to the human person, including the human body.

According to natural law theory, particular property arrangements are established by human convention as a way of implementing several general moral requirements. The most basic of these is the requirement of "common use." Property is not to be used simply for the advantage of the owner but also for the advantage of others as well – those whose needs might be fulfilled by the use or the fruits of the use of the item in question. But the most relevant of these requirements for present purposes is the idea that use of the things of the world best serves human needs if some individuals have responsibility for the use of some of these things and discretion over how they are to be used. Aquinas thinks that in the world as it actually is, some sort of determination of who has charge over various things is necessary for an orderly and just realization of the goods which can be realized by their use (*ST, Secunda Secunae*, q. 66, aa. 1-2; Boyle, 1989b, pp. 191-197).

So in dealing with property, and while maintaining the assumption that the actions involved are tightly bound by moral considerations, the natural law tradition holds that there should be room for individual discretion and freedom of action.

The natural law view about the discretion of owners is an application of the natural law view of the principles for allocating decision making responsibility and discretion within any group of people. According to this view, decision making authority on any given matter should be located in those people who are in a position to act effectively concerning that matter. Such people are ordinarily those who must perform the actions and who are in a position to know the facts, complications and possibilities for action. Those removed from the situation calling for action are not ordinarily in a position to act according to the requirements of practical wisdom, what Aquinas called *prudentia*. Perhaps more importantly, if those who are removed from the situation should impose their decisions on those who are more immediately involved, then the latter are deprived of the opportunity for actions which carry out their responsibilities and fulfill them as persons.

Recent natural law theorists discuss this general view under the heading of the principle of subsidiarity. The word is derived from *subsidium*, which means help, and the idea it expresses is that larger communities such as a political society should relate to smaller communities and individuals in such a way as to help them fulfill their responsibilities, not to take over these responsibilities. Another side of this principle is that political authorities do have under certain circumstances the authority to command actions which will contribute to the fulfillment of individual responsibilities (Finnis, 1980, pp. 144-147, 159).

This account of the principles for allocating decision making involves a move from a person's responsibility to the responsibilities of others to allow that person to carry out his or her responsibility. But it does not include the general thesis which Eric Mack criticizes: that for every duty a person has, others have a corresponding duty to allow the person to carry it out (1999). The natural law move is not underwritten by a supposed conceptual connection between duties and rights, but on more general considerations about the point of acting responsibly and the nature of human cooperation. These considerations plainly do not require that all a person's duties imply corresponding rights of non-interference.

It should now be clear why the discretion of property owners is an application of natural law's general view about the allocation of decision making authority. The use of some things is needed for people to carry out their responsibilities, meet their own needs and fulfill themselves. Discretion by persons capable of making good use of certain things is therefore appropriate. The claims of this discretion are clearly defeasible on the natural law conception, and the extent and character of the discretion is based on reasonable convention, but discretion by owners is necessary in the world as it exists.

The natural law view of the principles for allocating decision making responsibility is also part of the natural law account of bodily intangibility (the other part being the personal status of the human body). Most generally, a person's capability of moving around in an unhampered way within some limits is a necessary condition for fulfilling almost any responsibility and achieving anything at all through human action. Part of what property rights protect is this capability, but bodily intangibility protects it more radically.

The application of this consideration to the case of actions undertaken for the health and bodily welfare of a competent adult is especially compelling. Two related lines of reasoning clarify this application. Both depend upon the definition of health.

I assume that health refers to well integrated, harmonious, psychosomatic functioning. As such it is the perfection and flourishing of the bodily life of a human being. Thus, it is a condition of the person, not simply an extrinsic instrumentality. Clearly, health has instrumental value as a condition for many other worthwhile human activities. But this instrumental value is limited: some goods can be pursued in conditions of less than adequate health, and ill health provides a context for the pursuit of some goods; for example, developing the virtues of patience and courage. Still, as a condition of the person, in which a dimension of the human self is perfected, health has, at least in most circumstances, features of goods which are sought for their own sake. In addition to its instrumental value, it seems that being healthy is a goal worth pursuing.

These considerations about the nature and value of health underlie the application of the idea of bodily intangibility to decisions about the health care of competent adults because they are basic premises in an argument for the conclusion that only these persons are capable of knowing how health concerns reasonably fit into the whole set of commitments, obligations and concerns which make up their lives as human undertakings.

My comments about the nature and value of health suggest that health should have some place within the life plan of an upright person. But they do not, except in a very general way, suggest what that place should be. For the instrumental value of health in any person's life will be dependent on its relation to other valued activities, and will be limited by the play and conflict between these values, including those for which health is not especially important.

Even insofar as health is recognized as a non-instrumental benefit, its place in an upright person's life plan is not settled in a determinate way. For it is surely not the only thing which people seek as an ultimate benefit, and plainly can be accorded a status which is unreasonably high. People can become too concerned about feeling and looking good, about avoiding risks to life and health, and staving off old age and death.

It is useful to distinguish two kinds of relativity here. First, health is surely part of the good about which the virtue of temperance is concerned. This virtue, according to Aquinas, regulates desires in accord with "the necessity of this life" (*ST, Secunda Secundae*, q. 141, a.6). Desires for food, drink, rest and other bodily satisfactions, although related to life and health, are not necessarily in line with these goods, but can spontaneously move us to inadequate fragments or appearances of them, like feeling or looking good. So, temperance requires that desires relating to bodily life be moderated by consideration of their proper purpose. But other virtues come into play as well, most obviously courage. A person's emotional make-up, general state of health, and cultural circumstances obviously are relevant to determining what actions are temperate and courageous, and such factors as these vary considerably.

Secondly, even if temperance and courage provide rational structure to the pursuit of health and related bodily goods, there is the further question of how these goods, even rationally pursued, fit into the larger framework of a person's life. This more architectonic relationship is rationally structured by the virtues of prudence and justice. Here again the demands of the relevant virtues are relative to a number of personal and circumstantial factors. Since people reasonably have different commitments, life plans and opportunities for action, there are many patterns, compatible with virtuous living, for integrating health with the other goods of life.

Thus, responsible concern for health and bodily well being demands rationally directing our feelings about these things so that the genuine goods are promoted, and fitting our concerns about health into a larger pattern of human concerns. In both respects, the patterns of responsible action can vary greatly. It is very difficult, therefore, for others to say with confidence whether a person's placing of health within his or her overall life plan is reasonable or whether the actions a person chooses to take or refuse are in fact properly governed by the relevant virtues. Surely, there is no general norm or set of norms which might give others a basis for decisively criticizing a competent adult's judgment about such matters.

Thus, for example, perhaps health care concerns could reasonably rank rather low in the life of a person who is healthy by nature or luck, or in the life of someone who lacks immediate responsibilities for a family, or who feels called to some important but risky or unhealthy occupation; perhaps they should rank somewhat higher in the life of one having serious family responsibilities or one called by one's profession to provide a good example about health matters to other people. And perhaps actions which seem to others required by a rational concern for health are not in fact compatible with temperance or courage.

The second reason for thinking that the person whose health is at stake should be the one to judge what health care is appropriate depends upon another aspect of the nature of health. Since, as I am assuming, health is a harmony of organic and psychic functions, the person whose functions these are has a unique chance to affect them for good or ill. Furthermore, since these are functions which go on throughout life and which are always subject to disequilibrating factors, the protection and promotion of health are generally ongoing undertakings, not objectives which can be realized more or less definitively by a set of actions which can be completed and forgotten. These undertakings are closely connected to other activities which form most of the fabric of everyday social and personal life, such as one's eating, sleeping, working and recreation patterns.

In short, the fundamental responsibility for the health and health care decisions of competent adults is necessarily their own. Only the individual is capable of exercising the needed virtuous judgments. Furthermore, the individual's unique position of being the one who experiences his or her own good or bad health and his or her unique capacity to affect it for good or ill, together with the fact that steps to heal, protect and promote it are ongoing and so intimately tied up with the basic actions which form the fabric of a person's life make this conclusion inescapable. Analogous considerations suggest why parents, family members and close friends are appropriate proxies when people cannot make these decisions for themselves.

IV. THE LIMITS OF PATIENT AUTONOMY

The preceding account of the grounds for bodily intangibility and its application to the case of decisions about the health care of competent adults might appear insufficient because it seems to include such grounds for overriding or limiting patient autonomy that my proposition about patient autonomy is not justifiable. The objection might be developed as follows. According to the natural law conception, there appear to be grounds for defeating any of the normal presumptions about the allocation of decision making authority. This is clear in the case of property. There are a variety of morally defensible property arrangements, and the rights of owners are defeasible. The same appears to hold more generally, and so also in the case of patient autonomy as well. So, one must conclude, nothing as general or normatively powerful as modern conceptions of bodily intangibility and patient autonomy or their applications to health care decisions can be justified on natural law grounds.

But the analogy of patient autonomy to property rights is weak in two respects. The first is that the requirement of common use which is the ground for the defeasibility of property rights has no application to persons and their parts (which are what bodily intangibility protects). Property rights may rightly be overridden on the natural law account only when the use of the things owned plainly does not serve human needs in a reasonable way. This can occur in situations in which one person, faced with necessity, must take the property of another who has no similarly exigent need for the thing taken, and it can happen as a matter of public policy when a government changes property arrangements to better serve the common good. So, the basic ground for overriding property rights depends on the fact that they are in things which have a use. But persons and their bodily life and activity cannot be reasonably added to the category of things to be used; they are the ends for whose sake the use of nonhuman things should be made. Thus, even if one's decisions about one's health are tightly bound by moral norms, including norms demanding service to others, one's bodily life and health is not simply a thing to be used for other's purposes. It is one's own responsibility to judge how one is to fulfill these responsibilities.

The second disanalogy between property rights and patient autonomy lies in the fact that the contingency of property arrangements upon reasonable convention does not hold in the case of patient autonomy. Property arrangements depend upon reasonable convention because there is no unequivocally best way for all cultural and social conditions to allocate the responsibility for things of the world and the goods which can be developed from them. But this kind of contingency does not obtain when it comes to allocating decision making authority for health care. Technology plainly changes the possibilities for actions in pursuit of health, but does not affect the basic considerations which make the application of bodily intangibility to these matters so compelling.

So, the analogy between property rights and patient autonomy as conceived on natural law grounds does not provide convincing grounds for supposing that patient autonomy can be easily overridden. But perhaps there are other reasons for thinking so. One such reason is suggested by the fact that one of my arguments for applying bodily intangibility to a competent adult's health care decisions appeals to the rather ideal condition of a virtuous person who has a developed, rational life plan. Could not the autonomy of those less virtuous and integrated be rightly set aside in favor of some authority?

Two factors count against such an effort. First, people can have authority only where there is some real possibility that the actions they can command are likely to promote the goods at stake more effectively than if there were no authority. For the reasons indicated above in the second argument for locating this responsibility in the person whose health is at stake, it seems unlikely that another authority could be effective in significantly promoting the health of a competent but irresponsible adult. Good luck or unhealthiness seem to be the prospects for such persons. For similar reasons, moralistic motivations for the use of authority to coerce the irresponsible to act responsibly are unrealistic. Taking over health care responsibilities from individuals would seem to involve taking over the responsibility of much of their daily lives and that can hardly be thought to be helping a person to act responsibly.

Secondly, authorities are not generally in a position to distinguish between those who are virtuous and those who are not with respect to health care matters. The variability of possibly reasonable attitudes and decisions about health care matters is one of the things which makes personal responsibility for them so compelling, and this very factor puts putative authorities in a difficult epistemic position.

Still, these two factors do not exclude all attempts to introduce authority into decisions about the health care of competent patients. For there are some steps compatible with them which at least public authorities might take. There are some positive and negative norms which govern a person's health care which are of general application and sufficiently clear to enforce – for example, laws prohibiting some unhealthful activities, such as taking certain drugs, or requiring the use of seat belts in automobiles, or a requirement that everybody get an annual checkup (assuming that this service were readily and cheaply available as it is in Canada). Furthermore, if people have obligations towards the health of others, as natural law theory surely implies, then there may be other enforceable requirements which, although compatible with the limitations for which I argues in the previous two paragraphs, would constrain patient autonomy substantially.

The natural law account I have been developing does not exclude all such uses of authority, but admitting this adds little to the limitations on patient autonomy which are widely accepted; for example, that public authority can override the refusal of treatment if it is reasonably judged that the refusal will put others in danger or cause the person refusing to fail in fulfilling an important social duty.

A. Limits Based on Concern for Others

The widely accepted limits to patient autonomy are not based on a concern for the competent person's health but on the impact of the person's decisions on others. There is likely to be dispute about the cases in which these limits are invoked as a ground for overriding patient autonomy, but when it is clear that the exercise of patient autonomy will place others in real danger or cause important social responsibilities to go unmet, the use of public authority appears to be operating within its proper sphere and to be justified. For in both cases public authority is commanding what those commanded already have a responsibility to others to do, and does so in conditions under which the failure to command would be unfair to those who depend upon public authority to protect the common good. Thus, for example, requiring inoculations for everyone can be justified when there is strong reason to think that failure of some to be inoculated will put others at significant risk.

The most common and plausible limitations on individual freedom justified by this line of reasoning are those in which bodily intangibility and patient autonomy are not violated or are violated only in minimal ways, for example, activities which cause various forms of pollution, spread disease, or put others' life and health at risk. Still, as the inoculation example suggests there is here a rationale for some limitation of a person's discretion concerning medical treatment.

It is important to note that this line of reasoning justifies only the use of public authority to command actions contrary to a person's exercise of his or her discretion. This kind of authority cannot be exercised by those acting in a private capacity, because the common good which justifies the exercise of authority is the responsibility only of public officials acting in their public capacities. So, unless we are to suppose that health care professionals are public authorities, they cannot on their own override patient autonomy on these grounds. The supposition that health care professionals are public officials seems indefensible because of the kind of work they are expected to do. Thus the authority to command actions incompatible with patient autonomy is very restricted.

But public officials are not limited to protecting the common good from obvious harm; they can take actions to promote it. So, if people have obligations to the health of others, public authorities can surely command some of them. And natural law supposes that people do have obligations to other people. The obligations of parents bearing on the health of children and family members are perhaps clear enough, but there also are neighborly responsibilities towards the health of others. The common vulnerability and need for help by those who are sick, together with considerations of fairness based on the Golden Rule, establish an obligation on the part of all people to help others who are sick insofar as they can do so compatibly with other responsibilities.

The number of people each of us is obligated to help is potentially great. And this number can be enlarged by way of social cooperation through such mechanisms as socialized medicine and health insurance. But the kind of help we might be obliged to provide cannot be such as to take over people's responsibility for their health care. For that, as I argued above, is beyond our power. So, the assistance we are obliged to give is neighborly, not familial – that is, not like the obligations of parents towards their minor children. Thus, while we might have an obligation to pay for someone's health care, we are not thereby entitled to tell them what health care to get. Nor, for the same reason, do public officials have any such entitlement.

This general responsibility for the health care of others has some pointed implications which bear on our discretion towards our own health: we should be ready to give blood, to donate our organs when we have died, and sometimes even to donate organs while alive. Could any of these and similar responsibilities be proper objects for the commands of public authorities? If adequate provisions are made for those who conscientiously object to the use of cadaver organs, there might be justification for the use of public authority in this area. No person is touched; patient autonomy is not violated; and there is an obvious good at stake which might best be served by public action.

But the other cases involve actions which do involve touching people and overruling patient autonomy. The requirement that people give blood might be reasonable in emergency situations, but as a general regulation it would seem to be a potentially disruptive intrusion into people's lives. Its enforcement would be even more intrusive. Any sort of general social requirement that a person give his or her extra kidney, or part of a liver, or bone marrow appears to involve just the sort of ignoring of how such actions and their effects will fit into the person's life as only personal discretion can remedy.

Furthermore, if the distinction between persons and things is to be maintained in practice, and the status of persons as ends of use and not items for the use of others is to be maintained, such actions as these cannot be commanded. One who, quite literally, gives of oneself, either out of generosity or a sense of duty, is not being treated as a thing to be used, but is fulfilling himself or herself through responsible action. But one whose very self is invaded, without regard for his or her judgments about the matter, does appear as being treated as a mere source of materials for use by others.

In short, while natural law theory includes the idea that individuals have positive and negative responsibilities for the health of others, it seems that the only positive responsibilities which can be reasonably commanded by social authority are those which do not violate bodily intangibility and patient autonomy, and that the only negative responsibilities which can be reasonably commanded are either those which do not violate bodily intangibility or prohibit actions which clearly pose a danger or other harm to others. Responsibilities to others, including to the health care of others, do not provide grounds for significant limitations on patient autonomy.

B. More Paternalistic Limitations?

For many people, no doubt, a consideration of such direct responsibilities to others would be the end of any consideration of what might be legally commanded in the matter of health care. For many assume something like the harm principle, which would rule out any consideration of the use of political authority in self-regarding acts such as a person's decisions about his or her own health. But natural law theory does not contain a rationale, such as the harm principle, for a sharp and clean division between the public domain where legal enforcement is ruled out of bounds. According to Aquinas, some acts of all the virtues can be commanded by public authority, on the condition that they can be ordered to the common good of political society either immediately or mediately. He says:

> But law, as was stated above, is ordained to the common good. Therefore there is no virtue whose acts cannot be prescribed by the law. Nevertheless, human law does not prescribe concerning all the acts of every virtue but only in regard to those that are ordainable to the common good – either immediately as when certain things are done for the common good, or mediately as when a legislator prescribes certain things pertaining to good order (*bonam disciplinam*) whereby citizens are directed in the upholding of the common good of justice and peace (*ST, Prima Secundae*, q. 96, a.3; my translation).

This draws a line between the public and the private domain which is much less obvious, permanent and sharp (though not, I think less intuitively plausible) than that drawn by the harm principle. And it raises the question whether legal authority might command some actions which appear to benefit primarily the person doing the action, and prohibit some actions which appear to harm primarily the person who would do them. The cases, mentioned above, of requiring seat belts and regular medical checkups, exemplify the first possibility, and the case of prohibiting some unhealthy activities like taking certain drugs exemplifies the second.

Someone might suppose that my examples are all cases in which authority is justified, if at all, on exactly the same grounds as in the cases discussed above: because of the direct social impact of following or violating the regulation in question. But that seems implausible. For surely the people who will derive the greatest benefit and most immediately from such laws are the persons whose life and health are protected by obeying them. The significant impact on others, if there is any, will come as a result of this benefit. And, equally surely, the motivation for deliberating seriously about making such matters subject to public authority is at least partially paternalistic. So it is necessary to consider whether such paternalistically based legislation, which appears to go beyond what could be justified by the harm principle, can be justified on natural law grounds, presumably because of its indirect connection to the common good.

Plausibly, a person's concern for his or her health is related to the common good. For one's responsibility for one's health has a twofold social dimension: (1) health is instrumental to carrying out one's commitments and thus maintaining health affects others; and (2) ill health makes moral demands on others because of the responsibility of all to provide assistance to the sick. And the commanding of some actions might plausibly contribute to people's developing a responsible concern for these social aspects of their own health. So, personal health concerns plausibly have some connection to the common good of a political society, and so would seem to be subject to public authority.

But this is not sufficient to justify the use of public authority in this context: considerations in favor of bodily intangibility and its application in patient autonomy also need to be taken into account. Only one of my examples involves health care personnel and the need for one person to touch another. The others simply involve respectively a person's routinely doing a simple act and a person's refraining from doing certain kinds of things. And it is the case of the mandatory check up which is the most doubtfully justifiable; this requires some initiative, more disruptive than fastening a seat belt, and would seem to justify public scrutiny of a person's life which could prove to be even more disruptive.

Thus I think that on natural law grounds authoritative interventions for the sake of a competent person's health should be limited to prohibitions of certain actions and to the prescriptions of certain actions which are easily undertaken without any significant disruptions of one's life. The actions which might plausibly fall into either of these classes, but especially into the latter category of prescribed actions, are likely to be small. Consequently, such paternalistic interventions as might be justifiable will not simple swamp the idea of patient autonomy, and render the qualifier "generally" only a functional term in my formulation of people's considered judgments about patient autonomy.

Moreover, the reasons underlying bodily intangibility and patient autonomy themselves provide a direct limit on how many actions could reasonably be authoritatively commanded or prohibited. For if there were a great many actions commanded and prohibited, even if each command demanded doing only fairly simple and routine things and each prohibition avoiding some definite performance, there would be significant interruption of a person's life and a compromise of the discretion appropriate in living it.

A closer look at the relevant positive and negative duties makes clear how short the list of actions which might plausibly be commanded or proscribed by public authority actually is. The positive responsibilities of a person towards his or her health are to do actions or undertake projects which are described in fairly general terms — such things as taking appropriate exercise, learning and making use of up to date information about health matters, developing habits of personal hygiene, and so forth. Such generally described acts and projects are difficult to justly command, because it is not clear what counts as obeying the command. So the basic general responsibilities themselves are not good candidates for enforcement. The more specifically described and discrete actions by which people carry out these responsibilities are a more plausible subject for command – things like getting an annual checkup or inoculations. But none of these more specific and discrete actions is sufficient for fulfilling the general obligation it implements, nor, in many cases, is it strictly necessary. These facts put in question both the utility and the fairness of commanding actions to implement people's positive obligations towards their health. Furthermore, many of the discrete actions which might be commanded, such as brushing one's teeth, bathing, or eating regularly, would be actions within the flow of everyday life whose disruption public authority can justify with great difficulty. The plausible candidates here are, on examination, very few.

With respect to the negative responsibilities to one's health there is an analogous difficulty also rooted in the character of the general obligations. In Donagan's language, we are not to impair our health or mutilate our bodies *at will*. But the actions which this norm excludes are not easily distinguished from morally justified actions. The actions which are excluded are much less easy to identify by public criteria than actions like murder, adultery, theft and perjury. For example, we should, no doubt, avoid eating foods which are harmful to health, but social responsibilities or other good reasons may require us to eat what we know is harmful and accept that harm as a side effect. So, when the issue is avoiding harms to health, or risks of such harms, the norm must prohibit avoidance when there is not good reason to perform the action. And it is generally difficult for outsiders to say when this condition obtains. Thus, legal prohibition of taking some drugs, to the extent that is based on a concern for the health of the drug taker, must assume that there is no good reason for taking them. I think that condition can be met, but not for many activities which affect one's health negatively.

Mutilations might seem to be an exception to this, and perhaps would be if anything as general as Kant's rejection of mutilation were justifiable. But natural law theory in the twentieth century has rejected the older absolutism of common morality about this. This rejection is expressed in the so called "principle of totality," which allows the removal of an organ which is dysfunctional, or even the removal of an organ which is itself healthy but whose removal is necessary for proper biological function. This principle made clear that removal of the organ as such was not the morally decisive factor. Proper respect for the human body is preserved by doing what one can to preserve organic functioning of the whole human body (Gallagher, 1984, pp. 217-242).

This focus on organic function as distinct from considerations of simple bodily integrity became the basis for justifying the removal of organs for transplant. This can be justified, according to the natural law account, if functioning is not compromised. So, one who gives up a kidney which is not needed is not sacrificing function but merely accepting a risk to good functioning in the future for the sake of helping another. The case would be different if one wished to donate a cornea while still alive, since both eyes are needed for proper sight. This, it seems to me, is not justified on natural law grounds. Still, this line of analysis vindicates Donagan's judgments about mutilation that "... the only conditions on which most people would be inclined to submit to it coincide with those in which it is permissible according to the first principle to do so" (1977, p. 79).

There is a notorious exception to Donagan's observation in the natural law tradition: the rejection of contraceptive sterilization. It seems to me that the traditional Catholic position, though increasingly disputed within Catholic circles, draws the correct implications from natural law principles. Nevertheless, given that this judgment concerns so personal and so controversial a matter, there is no likelihood that it could become a matter for legal prohibition.

In short, the possibility that the list of negative responsibilities toward a person's health which could reasonably be prohibited by public authority might become oppressively long is not significant.

C. Limitations Based Upon Commercialization of the Body

It is possible to sell body parts and to otherwise make use of the human body commercially; for example, surrogate mothers who on the basis of a contract help a couple to bear a child in one or another way, and people who under exigent circumstances mutilate themselves to obtain a job or beg. Some aspects of such activities have already been considered, but it is worth considering some of these cases separately because they raise difficult questions for any approach which recognizes the discretion which competent adults should have concerning their own health and bodily life. I assume that whenever commercialization of the human body is morally wrong there is good reason for legal prohibition, since buying and selling are public activities which need some form of regulation. But even here the reason is not necessarily decisive. Authorities might reasonably judge that the side effects of prohibiting some commercial transactions are too high – for example, that enforcement costs are too high, or that unacceptable black markets are created.

For our purposes the most interesting cases of commercialization of the human body are those in which the action involved is morally objectionable, if at all, because of the commercialization. Selling blood, bone marrow or a kidney are examples. Many people think that commerce in these items is a bad idea, even though giving blood and donating organs is or can be a very good thing to do. And very many appear to think that it would be preferable not to commercialize these activities even if there are no decisive moral objections to doing so.

There are specific objections to selling one or another of these things: that the chances of contaminated blood are too great if it is treated as a commodity, that commercializing organ donation provides an undue inducement for the poor to take health risks that they can least afford, or that commercialization is simply a way for the rich to exploit the poor. Questions of these kinds are difficult to answer and need to be addressed before social decisions about selling body parts are settled. Much of the thinking needed to answer them is not moral analysis. Still, the issue of undue inducements and exploitation calls for some philosophical comment.

First, let us consider the matter from the point of view of the potential donor in such an arrangement. I see no ground for thinking that it would necessarily be wrong for such a person to sell his kidney (or perhaps a slice of his liver or some bone marrow). There is nothing wrong with such a person's donating his kidney to his son to save the child's life: he chooses to have his kidney removed not to impair his functioning but for a good purpose, and he accepts the risk to his health as a side effect. Now let us consider a different case: suppose his son who does not have renal problems will starve to death unless he gets some money, and can get it only by selling his kidney. The structure of the action is the same as that of the first scenario except that there is one more step in the instrumental reasoning: it's getting the money which saves his son's life. Furthermore, he is still saving someone's life more immediately by his donation – the purchaser's. And this, as we have seen, is something he might well have reason, indeed morally compelling reason, to do. This suggests that commercializing the donation of organs might facilitate the cooperation in health care matters. So commercialization as such cannot fairly be designated as an inducement for people to act irresponsibly in regard to their health.

Now let us consider the matter from the perspective of the purchaser. This person can surely have good and morally compelling reason for acquiring an organ by way of a transplant. Presumably he can seek to get it in any way compatible with his whole set of moral responsibilities. So using financial resources available for such a purpose is not necessarily wrong. But is he exploiting the donor? Surely, not necessarily. For the donor can consider the proposition and, unless we make the elitist supposition that those ready to sell organs are not capable of exercising rational discretion, can consider it responsibly. Among the things to be considered is his own obligation to take some risks to help others.

Natural law theory, like common morality more generally, is not indifferent to the exploitation of the poor by the wealthy, but it cannot underwrite an absolute objection to a social arrangement based on abuses which that arrangement makes possible. Thus, it is difficult to see how the objection to selling organs based on the dangers of undue inducements and exploitation can simply rule it out.

But there is a deeper and more general objection to selling blood and organs, which surely underlies Kant's horror at the idea of selling even one's finger. That, of course, is the idea that the human body is different from other things, that as a part of the human person it has a dignity which makes putting a price on it a kind of defilement and falsification of its status. But does selling blood or organs imply that they are property and so things to be used, not parts of the person?

Once they are distinct from the person's body, they are very like other things. There does not seem to be moral objection to using hair that is shorn from a person for various purposes, and even to putting a price on it, so also, it would seem in respect to blood or organs. So the problem must be in the taking itself. And taking such things is not itself incompatible with respect for the human body, since organs and blood can rightly be taken. So, the objection must lie in their being taken to be sold. Is taking blood or an organ from the body so that it can be sold treating that part of the person as a mere thing? It is plausible to think that if such things are taken against the will of the person from whom they are taken, the answer must be affirmative. But that is not the case under consideration. For the issue concerns any example of voluntarily selling one's organs, and the best of these cases are more like the situation in which a person gives of himself or herself to benefit others out of a sense of duty or generosity. In these best cases, the seller has morally compelling reasons to sell his or her organ, and so the only difference from donation without payment seems to be that some of the benefits are mediated by the payments received.

Thus, I think that if selling organs necessarily shows disrespect for the human person by treating a human part as if it is a subhuman thing, then the argument has not yet been made out. Selling organs sometimes might involve this sort of disrespect, for example, when done without any regard for one's bodily welfare, but it is hard to see how giving organs can be permissible and selling them always wrong.

The case of surrogate motherhood is somewhat different and more difficult. No doubt, one's views about the legitimacy of contractual relationships in this area will depend on one's moral evaluation of such things as artificial insemination and third party involved procreation. But even if we set these opaque matters aside, the use of a woman's body to carry a baby under contract raises special problems. For there are imaginable cases in which it seems that one woman could rightly carry the baby of another. Thus, if techniques were developed which would allow the transferring of an embryo from one woman's womb to another's, there would seem to be cases in which performing the transfer would be morally justifiable – for example, to save a fetus from a spontaneous or deliberate abortion. Could such a transfer reasonably be the object of a contract for money?

Here the thing being used is not what was formerly a part, but has been made distinct from one's bodily self and its vital activities which are one's own functions. In other words, here what is being sold is not a thing like an organ but a service; it is more like contracting for a job than selling something. And the service a woman sells here is not a definite action or a set of performances which can be done like the tasks which make up a day's work. The service seems rather to be a portion of her living over a period of time. So any such contract would seem to give others a claim over the woman's bodily life that is potentially quite intrusive. An analogy to slavery, admittedly of short term and somewhat limited scope, comes to mind. For the woman is putting much of her living under the discretion of others. But these considerations are not obviously decisive: if one can perform such a service legitimately without pay, then the very business of putting one's life under this kind of discretion of others need not be wrong and that surrender of discretion is central to the analogy to slavery. Think of an altruistically minded woman who wants to provide this service on a regular basis on the analogy of running a very early foster home for babies, but who needs compensation to support herself and do the job. If the resources of natural law do not run out here, my ability to make use of them does.

Finally, there are the uncommon but unsavory cases in which people are tempted by financial gain to mutilate themselves to make themselves appear freakish or pitiable. It seems wrong for people to mutilate themselves for such reasons, but at the same time it is easy to excuse them. For people would not consent to such mutilations unless they were in dire circumstances. It is also clearly wrong for others to induce them to do such things by paying them. So, such inducements are reasonably prohibited by law. But in some cases the financial inducement need not be provided by another: consider the case of the desperately poor person who believes that crippling himself to beg more efficiently will greatly increase his changes of survival. Perhaps political society has a duty to keep people from this kind of desperate situation, but if it cannot do this, then it is hard to see how it can justly prohibit people from doing what they to do to stay alive.

In short, the limitations on the discretion of competent adults concerning their health which is provided by considerations of commercialization seem to be rather fewer than one might expect.

V. CONCLUDING COMMENTS

The argument of the first three parts of this paper can be summarized as follows. The presumption in favor of the discretion of competent adults in decisions concerning their own health care is a very strong one. Indeed, at least one aspect of this discretion can be formulated in a proposition which is plausibly taken as a considered judgment of reasonable people: that there is generally no authority governing the decisions of a competent adult concerning his or her own health care. Natural law theory provides an account of this judgment, an account which allows us to understand the point of bodily intangibility and patient autonomy within a larger framework of moral obligations and social relationships. This account does allow that there are limitations on bodily intangibility, but these limitations do not go beyond what people are generally willing to allow, and they surely do not contradict relevant considered judgments. These limitations are not established by the application of a general principle but by casuistry which brings a host of moral considerations to bear on proposals for limiting patient autonomy.

This line of argumentation does not, of course, show that the natural law approach to these matters provides the best account of them. To determine that two very large questions would need to be addressed: (1) whether other normative approaches can provide an account of patient autonomy, and (2) how natural law and alternative moral theories would stand up in the larger dialogue about ethical theory. I cannot enter into these complicated matters here.

Suffice it to say that it is not clear that natural law will fare badly in either of these comparisons. For example, it is not clear that utilitarianism does provide an account of patient autonomy. More generally, it is not clear that anything like the harm principle can be justified on utilitarian grounds or that the harm principle marks the correct line between the area of private action and the area where legislation and public sanction is appropriate.

Locke, by way of contrast, does seem to have an account of patient autonomy, but his account includes the notion of natural rights as side constraints, and sharply separates considerations based on the good and on individuals' personal responsibilities from considerations about rights and social relationships. The sharp dichotomies which these claims introduce into moral and social life seem to many to be problematic and surely raise fundamental issues in moral theory. The more integrated approach of natural law theory avoids these difficulties, though, of course, it has plenty of its own to face. So in the dialectic between moral traditions and theories, natural law is not without resources.

One of these resources which emerges clearly in the effort to deal with questions of individual discretion in health care matters is the capacity of natural law theory to deal with the important objection that it is blind to contemporary concerns about personal freedom. As noted at the start of this paper, natural law theorizing includes a highly moralized view of social relationships and moral life, which many believe to be overly moralized and authoritarian. It is true that natural law leaves little room for action based simply on personal desire or preference. But if natural law does provide a reasonable justification for patient autonomy within an account which makes room for significant individual discretion and limitation upon the authority of political leaders and others having social status, then at least this objection to its relevance to contemporary discussions can be set aside. Ethical approaches which emphasize the moral perfectionism required by a life of virtue are surely in conflict, at a deep level, with those which assume that there must be considerable room for people to do just as they please. But the argument of this paper is on the right track, these perfectionist approaches cannot be dismissed as simply authoritarian or paternalistic.

St. Michael's College, Office of the Principal
University of Toronto
Canada

NOTE

[1]   For a discussion of Aquinas' analysis of suicide, see Boyle, 1989a, pp. 221-250.

BIBLIOGRAPHY

Aquinas, T.: 1941, 1942, *Summa Theologiae*, Volumes two and three, Institute of Medieval Studies of Ottawa, Ottawa.
Aquinas, T.: 1935, *Super Primam Epistolam ad Corinthios Lectura* in *Super Epistolas S. Pauli Lectura*, Vol. 1, Marietti, Editori Ltd., Turin, pp. 231-495.
Boyle, J.: 1987, 'Natural law,' in J. Komonchak, M. Collins and D. Lane (eds.), *The New Dictionary of Theology*, Michael Glazier, Inc., Wilmington, Delaware, pp. 703-708.

Boyle, J.: 1989a, 'Sanctity of life and suicide: Tensions and developments within common morality,' in B.A. Brody (ed.), *Suicide and Euthanasia*, Kluwer Academic Publishers, Dordrecht.

Boyle, J.: 1989b, 'Natural law, ownership, and the world's natural resources,' *The Journal of Value Inquiry*, 23, 191-207.

Cardozo, B.: 1914, *Schloendorff v. Society of New York Hospital*, 211, N.Y., 125, N.E., 92.

Donagan, A.: 1977, *The Theory of Morality*, University of Chicago Press, Chicago.

Finnis, J.: 1980, *Natural Law and Natural Rights*, Oxford University Press, Oxford.

Gallagher, J.: 1984, 'The principle of totality: Man's stewardship of his body,' in D. McCarthy (ed.), *Moral Theology Today: Certitudes and Doubts*, The Pope John Center, Braintree, MA, pp. 217-242.

Hegland, K.: 1986, 'Unauthorized rendition of lifesaving medical treatment,' in T. Mappes and J. Zembaty (eds.), *Biomedical Ethics*, second edition, McGraw Hill, New York, pp. 354-359.

Mack, E.: 1999, 'The alienability of Lockean natural rights,' this volume, pp. 143-176.

Pius XII, Pope: 1960, 'The intangibility of the human body (Allocution to the First International Congress of Histopathology, September 13, 1952), ' in The Monks of Solesmes (eds.), *Papal Teachings: The Human Body*, Daughters of St. Paul, Boston, pp. 194-207.

Pius XII, Pope: 1958, 'The prolongation of life,' *The Pope Speaks* 4, 393-398; *Acta Apostolicae Sedis* 49, 1027-1033.

Simon, Y.: 1951, *Philosophy of Democratic Government*, University of Chicago Press.

ERIC MACK

THE ALIENABILITY OF LOCKEAN NATURAL RIGHTS

This essay is about natural rights over one's person, one's life, and one's liberty.[1] The possession of such basic moral rights has profound implications for one's legitimate freedom with regard to medical treatment and procedures, and may have similarly profound implications for one's responsibilities with regard to medical treatment. Yet what the precise implications of these rights are crucially depend upon how these natural rights are understood. Perhaps the key interpretative issue here is whether these rights are to be understood as inalienable. A right is alienable if and only if the bearer of that right can, by his choice, waive or transfer that right.[2] If the moral rights of each person over himself, his life and his liberty are alienable and are not otherwise compromised by significant countervailing moral considerations, then these rights vindicate a sweeping doctrine of informed consent and mutually voluntary interaction in the biomedical sphere. Suicide and the assistance of suicide, the donation and the sale of bodily parts, surrogate childbearing, and broad freedoms of self-medication are only the most obvious libertarian implications of such a conception of natural rights.

On one level, the question of whether natural rights are alienable has, I believe, an easy and obvious answer, viz., yes. This is to say, if natural rights are understood as defining for each person a basic sphere of moral jurisdiction within which each rightholder may do as he sees fit, then it seems obvious that no feature or aspect of a person's rights can block the moral permissibility of chosen actions within his protected domain. However, this argument may rely too much, or at least too readily, upon assuming this particular domain of choice conception of natural rights. For there may be alternative conceptions of basic moral rights according to which some or all of these rights cannot be waived or transferred. Under such an alternative conception of rights, the bearers of these rights do not have the moral power to choose not to be protected by them. They cannot, by their choice, open themselves to forms of treatment, e.g., to being euthanized or to having bodily parts removed to be transplanted in others, which absent their consent would profoundly wrong them.

In this essay the key issue of the alienability of basic moral rights is cast as an exercise in interpreting John Locke's views on natural rights.

M.J. Cherry (ed.), Persons and their Bodies: Rights, Responsibilities, Relationships, 143–176.
© 1999 Kluwer Academic Publishers. Printed in Great Britain.

Locke is the historically ensconced exemplar of the natural rights tradition to which we so often appeal in contemporary debates within medical ethics – especially in debates about the permissibility of imposing certain forms of treatment upon individuals and in debates about the capacity of the consent of individuals to make otherwise impermissible forms of treatment morally legitimate. Given this exemplary role, a comprehension of the nature of Lockean rights and their status as alienable or inalienable will importantly inform the direction of these debates. Moreover, an examination of Locke requires that we directly confront the issue of the alienability of rights because Locke himself is commonly thought *both* to have endorsed the inalienability of the natural rights over one's person, life, and liberty, and to have articulated a conception of natural rights that does not permit those rights to be inalienable.

That Locke endorsed the inalienable character of rights seems to be strongly supported in well-known passages in which Locke denies that a person may destroy his own life and denies that a person may enslave himself.[3] Though in the state of nature a man has "an uncontroulable liberty to dispose of his person or possessions, yet he has not a liberty to destroy himself ..." (sect. 6).[4] And since entering into slavery is itself life-threatening,

> ... a man, not having the power of his own life, cannot by compact, or by his own consent, *enslave himself* to any one, nor put himself under the absolute, arbitrary power of another, to take away his life, when he pleases (sect. 23).

Yet Locke's apparent belief in the inalienability of these core natural rights is at odds with the sense of absolute or near absolute control over oneself which in conveyed to us by Locke's claim that the "state that all men are naturally in" is:

> ... a *state of perfect freedom* to order their actions, and dispose of their possessions and persons, as they think fit, within the bounds of the law of nature, without asking leave, or depending upon the will of any other man (sect. 4).

As A. John Simmons points out in 'Inalienable rights and Locke's *Treatises*':

> the very idea of an inalienable right, with its paternalistic air, [does] not sit comfortably with the radical voluntarism Locke routinely

espouses. The concern to protect persons from the consequences of their own *voluntary* choices, on which inalienability theses seem to rest, is a concern apparently inconsistent with [Locke's] arguments for the basic value of freedom (and free choice and execution of a life plan) (Simmons, 1983, p. 186).[5]

This incongruity within the Lockean text reflects a philosophical tension between the constraining character of claims about inalienability and the liberty sanctioning quality of ascriptions of rights. Rights, Locke seems to believe, mark off spheres of jurisdiction. A given person's rights define the area within which he is morally at liberty to do as he chooses. Thus to ascribe to William the right to life, the right over his life, while at the same time denying to William the moral power to decide what is to happen to his life – in particular and most dramatically, to decide whether that life will continue or not – seems to retract with one breathe what has been asserted with another.[6]

There are essentially three ways of reading the situation: (1) Locke is fundamentally confused, i.e., he endorses the inalienability of rights which, by their very nature, cannot be inalienable; (2) Locke is not confused because he does not really believe in the inalienability of these core rights; and (3) Locke is not confused because his belief in the inalienability of these rights reflects a conception of these rights that allows, indeed sustains, their inalienability. I defend alternative (2): the natural rights that Locke asserts do have the domain-defining character which precludes their inalienability and Locke is not committed to the inalienability of those rights. If we understand our basic moral rights as Locke did, and we should, those rights do vindicate a broad and radical doctrine of informed consent within medical ethics.

In section I of this paper, I defend the claim that Locke is not committed to the inalienability of natural rights by separating what *appears to be* Locke's assertion of this inalienability from Locke's theory of human rights. In section II, I support the ascription to Locke of the domain-defining conception of rights. The discussion there proceeds by way of a critical rejection of Simmons' understanding of Locke as a rule utilitarian of human preservation. This rejection allows one to reaffirm that Lockean rights assign spheres of authority to their bearers within which they may act as they see fit. Locke's theory of human rights does not encompass, as the rule utilitarian interpretation implies, even a *prima facie* duty to preserve oneself and, therefore, does not mandate particular ways in which persons must exercise their authority over themselves.[7] In section III, I briefly and critically assess certain additional arguments against the alienability of natural rights. While their consistency with a Lockean framework make these arguments available to Locke, they turn out to be arguments the availability of which is not greatly to be prized.[8] In section IV, I briefly comment on the viability of Lockean natural rights as fundamental norms for the evaluation of legal institutions and social policies and indicate some of the main implications of those norms in the bioethical arena.

I. ACCOUNTING FOR THE APPEARANCE OF INALIENABILITY

Any plausible denial of Locke's commitment to the inalienability of rights must offer some explanation for the *appearance* of his belief in their inalienability. I address this demand in two ways. First, drawing on arguments advanced by Simmons in his essay on inalienable rights in Locke, I maintain that what is primarily at work in Locke and *mistaken* for a commitment to the inalienability of rights is Locke's belief in each person's duty to God to preserve his life and his liberty. I seek to highlight the distance between Lockean duties to God and Lockean rights against others by drawing attention to Locke's appeal to precisely this distinction in his *Letter Concerning Toleration* (Locke, 1983).

Second, I discuss Locke's appeal to the psychological impossibility of rational men waiving or transferring their natural rights to life or liberty. I contend that Locke advances a much narrower psychological impossibility thesis than is often thought, and that, in any case, empirically falsifiable theses about what consent is possible or impossible among rational men are ill-suited to support the philosophical assertion of the inalienability of rights.

However, before turning to the major tasks of this section, it may be helpful to put aside a simple, but misleading, basis on which belief in the inalienability of natural rights might be ascribed to Locke. In doing so, I will also clarify what is and what is not precluded by the inalienability of a right, e.g., the inalienability of an individual person's right to life. It is obvious that Locke asserts both a duty to preserve oneself and a personal right to life. Especially since each of these assertions are tied to the importance of the preservation of human life and each are in some way expressive of the law of nature, it is only natural to surmise that they constitute two sides of a single normative coin – an inalienable right. On one side is the right as such; the moral immunity against being killed. On the other side is the right's inalienability; which is taken to be the impermissibility of allowing oneself to be killed.

But this surmise is based on a mistaken sense of what it would be for a right to be inalienable. For a right to be inalienable it must be that its bearer cannot, by his choice, waive or transfer it. For William to waive the right with respect to Mary would be for him, by his choice, to make it permissible for Mary to kill him. For William to transfer this right to Mary would be for him, by his choice, to make it permissible for Mary to kill him. Thus, the inalienability of William's right to life would not as such, speak to the issue of whether William may take his own life. [9] William's right not to be killed, like all rights, is a claim against others. He does not have to be able to waive that claim against others in order to be morally at liberty to take his own life.[10] Thus, Locke's belief in the impermissibility of suicide cannot be captured by attaching inalienability to the right to life. That impermissibility reflects, as we shall see, not a feature – or at least not any inherent feature – of any human person's rights, but rather God's right over human persons. We now turn to the question of whether the impermissibility of allowing another to take one's life, which derives from one's duty (to God) to preserve one's life, plausibly supports ascribing inalienability to one's right to life.

A. The Duty of Self-Preservation Versus the Right to Life

A key and illuminating contention of Simmons is that, for Locke, it is the duty to preserve one's own life and not any inalienable right as such which bars self-destruction and, in turn, prohibits each person from placing himself under the inherently dangerous institution of arbitrary government. That this line of argument exists in Locke is simply undeniable. Locke explains that each of us "is *bound to preserve himself*, and not quit his station wilfully" (sect. 6). Locke repeatedly connects the *denial* that a person has a right over his own life with the claim that there cannot have been any valid consent to arbitrary government. For such a consent, Locke says, would require that we possess a power, i.e., a right, to dispose of our lives which, in fact, we lack.

> ... [A] man, not having the power of his own life, cannot by compact, or by his own consent, *enslave himself* to any one, nor put himself under the absolute, arbitrary power of another, to take away his life, when he pleases. No body can give more power than he has himself;

and he that cannot take away his own life, cannot give another person power over it (sect. 23).

It is in virtue of the "fundamental, sacred, and unalterable law of *self-preservation*" that

> ... no man or society of men [has] a power of delivering up their *preservation*, or consequently the means of it, to the absolute will and arbitrary dominion of another (sect. 149).

Since William has a duty to preserve himself, and this is a *duty* he cannot shed, he is not at liberty to destroy his own life or to allow another to threaten his life. It is not that William has a right to life that is inalienable. Rather, in some crucial respect, *viz.*, *vis-á-vis* God, William has no right to life and, hence, in that respect, possesses no right which he might alienate (Simmons, 1983, especially pp. 190-193).[11] (Thus, ironically, the most persistent and forceful argument in Locke against the legitimacy of a tyrannical and human rights violating government is based on the claim that, at least *vis-á-vis* God, people do not possess enough authority over themselves to place themselves under such a regime.)

The only defect in Simmons' account of the duty to God strand in Locke's thought is that he continues to construe this strand as a doctrine of inalienable *rights* – albeit of admittedly odd "mandatory claim rights." A mandatory claim right is possessed by one, not as a correlative of someone else's duty, but as a derivation from one's own duty. It is a right against others that they (at least) allow one to do what one has a duty to do. According to Simmons, "This kind of right is, for Locke, a trivial consequence of possessing a duty; for it seems obvious that if something is my moral duty, I should be left free by others to perform it" (1983, p. 191). And, indeed, Locke does make this inference. For he declares of people that "they will always have a right to preserve, what they have not a power to part with" (sect. 149).

But, as Simmons (1983, pp. 195-196, n.52) himself recognizes, the inference from William's duty to do X to William's right against Mary that Mary (at least) allow him to do X is mistaken. This is easily seen by imaging cases in which, although William has a duty to do X (e.g., feed his child), Mary has a duty to do Y (e.g., feed her child), and Mary's doing Y precludes William's doing X (since there is only enough food for one child). Were Locke's inference correct such cases would be impossible. Either William's duty would imply an obligation on Mary to allow William to feed his child and, hence, the absence of a duty to feed her own child or Mary's duty would imply an obligation on William to allow Mary to feed her child. But such cases are perfectly, albeit tragically, possible. In fact, William's duty to feed his child no more implies a duty of others to allow him to do so than soldier A's duty to capture the strategic hill implies a duty on the part of the defender to surrender it. Similarly, William's duty to preserve himself implies no duty on the part of others to allow his preservation and, hence, implies no right to preserve himself. Thus, it is puzzling how, after some hesitation, Simmons reaches the conclusion that, "Locke, must ... be committed to inalienable *mandatory* rights – that is, inalienable rights to do what is required by the 'eternal' duties of natural law" (1983, p. 198). It would be far better, in light of this lack of implication, for us clearly to divide the Lockean beliefs that center on the duty of each to preserve himself from our account of his theory of rights.

If the rule against self-destruction is not among the dictates of the law of nature – at least insofar as the law of nature is identified with a specification of *people's* natural rights – then the secular right of enforcing the law of nature ought not to extend to forcibly preventing suicide. Were Locke to maintain that, although profoundly contrary to a duty to God, even suicide ought to be politically tolerated, we would have a strong confirmation of the separation of the duty to God and the natural right strands of Locke's thought. Unfortunately, I do not know of any direct discussion by Locke of the permissibility of forcibly preventing suicide.[12] But there are passages in Locke's first letter on toleration that come pretty close to the point.

A constant theme throughout this letter is that the Magistrate's power does not extend to the prevention of error. The right to err, indeed to make costly mistakes in one's own affairs, is and ought to be acknowledged. This right even encompasses a right to secure one's own damnation. Locke is at pains to emphasize both the extent to which a person's self-harming activity simply is not the business of the Magistrate and the extent to which it is appropriate to allow even the most self-injurious of religious activities.

> [T]he Care of Souls does not belong to the Magistrate ... The Care ... of every man's Soul belongs unto himself. But what if he neglect the Care of his Soul? I answer, What if he neglect the Care of his Health, or of his Estate, which things are nearlier related to the Government of the Magistrate than the other? Will the Magistrate provide by an express Law, That such an one shall not become poor or sick? Laws provide,

as much as is possible, that the Goods and Health of Subjects be not injured by the Fraud or Violence of others; they do not guard them from the Negligence or Ill-husbandry of the Possessors themselves. No man can be forced to be Rich or Healthful, whether he will or no. Nay, God himself will not save men against their wills (Locke, 1983, p. 35).

Moreover, the reason that Locke gives for this limit on the province of the Magistrate is that even eternally self-harming choices and actions do not trespass upon the rights of others.

... [O]ne Man does not violate the Right of another, by his Erroneous Opinions, and undue manner of Worship, nor is his Perdition any prejudice to another Mans Affairs; ... Every man ... has the supreme and absolute Authority of judging for himself. And the Reason is, because no body else is concerned in it, nor can receive any prejudice from his Conduct therein (1983, p. 47).

And Locke explicitly denies that the sinfulness of a choice or activity makes it the concern of others and, thereby, the subject to the Magistrate's authority.

Covetousness, Uncharitableness, Idleness, and many other things are sins, by the consent of all men, which yet no man ever said were to be punished by the Magistrate. The reason is, because they are not prejudicial to other mens Rights, nor do they break the publick Peace of Societies (1983, p. 44).[13]

Of course, neither the neglect of one's "bodily health," nor even the mismanagement of one's spiritual health, is the same as intentionally taking (or allowing the taking of) one's life. Nevertheless, these strongly anti-paternalistic and anti-moralist pronouncements against political interference with persons' choices affecting either their bodily or spiritual well-being strongly suggests that even intentional suicide would not count as "prejudicial to other mens Rights," and in this crucial sense would not be contrary to the law of nature. While, according to Locke, suicide is indeed a sin against God, it seems that Locke should maintain about suicide that:

... it does not follow, that because it is a sin it ought therefore to be punished by the Magistrate. For it does not belong unto the Magistrate to make use of the Sword in punishing every thing, indifferently, that he takes to be a sin against God (1983, pp. 43-44).

Like all of us, however, Locke was sometimes unwilling to accept the implications of his own theory. We can see in his refusal to accept fully the anti-moralistic implications of his doctrine on tolerance an indication that he would also have resisted the implied conclusion that persons have a right of suicide, i.e., that no one's voluntary suicide is prejudicial to other men's rights. Locke considers the objection that his doctrine of tolerance would require Magistrates to tolerate congregations that "sacrifice Infants, or ... lustfully pollute themselves in promiscuous Uncleanness, or practice any other heinous Enormities." Rather than distinguishing between the rights violation involved in the sacrifice of infants and the (mere) sinfulness involved in lustful self-pollution, he asserts that *all* these activities "are not lawful in the ordinary course of life" and are not made lawful by religious motivation (Locke, 1983, pp. 41-42). He might well have likewise declared suicide, and the alienation of rights needed to secure assistance in suicide, to be "not lawful in the ordinary course of life." But, given Locke's endorsement of a firm distinction between acts that are (merely) sins against God and acts that are (also) prejudicial to other men's rights,[14] it is hard to see how Locke could support either his actual contention that lustful self-pollution in promiscuous uncleanness violates others' rights or his probable contention that suicide trespasses on those rights and is, therefore, subject to the Magistrate's authority.

In ascribing to Locke both belief in a duty to God not to commit suicide and a belief in a right against others to pursue one's own death, we are not shackling Locke with an inconsistency. Indeed, Locke has almost at hand the materials to explain the consistency of these two claims. For, in the *First Treatise*, Locke draws the following distinction with regard to property in creatures, *viz.*,

> ... in respect of one another, Men may be allowed to have property in their distinct Portions of the Creatures; yet in respect of God the Maker of Heaven and Earth, who is sole Lord and Proprietor of the whole World, Mans Propriety in the Creatures is nothing but that *Liberty to use them*, which God has permitted ... (sect. 39).

If we apply this distinction to persons, we can say: (a) "in respect of God," William's right in his own person is only such as God permits and, therefore, William's suicide sinfully contravenes God's permission; but (b) in respect of all other human beings, William has property in his own person and, therefore, with respect to them, he is morally at liberty to dispose of his own life as he sees fit. Thus, the existence of the duty to God strand still allows that, within his theory of *human* rights, Locke remains an avid promoter of the authority of each over his own life and liberty.[15]

B. The Argument from Psychological Impossibility

Although the chief argument against consent to arbitrary government is the argument from lack of moral power (i.e., lack of right) to offer this consent, Locke also seems to embrace an argument from lack of psychological capacity. It would be so irrational to surrender one's rights to life or liberty that no such surrender can be understood as the act of a competent rational person. Since a right can effectively be waived only through the act of a competent rational person, the rights to life and liberty cannot be surrendered by voluntary choice. To posit such a surrender is to posit what is psychologically impossible. "[A] rational creature cannot be supposed, when free, to put himself into subjection to another, for his own harm" (sect. 164). No rational person would place himself under the authority of one who was not himself subject to "the restraint of laws." To believe that men might do so,

> ... is to think, that men are so foolish, that they take care to avoid what mischiefs may be done them by *pole-cats* and *foxes*; but are content, nay, think it safety, to be devoured by *lions* (sect. 93).

Only a non-competent, non-rational, being would waive his right to life or liberty. But such a being would already be bereft of the status and rights of a rational being.[16]

We ought not, however, to conclude too quickly that Locke actually asserts the irrationality of *every* waiver of a right to life or a right to liberty. All the passages from the *Second Treatise* most likely to be read in this way (sects. 93, 131, 137, 164) explicitly focus on the waiver that would be necessary for establishing arbitrary government. It would be entirely consistent to maintain that all such waivers signify mental incompetence and also to assert the rationality of other waivers of one's rights to life or liberty. Moreover, some passages, which on first reading might suggest a psychological inalienability thesis, do not do so on further inspection. Consider, for example, the argument that,

> the preservation of property being the end of government, and that for which men enter into society, it necessarily supposes and requires, that

the people should *have property*, without which they must be supposed
to lose that, by entering into society, which was the end for which they
entered into it; too gross an absurdity for any man to own (sect. 138).

Here Locke seems to be pointing, not to the psychological impossibility
of people agreeing to a society which threatens their property,[17] but rather
to the conceptual oddity of supposing that a social order which threatens
property can satisfy the end for which society exists.

There is one passage in the *Second Treatise* – a passage I have already
taken note of – in which Locke does argue "a man ... *cannot* by compact,
or his own consent, *enslave himself* to any one" (sect. 23). But here Locke
is advancing, not the psychological impossibility argument, but rather the
duty to God argument. A man cannot enslave himself to another because
he does not have "the power of his own life." Locke explains that, "No
body can give more power than he has himself; and he that cannot take
away his own life, cannot give another power over it."

In fact, the very same section of the *Second Treatise* points indirectly
to the possibility of the *rational* waiver of one's right to liberty. For, in
discussing the enslavement of an individual who through aggression has
forfeited his right to life, Locke implies that such a person might well be
better off enslaved than executed. He implies that it is only when "the
hardship of his slavery outweigh[s] the value of his life" that it will be
rational for him "by resisting the will of his master, to draw on himself
the death he deserves." But, if slavery can sometimes be better than death
for a person who has forfeited his life and freedom, it seems quite
possible that slavery can sometimes be better than death for someone
whose continued life depends upon his alienating his freedom. In such
cases, Locke should grant the rationality of enslaving oneself to another.
So there are serious problems for the view that Locke did maintain or was
committed to maintaining the broad psychological impossibility thesis.

Had he maintained this thesis, then, as numerous commentators have
noted (Simmons, 1983, pp. 203-204), it would be in virtue of his
acceptance of a highly implausible empirical claim, *viz.*, that it is always
so contrary to a person's good to forego either is a sure mark of
irrationality. The problem, of course, is that it is rather easy to think of
circumstances in which it would be rational for a person to waive his right
to life or to liberty. William might well find himself in such dreadful and
untreatable pain that it would be rational for him to waive his right to life
so as to grant another the moral liberty to end that unbearable existence.
Mary might well find herself in circumstances such that only a period of

confinement under the supervision and direction of others holds out any reasonable hope for a recovery of her mental health and, so, find it rational for her to waive important aspects of her right to liberty at least for some specified period. If Locke did believe in the broad psychological impossibility thesis, he ought to be freed from it. And he can be freed from it merely by jettisoning the implausible empirical claim that is crucial to it. If, on the other hand, Locke only believed something like the narrow psychological impossibility thesis, then Locke is perfectly well positioned to allow that people can rationally waive even their core natural rights and that nothing about the inner character or source of those rights precludes their being alienated.

Thus, on the one hand, the appearance of inalienability that derives from Locke's duty to God doctrine is not to be tied to Locke's theory of rights or treated as a sign of his belief in the inalienable character of those rights. On the other hand, Locke's empirical claims about the psychological impossibility of individuals' exercising their rights in ways that are normally self-destructive are highly contestable in themselves and of dubious relevance for establishing the inalienability of those rights.

II. THE LOCKEAN THEORY OF RIGHTS

It remains to be shown, however, that Lockean natural rights are of the domain of choice sort which by their very nature preclude the inalienability of those rights. To investigate whether they are, we proceed to a critical consideration of Simmons' account of the Lockean theory of rights. What is the moral principle underlying each person's ("optional") claim rights to life and liberty? Simmons' answer (1983, p. 198) is "the fundamental law of nature that mankind is to be preserved" – which proceeds from God's willing man's preservation. According to Simmons,

> Locke's moral theory is ... a kind of rule utilitarianism, with the preservation of men serving as the ultimate end to be advanced. The fundamental law specifies this end, and all other rules of natural law are members of that set of specific rules which best promotes the preservation of mankind. Our duty is to follow these rules, respecting the rights which they imply. When, occasionally, the rules yield conflicting claims ..., the conflicts are to be resolved by appeal to the fundamental law (which directs that whatever solution best preserves all involved is to be pursued ...) (1983, pp. 199-200).[18]

If this sort of utilitarianism were the basis for Locke's theory of rights, then the very basis of that theory would imply that each person was bound by at least a *prima facie* duty to preserve himself and, hence, would be subject to at least a *prima facie* moral necessity not to waive his rights of life and liberty. Thus, if this utilitarianism were the basis of Locke's doctrine of rights, it would not be possible to deny Locke's commitment to inalienable rights by separating as sharply and deeply as I would wish Locke's belief in a duty of self-preservation and Locke's rights theory proper.

More fundamentally, if Locke's moral theory is a type of utilitarianism, even a type of rule utilitarianism, the rights it generates will not be moral side-constraints which, by defining a sphere of moral jurisdiction surrounding each individual, specify for each individual a realm in which he may do as he thinks fit "without asking leave or depending upon the will of any other man." No utilitarian basis for rights will generate a theory of rights with the quality of "radical voluntarism" that Simmons himself ascribes to Locke's theory. For on some level, each person's rights or at least his claim to exercise his rights would be conditional upon and hostage to their exercise contributing to the common goal, e.g., the maximum preservation of mankind, for the sake of which those rights are ascribed to people. Thus, if Locke's underlying moral theory is a type of utilitarianism, people's basic rights will not confer on them the sort of absolute or near-absolute authority over themselves contemplated by the combined assertion of Lockean rights to life and liberty and of their alienability. For these reasons, to maintain the alienability of Lockean natural rights, one must reject Simmons' account of Locke's theory of rights.

There are two main bodies of evidence that Simmons can draw upon in support of his utilitarian interpretation of Locke. The first consists in Locke's repeated characterization of the natural law as willing the "*preservation of all mankind*" (sect. 6). The second is Locke's disposition toward describing certain cases as involving conflicts of rights which ought to be resolved by maximizing the preservation of all involved.[19] Each of the next two subsections offers a reading of the evidence that supports the ascription to Locke of belief in rights as moral side-constraints, as morally protected domains of choice.

A. *Locke's Argument for Rights*

While God may will the preservation of mankind, He pretty clearly does not will that each person take upon himself the task of preserving all of mankind. Indeed, had God prescribed this task for each person, He could not have *absolutely* proscribed suicide. For circumstances might easily arise in which William's suicide would be the course of action that would maximize human preservation. We misread Locke if we adopt, as Simmons does, an aggregate rather than a distributive reading of God's willing mankind's preservation. By a distributive reading, I mean one according to which God specifically entrusts each human being with the task of his own preservation. To recognize and act in accordance with God's will on the matter of human preservation, William must preserve himself and " ... be restrained from invading others rights, and from doing hurt to ... another" (sect.7). Others' preservation is not part of William's mandated goal; rather the existence of other persons with like natures and like goals imposes side-constraints upon how William may pursue his mandated role.[20]

Locke's entire explication of the basis of the law of nature – at least insofar as the law of nature dictates man's relationship to his fellow man – focuses, not on any moral demand to advance the preservation of others, but rather on the constraints on interpersonal behavior that co-exist with, and make possible, the *"state of perfect freedom"* (sect. 4). Since our condition of natural freedom is also a condition of natural equality, since we are "creatures of the same species and rank" and there is no "manifest declaration" to the contrary by "the lord and master of ... all," we "should also be equal one amongst another without subordination or subjection." In this state of equal freedom, "all the power and jurisdiction is reciprocal" (sect. 4). If a person has jurisdiction over his own person, life, actions and possessions, then each other person must have a comparable jurisdiction over his own. Given this *"equality* of men by nature,"* whatever one (rationally) claims for oneself against others, one must grant others as an equally legitimate interpersonal claim against oneself (sect. 5). Since for the sake of his own preservation, each person rationally claims for himself freedom from others (see, for example, sect. 17), each must grant a like freedom to others.

This line of argument, which focuses on the likeness of the claims that equal beings, each oriented toward his own preservation, can make against each other, is carried forward in the crucial section six. "[R]eason, which is [the law of nature], teaches all mankind, who will but consult it, the being all *equal and independent*, no one ought to harm another in his life, health, liberty, or possessions." Similarly, Locke declares that " ... sharing all in one community of nature, there cannot be supposed any such *subordination* among us, that may authorize us to destroy one another." In both cases, Locke infers a prohibition on harming or destroying others, not the inclusion of others' preservation within each person's obligatory ends. Nor is there any basis for believing that this restraint on the manner in which each may advance his own preservation is ultimately justified by its consequences for the aggregate preservation of mankind. There is no indication in Locke that the rule against injuring others is to be adopted on the basis of a calculation of the social benefits which respect for it will bring.

Similarly, the argument from the latter portion of the crucial section six is directed toward a side-constraint conclusion:

> Everyone, as he is *bound to preserve himself*, and not quit his station willfully, so by the like reason, when his preservation comes not in competition, ought he, as much as he can, to *preserve the rest of mankind*, and may not unless it be to do justice on an offender, take away, or impair the life, or what tends to the preservation of the life, the liberty, health, limb or goods of another.

One's being bound to preserve the rest of mankind consists in one's being bound "not ... to take away, or impair the life, or what tends to the preservation of the life, the liberty, health, limb or goods of another." Reason dictates and God wills that each preserve himself. The implication of this dictate for William's treatment *of his own life* is that William ought to preserve that life while the implication of this dictate for William's treatment of Mary is grounded in "the like reason" that binds William to sustain his own existence (see, for example, Mack, 1980).

It is striking that, within this account of the dictates of the law of nature regarding the treatment of man by man, even the claims that we are "all workmanship of one omnipotent, and infinitely wise maker" and are "his property ... made to last during his, not one another's pleasure" (sect. 6) are *not* employed on behalf of the conclusion that we each ought to nurture the whole of this workmanship. Rather, these claims are used in service of the prohibition on taking the lives of others. For such a taking of lives, Locke argues, supposes that we are authorized "to destroy one another, as if we were made for one another's uses, as the inferior ranks of creatures are for our's." Our each being God's workmanship merely serves, at this point, to rebut any supposition that we were made for, or exist for, one another's uses.[21]

It seems, then, that for Locke the assertion of God's ownership of human beings has two distinct functions. It functions positively to establish a duty each owes to God to preserve his own life. And it functions separately and negatively to exclude any human's ownership by another human and, thereby, to rebut any assertion of the natural mastery of one human being over another. In this second respect, God's ownership of mankind operates in a fashion that parallels Locke's assertion that God gave the earth to mankind in common. This claim functions negatively to exclude any particular individual from having, through God's specific grant to him or his ancestors, a special claim on some portion of the earth.[22] In both cases, the negative use of God's original authority serves to defend an original moral equality among men. In short, both the character of Locke's arguments for the rights defined by the law of nature and their repeated expression in the form of negative side-constraints – e.g., "no one ought to harm another in his life, health, liberty, or possessions" (sect. 6), "Subjects [are] not [to be] injured by the Fraud or Violence of others" (Locke, 1983, p. 35) – rather than as guides toward the maximization of human preservation are strong evidence that Simmons is incorrect in seeing Locke's doctrine as a type of utilitarianism of preservation.

What, then, can be said about Locke's repeated description of the goal of the fundamental law of nature as *"being the preservation of mankind"* (sect. 135)? How strongly does this sustain Simmons' utilitarian interpretation? Locke certainly uses *"the preservation of all mankind"* to refer jointly and indiscriminately to the duty each has to preserve himself, to the right to preserve oneself which this duty is supposed to imply, and to the duty each has "by the like reason" not to destroy the equal and non-subordinated lives of others. The common theme of human preservation and the common ground of these moral rules in Locke's God and in Locke's conception of human reason make it unsurprising that Locke sees these rules as being all of a piece and brings them under a common phrase. Given this explanation for the grouping of these elements of Locke's moral thought under the common umbrella of preservation, this broad use of the idiom of human preservation does not, in itself, provide much evidence for Simmons' interpretation of Locke as a sort of utilitarian.

B. Resolving Conflicts Among Lockean Rights

More apparently supportive of the utilitarian reading, however, are several passages in which Locke seems to contemplate conflicts of rights and to assert that such conflicts are to be resolved in the ways that overall would most preserve mankind. In the first of three such passages, Locke defends injurious self-defense by arguing:

> for, *by the fundamental law of nature, men being to be preserved* as much as possible, when all cannot be preserved, the safety of the innocent is to be preferred ... (sect. 16).

By describing the rationale for self-defense as human preservation "as much as possible," Locke certainly suggests a utilitarian orientation. But further inspection reveals that this suggestion is misleading. It is not *that* the innocent is to be preferred to the aggressor because there is more preservation to be had by favoring the innocent. This would not be the case if, for example, all the members of an aggressive mob had to be killed in order for a single innocent to be defended. Yet, it is hard to believe that Locke would not endorse defensive killing in this case as well. Nor is there any hint that Locke is thinking that following the rule, "prefer the innocent," is justified by the calculation that this more preserves mankind than adherence to any alternative rule. Rather, for Locke, defense is justified because the attacker puts himself in a state of war with respect to his target who, therefore, has "a right to destroy that which threatens [him] with destruction" just as "he may kill a *wolf* or a *lion*" (sect. 16). Again, the type of justification offered by Locke, this time for the use of deadly force against the attacker, shows that his position is not a species of utilitarianism.[23]

In a second passage in the *Second Treatise*, Locke points to the need for discretionary power in the executive with regard to the enforcement of municipal laws – a discretion which would take the form of recourse back to "this fundamental law of nature and government, *viz.*, that as much as may be, *all* the members of the society are to be preserved." As an example of a justified transgression of the letter (but not the spirit of the municipal law, Locke cites the need "to pull down an innocent man's house to stop the fire, when the house next to it is burning" (sect. 159). Without doubt Locke's language strongly suggests that the justification of the pre-emptive destruction of the innocent's property is the prevention of the greater loss of property – just as the similar employment of the language of preservation "as much as possible" suggests a utilitarian vindication of self-defense.

Nevertheless, one must avoid reading much of a utilitarian message into Locke's language. After all, is it plausible to project that Locke would have favored the destruction of *any* innocent's house the destruction of which would prevent a greater property from being destroyed? Would Locke, e.g., have favored the dismantling of William's house, which, let us suppose, is quite distant from the fire, for the purpose of constructing equipment with which to extinguish the fire and thereby save Mary's endangered and more valuable house? While this dismantling would more preserve mankind or, at least, mankind's property, it is doubtful that this would suffice to convince Locke of its propriety. One can hypothesize that what does convince Locke of the propriety of pre-emptive destruction in the case he has in mind is that the targeted house will otherwise serve as a conduit bringing the fire to yet other dwellings. What justifies its destruction is its dangerousness, under the circumstances, to the neighboring properties (plus, perhaps, the fact that it will be destroyed anyway). If this is what Locke had vaguely in mind or would "really" have said if pressed on the matter, then this passage does not support the utilitarian interpretation.

In a third passage, Locke discusses a case in which there is an (apparent) oversupply of just claims against a given person's estate. An unjust aggressor has been conquered and the conqueror's just reparation may call for his acquisition of the entirety of the aggressor's estate. Yet, the aggressor's innocent children also have just titles in portions of that estate. Locke asks, "What must be done in the case?" Then, as he often does, Locke casts his answer in terms of most preserving mankind.

> ... [T]he fundamental law of nature being, that all, as much as many be, should be preserved, it follows, that if there be not enough fully to *satisfy* both, *viz* for the *conqueror's losses*, and children's maintenance, he that hath, and to spare, must remit something of his full satisfaction, and give way to the pressing and preferable title of those who are in danger to perish without it (sect. 183).

In this passage, it seems that Locke is saying that the children's titles (at least to maintenance) are "preferable" *because* the children are in danger of perishing. Fortunately, we do not have to address here the complicated question of whether according this sort of preference to the children's claims over those of the just conqueror manifests an underlying utilitarianism. For, contrary to appearances, Locke does not really view this case as one of conflicting rights the priority of which needs to be ascertained – perhaps by determining which conflicting right most promotes human preservation.

Instead Locke's actual position is that the children have title to a portion of the father's estate – a portion at least large enough "to keep them from perishing" – and that nothing the father has done toward forfeiting the portion of his estate which is unencumbered can diminish the title of the children.

> ... [T]he father, by his miscarriages and violence, can forfeit his own life, but involves not his children in his guilt or destruction. His goods, which nature, that willeth the preservation of all mankind as much as is possible, hath made to belong to the children to keep them from perishing, do still continue to belong to his children (sect. 182).

Locke then seems to waiver a bit, wondering whether the just conqueror's right to reparation may extend into the portion of the estate that belongs to the children. But he concludes that, although the conqueror can take reparations from the aggressor's estate for "the damages received, and the charges of war," this may be done only "with reservation of the right of the innocent wife and children."[24] Precisely the same doctrine is repeated immediately before Locke's question, "What must be done in the case?" and his invocation of the idiom of preserving all as much as possible. What this does show, and rather strikingly, is that Locke's frequent recourse to this idiom is merely his way of making reference back to theory of rights based ultimately on God's desire for human preservation or to some segment of application of that theory, e.g., the rights of children against those who have begotten them (see, for example, *First Treatise*, sect. 88, and *Second Treatise*, sect. 56).

I conclude that these intriguing conflict of rights passages are not significant evidence that Locke intended a utilitarianism of preservation and provide even less reason to believe that Locke is committed to such a utilitarianism. Thus, neither these passages nor an examination of Locke's grounding for natural rights supports Simmons' understanding of Locke as a utilitarian of preservation. Lockean natural rights are domain of choice rights which impose moral side-constraints on others' treatment of the rightholder. The liberties and moral immunities associated with these rights are not limited to those conducive to the satisfaction of rightholders' moral goals or moral duties; in particular, these liberties and immunities are not restricted to those ministrative to the duties of self-preservation or the preservation of others.

III. FURTHER ARGUMENTS FOR INALIENABILITY

Still, we have not disproved the possibility that, somewhere and somehow, there is an argument made by or available to Locke for the inalienability of these side-constraining natural rights. Perhaps such an argument can be grounded on the very character of natural rights or on the character of rightholders as such. As I read him, Simmons offers one such argument. We are concerned here, not with the duty against self-destruction, but with the right not to be killed – the inalienability of which would consist in the rightholder being unable to waive this right and, thereby, being unable to make it permissible for another to kill him.[25] Simmons argues that:

> This right not to be killed is the logical correlate of other's duty not to kill me, and must be "eternal" if that duty is to be eternal ... Given that [Locke] is committed to an eternal natural duty not to kill (arbitrarily) another, he is obviously committed to an *inalienable* right not to be killed, as well (1983, p. 198).

Unfortunately, this is a weak argument. There are two ways to understand the claim that eternally people are bearers of rights to life and liberty. If one takes the claim to mean that people eternally bear these rights independent of any choice about continuing to possess (or be possessed by) them, then the inalienability conclusion follows most quickly. But the claim can also be understood to assert that eternally every person is "born to" certain rights, that very person does and will enter the moral community as a possessor of these rights. This second understanding does not at all imply that at every moment of time these rights must continue to attach to each and every person. Locke's arguments for the core natural rights point only to the second and more modest reading of the assertion that eternally people are bearers of rights. Each person's being born to a jurisdiction over himself equal to the authority others possess over themselves is consistent with each person's having the capacity to diminish his sphere of jurisdiction (and/or enlarge that of others) by his choice. Thus, from the "eternality" of Lockean natural rights, we cannot infer their inalienability.

In his, 'Contract remedies and inalienable rights,' Randy Barnett forcefully develops four arguments for the inalienability of natural rights (1986, pp. 185-195), two of which require explicit further consideration here.[26] Barnett's first argument focuses on limitations upon A's capacity to bind himself "always unquestioningly [to] follow the commands of B" (1986, p. 186). The relevant limitations here consists in C's rights against A – rights that A cannot abrogate by means of an agreement with B that is designed to render A bound to obey B's order to kill C.[27] Yet, this incapacity of A and B jointly to arrange for A to slough off his duty not to kill C hardly shows that some *right* of A's is inalienable. What, after all, is the right of A which is supposed to be inalienable?

Anticipating this question, Barnett suggests that the key right here is A's right to respect C's rights.

> Is A's right to control his own person and respect C's rights – a right that A indisputably starts with – a right that can be transferred? ... [N]o. Notwithstanding any agreement he may have made to B, *A is still under a duty to respect C's rights*. A cannot, therefore, alienate such complete control of his future actions to anyone (1986, p. 187, n. 33, emphasis added).[28]

But surely it is A's *duty* not to kill C which must bear all the argumentative weight here. Suppose that A merely had a right not to kill C (but *no duty* correlative to C's right not to be killed) so that it was permissible for A not to kill C and it was impermissible to force A to kill C. A would be able to waive *this right*. That is, A would be able to bind himself not to exercise this right to not kill C and A would be able to release others (e.g., B) from their duty against forcing A to kill C. A's waiving his right not to kill C would be no different from my waiving my right not to hop on my left foot, thereby making it permissible that I be forced to hop on that foot. Hence, the limitation upon what A can validly agree to is anchored in A's duty not to kill C, not his right not to kill C.

For A to be the bearer of an inalienable right is for there to be some area of control over himself, some realm of liberty, which ineluctably attaches to A. That there is some realm of liberty is surely what Barnett means to convey when he concludes that "A cannot, therefore, alienate such complete control of his future actions to anyone." But, because what in fact ineluctably attaches to A (absent C's agreement) is a duty (to C), there is no such irreducible residue of control of A over himself. There is only the moral necessity that he not kill C either under his own or B's command.[29]

Barnett's second argument for inalienability turns on the impossibility of A's transferring direct control of his body to B. Barnett argues that,

> a right to control a resource cannot be transferred where the control of that resource itself cannot in fact be transferred ... But bodies are different from other kinds of things. What is *my* body cannot in any literal sense be made *your* body. Because there is no obstacle to transferring control of a house or car (of the sort that is unavoidably presented when one attempts to transfer control over one's body) there is no obstacle to transferring the right to control a house or car. But if control cannot be transferred, then it is hard to see how a right to control can be transferred (1986, pp. 188-189).[30]

Certainly it is plausible that some transfers simply cannot take place and that, especially when that impossibility is known to both parties, there can be no valid contract effecting that transfer. This may, indeed, be a barrier to A's successfully transferring his agent-type control over A's body to B.[31] But no one asserting the alienability to B of A's right to liberty has such a transfer of agent-type control in mind. What is imagined is simply that A, due to his consent, no longer has a legitimate complaint when B treats him like a slave, i.e., when B orders him around, enforces those orders with threats, physically manipulates A, and so on. We know all too well that people can be subjected to this sort of treatment – though the subjection is almost always non-consensual. Why cannot this sort of initially impermissible, but perfectly possible, treatment be rendered permissible (even if not morally attractive) by the consent of its subject? A can waive his right to his 1987 Yugo, thereby making it permissible for B to demolish it with his sledgehammer. There seems to be no reason provided by Barnett that A cannot in parallel fashion waive his right not to be subject to B's coercive, albeit not agent-type, control. Thus, neither of these arguments suggested by Barnett offer a plausible route for asserting the inalienability of Lockean natural rights.

IV. THE STANDING OF NATURAL RIGHTS AND THEIR BIOLOGICAL IMPLICATIONS

This essay is primarily devoted to defending a particular understanding of Lockean natural rights and their inalienability – an understanding which seems to have many direct and robust libertarian bioethical implications. Nevertheless, it seems worthwhile to address explicitly two dimensions of the contemporary relevance of Lockean natural rights. The first dimension is the plausibility of the affirmation of such rights as the fundamental principle for evaluating legal institutions and social policies. The second dimension is basic structure and character of the bioethical implications of these rights and their alienability.

A. The Philosophical Standing of Lockean Natural Rights

It is, of course, impossible in a philosophical sidebar to provide either a compelling defense of one school of moral philosophical thought or a definitive critique of its major competitors. One can, however, do two things in a few paragraphs. First, one can draw attention to the extent to which the Lockean foundational arguments for rights which I described above and the force of these arguments are separable from the theological content that they have for Locke. Thus, one can see how the force of these arguments is independent of anything like Locke's own theological commitments or historical setting. Second, one can describe in more contemporary terms the general character of the case for Lockean-style natural rights as it has been presented by recent friends of such rights.

I have argued that God's role within Locke's theory of people's natural rights with respect to one another is fundamentally negative. God is invoked to deny that persons are bound to sacrifice their respective self-preservation and happiness for anything like the utilitarian aggregate. God is invoked to deny any natural (or divinely ordained) inequality of rights. God is invoked negatively in the sense that Locke's point is primarily to deny that there is an acceptable basis for others' claims that God has set us the task of directly serving all mankind or has established a natural order of mastery and subordination among us. Absent a context in which theologically based claims of this latter sort will be forceful, there is no need for the defender of Lockean rights to appeal even to Locke's negative claims about God's moral design.

Rather what the friend of Lockean natural rights needs to appeal to – what, I contend, Locke himself ultimately was appealing to – are the medley of facts associated with the separate, equally ultimate, moral importance of each individual. This separate value and importance makes it rational for each person to pursue the fullest realization of his own values, to advance his own (true) good. Yet, the comparable existence of others, each of whom pursues his or her own good with an equal propriety, imposes constraints on the manner in which each may promote his ends. Genuine recognition of the existence of these other, co-equal, value-pursuers requires that one not treat these others as mere resources at one's own disposal but rather as moral ends-in-themselves. Moreover, the existence of other value-pursuers makes it rational for each of us to assert rights that protect us from others' exploitation and, given the moral equality of persons, the rights one rationally asserts for oneself cannot rationally be denied of others. Within the kingdom of value-pursuers, each of whose life and happiness is of ultimate incommensurable value, the fundamental *interpersonal* principles protect each individual from his subordination to others and their purposes by ascribing to each a fundamental claim of self-sovereignty. The details of the concrete manifestations of these basic claim-rights are deeply affected by convention, social understanding, and institutional history. Nevertheless, these basic claim-rights to themselves stand as fundamental constraints upon conventions, social understandings, institutions, and policies – constraints that are protective of the individual's choices about what he shall do or what he shall permit to have done to his person.[32] Of course, the moral boundaries drawn by an individual's initial rights, by historical concretizations of her basic rights, can be redrawn by that individual's voluntary choices or by voluntary agreements between that individual and others.

B. Fundamental Bioethical Implications

There are two fundamental immediate implications of each person's alienable self-sovereignty. The first is that, absent some extraordinary and compelling countervailing consideration (such as, a forfeiture of rights because of heinous criminal activity), subjecting any standard (i.e., adult, competent) rightholder to unchosen medical treatment violates that individual's rights and is impermissible – whether or not that treatment is of, for, or by private individuals, non-governmental associations, or governmental agencies. The second implication is that any individual who qualifies as a standard rightholder may subject herself to any medical procedures, i.e., to any form of self-treatment, she chooses; and she may secure for herself any form of treatment by others that others are freely willing to provide. An individual need not waive her rights in order permissibly to subject *herself* to even foolishly dangerous or bizarre treatment. For her rights over herself constrain *others'* treatment of her; they do not constrain her treatment of herself. And, an individual may waive (alienate) her rights and, thereby, open herself up to treatment which otherwise would be impermissible – to, e.g., the surgeon's opening of her patient's thoracic cavity or the oncologist's injecting life-threatening chemicals into his patient. Unless they have been waived or contracted away, an individual's rights over herself protect her against moralistic or paternalistic interference with her subjecting herself to such treatment as she chooses or with her non-coercively securing such treatment from others.

Of course, an individual may bind herself not to pursue certain forms of treatment and, more generally, not to make certain "lifestyle" choices as part of a package whereby she secures others' commitment to pay for or provide her with other courses of medical care. She may, e.g., only be able to purchase a right to open-ended care for heart disease or cancer if she foreswears heavy drinking or medicating herself with quackcillin. In order to elicit Mary's binding agreement to forego certain choices, the prospective provider of health care may threaten to withhold from Mary *any and all care to which Mary does not already have a right.* Suppose, as I believe is certain on Lockean grounds, that Mary has no rights against William that he (or he in conjunction with other members of society) provide her with enormously expensive coronary or cancer care. If Mary lacks this right, then William (or William in conjunction with the other members of society) may offer to vest Mary with a right to that care on the condition that Mary foreswear certain choices which otherwise she would have every right to indulge. Mary, of course, may refuse the offer or search for a more pleasing set of terms from some other prospective supplier. But, should Mary agree to forego her right to imbibe a fifth of Old Grand Dad bourbon a day in exchange for William's making various medical care available to her, she will be bound to eschew this unhealthy habit and she will forfeit her contractual right to the care if she breaks her part of the agreement. Private medical insurers or health care organizations, who (we are supposing) have no initial obligation to provide health care to Mary, may propose to Mary whatever terms they think best in light of their own paternalistic or moralistic preferences or in light of their calculations of the cost-savings to be achieved through discouraging unhealthy choices and competitive pressures. Through agreements with such insurers and organizations, individuals may acquire additional motives or even obligations to act responsibly with regard to their own health. From the perspective of Lockean rights, of course, individuals must remain entirely free not to enter into any such agreements and to bear the full costs of their own decisions and fortune or misfortune.

Two major complications exist if the government is the provider of health insurance or health services. The first concerns the legitimacy of state taxation which, in effect, forces some individuals in part or in whole to pay for the medical care of others. The second concerns the requirement that state policies, be in some significant sense, neutral between competing conceptions of the good life. Is redistributive governmental financing of health insurance or health services through taxation mandated by, permitted by, or in violation of defensible principles of economic justice? My own view is that it violates defensible Lockean principles of economic justice, *viz.*, principles of ownership in extra-personal things which allow for the division of the extra-personal world into protected individual domains in a way that parallels the Lockean division of the world of persons into protected individual domains. My view is that redistributive governmental financing of health insurance or services, at the least, violates the property rights (where such property rights are permitted to arise – as they should be) of those on whom it imposes net losses. (If participation is mandatory, it violates the rights of anyone who would opt out except for the threatened sanctions.) However, this objection to the coercive socializing of the financing of medical treatment takes us far beyond Lockean *self-ownership* and the alienability of rights over one's person. It cannot be pursued or defended here.[33]

Nevertheless, another special problem arises if the state specifies the terms under which individuals will receive medical insurance or care; i.e., if the state decides what conditions individuals have to agree to and abide by in order to qualify for what forms of medical treatment. A private insurer or provider may legitimately insist that a prospective client/patient agree to forego this or that particular activity, e.g., smoking, mountain climbing, the consumption of rich desserts, or surrogate child bearing, in order to be vested with a right against that insurer or provider to so-and-so medical care. And a private insurer or provider may legitimately refuse to cover or to offer certain medical services, e.g., abortion, assisted euthanasia, cosmetic surgery, or primal scream therapy. In setting these terms or making these exclusions such private insurers or providers are not being neutral among the diverse (albeit rights-respecting) conceptions of the good life entertained by their prospective clients. This, as I have said, is no problem. But, within the liberal Lockean tradition, the state is supposed to remain neutral between competing (rights-respecting) conceptions of the good (Mack, 1988). It fails to be neutral if it sets any substantive conditions on persons' being vested with a right to insurance, or care, or if it substantively restricts what sorts of medical care is covered or offered. Yet the absence of conditions and restrictions is a recipe for societal financial disaster. The state could put in place diverse sets of conditions and restrictions while preserving some degree of neutrality by offering individuals many different packages – each of equal market value – within each of which some set of some restrictions upon persons' conduct would be linked with rights to a certain range of medical coverage or treatment. Better yet, the state could issue vouchers for medical insurance or care with which each individual could purchase a medical insurance or care package of her own choosing from among the diverse packages that would be offered by private insurers or providers. Such a system would have all the advantages with respect to preserving governmental neutrality and encouraging a wide range of packages responsive to consumer choice that a well-designed educational voucher system would have. Under such a system, aside from the provision of the vouchers, the role of government would be restricted to insuring that all parties comply faithfully with the terms of their voluntary agreements, that all parties remain free to enter into any rights-respecting agreement including ones that involve the alienation of rights, and that all individuals remain free to exercise their natural (and unalienated) rights and their duly acquired contractual rights.

V. CONCLUSION

All that remains is to review the reasons for denying that Locke asserts or is committed to the inalienability of the natural rights to life and liberty. I noted, following and consolidating points developed by Simmons, that what often passes for the inalienability of rights is actually the assertion of an inescapable duty to God to preserve one's own life. I have stressed the separateness of this assertion of duty and Locke's theory of natural human rights and supported this reading by appeal to the firm distinction between sins against God and violations of rights against men present in Locke's *A Letter Concerning Toleration*. I have also investigated the extent and character of Locke's belief in the psychological inalienability of the rights to life and liberty. It appears that Locke did not maintain that every full waiver of these rights would be impossible for a rational agent. And, were he to have made this broad claim, this would merely signify adherence to implausible empirical beliefs and not to any inherent feature of Lockean rights which point to their inalienability.

A reaffirmation of the domain of choice and side-constraint character of Lockean rights shows that these rights protect their bearers from certain forms of treatment at the hands of others and do not, themselves, prescribe goals, e.g., the goal of self-preservation, to the holders of these rights. The assertion of such rights does not imply that moral theory and moral insight are incapable of providing individuals with sound *advice* about how to exercise their moral jurisdiction over themselves, their lives, and their liberties. It merely implies that the unsoundness or even sinfulness of their chosen actions does not render them impermissible and subject to forcible prohibition as long as those actions are within the spheres protected by individual rights.

Thus, an endorsement of natural rights as domains of choice is fully compatible with a moral objectivism about which choices among a person's permissible set are proper, commendable, and duly responsible. Or, more precisely, belief in such rights is compatible with the forms of moral objectivism that do not so morally conscript everyone into the service of common ends as to leave no room or point in the basic structure of the normative universe for individual rights. Domain of choice basic rights generative of an ethic of informed consent and mutually voluntary interaction need not be seen as a morally minimalist solution to the post-modern demise of the objectively good life.[34]

Such a non-imperialistic moral objectivism is fully compatible with natural rights which serve as "eternal" fences against one's subjection by others – even when those others have one's well being at heart. These eternal fences do not, however, stand as inalienable barriers to one's own choices about one's person, life, or liberty or to one's acquired rights to medical insurance or care being subject to a variety of conditions and restrictions. Thus, if our fundamental Lockean heritage lies in the still plausible affirmation of natural rights to our persons, lives, and liberties, it is the moral power to choose what becomes of our lives and liberties that is essential to that worthy heritage; not duties to choose in certain externally mandated ways.

Department of Philosophy
Tulane University
New Orleans, Louisiana

NOTES

[1]   This essay was originally prepared for a Liberty Fund Conference on personal responsibility and personal freedom in health care organized by Baruch Brody. I am grateful to John Simmons and Mary Sirridge for their critical comments on earlier drafts and to my co-participants at the conference for their stimulating discussion.

[2]   We must distinguish three different ways in which a person may be thought to lose a right: alienation in which the right is transferred or waived by the choice of the rightholder; forfeiture in which the right is lost through the rightholder's wrongdoing; and prescription in which the right is taken or waived through the choice of some other party. Moreover, it is a separate issue whether a right is absolute or subject to being overridden. Often when people say that a right is inalienable what they really mean is that it is imprescriptible and/or absolute. See Simmons (1983, especially pp. 176-180).

[3]   See the references in Simmons (1983, n. 4, pp. 175-176).

[4]   Bracketed citations in the text are to sections of Locke's *Second Treatise.*

[5]   Note that Simmons argues that the inalienability theses only "seem" to rest on paternalistic concerns. On Simmons' reading of Locke, they actually rest on beliefs about duties to God. See also McConnell (1984). McConnell denies the need to appeal to paternalistic concerns to vindicate the classification of certain rights as inalienable. Instead he appeals to the social costs of determining whether certain apparent waivers or transfers of rights are genuinely voluntary.

[6]   In an important essay on inalienability, Margaret Radin provides an extended argument, in the name of inalienability, for restricting persons in the sale of their time, bodily parts, services, and property – restrictions that are all contrary to the "radical voluntarism Locke routinely espouses" (Radin, 1987). As a partial response to Radin, see Mack (1989a). Also, see Boyle (in this volume) and Mack (1991).

[7]   The conception of rights that Simmons reads into Locke is only one of many competitors to the domain of choice conception. Another would be the conception of rights congruent with the natural law morality outlined by Joseph Boyle. In both cases, individual rights are not morally fundamental. Rather they codify moral rules compliance with which is valued because it most advances either general preservation (in the case of Simmons) or basic goods and the fulfillment of human duties with regard to them (in the case of Boyle).

[8]   In particular, I assess the arguments drawn together in Barnett (1986, especially pp. 186-195).

[9]   See, for example, McConnell (1984, p. 43): "The inalienability of the right to life need not preclude the permissibility of suicide." Thus, the advocate of the inalienability of this right need not allow paternalistic preventions of self-reliant suicide.

      See also, Barnett (1986, pp. 192-193, emphasis added): "Surely the account just provided is not paternalist. No one may rightfully interfere in the consensual sacrificial conduct of a competent adult ... simply because the intermeddler knows what is truly best for the individual making the sacrifice. Rather, this argument against the enforceability of agreements to transfer control over one's body is that the law should not specifically enforce certain commitments *when the party who made the commitment thinks better of it.*"

      Notice, in fact, that Barnett seems only to disallow the *permanent* waiver of the right not to be killed. The religious leader who conducts the consensual sacrifice of a member of his flock before that member "thinks better of it" does not infringe upon the member's rights. Compare this with Feinberg's distinction between the inalienability that disallows the permanent and irrevocable surrender of a right and the inalienability that also disallows temporary waivers in which one retains "the right to change one's mind *at any point* and thereby nullify the transaction. Feinberg endorses only the former inalienability (1978, p. 118).

[10] But he does have to be able to waive that right in order to arrange for others to be at liberty to assist his suicide. That is the way in which the inalienability of William's right not to be killed would constitute a serious barrier to William's voluntary suicide.

[11] See, for example, McConnell's interpretation of these passages, *viz.*, "According to Locke, people may not transfer their rights to life and liberty because in some sense these are not theirs to transfer" (1984, p. 45).

[12] And had Locke explicitly addressed this issue many factors other than the inner logic of his theory of rights might have affected his actual pronouncements.

[13] That uncharitableness is a sin but not a violation of rights either strongly contrasts with, or requires a rereading of the famous passage from the *First Treatise* according to which, "Charity gives every man a title to so much out of another's plenty, as will keep him from extreme want, where he has no means to subsist otherwise ... " (Locke, 1967, sect. 42). One should also remember the context of section 42 of the *First Treatise*. Locke is arguing that even if one man, in the direct line of Adam, were the owner of all land, he would not be justified in withholding everything from other people in order to secure their consent to his political authority.

[14] In *An Essay on Human Understanding* Locke asserts that, "the true ground of morality ... can only be the will of God, who sees men in the dark, has in his hand rewards and punishments and power enough to call to account the proudest offender" (1959, p. 70). Similarly, "what duty is, cannot be understood without a law; nor a law be known or supposed without a lawmaker, or without reward and punishment" (1959, p. 76). It is difficult to see how, on this theory, any moral knowledge would be available to someone who lacked a detailed knowledge of God's existence and intentions. (Locke seems eager to reach this conclusion for the sake of demonstrating that there are no *innate* practical principles.) Putting that massive problem aside, it can be noted that the dependence of all duties on God, need not require that all duties be *to* God, i.e., that the bearer of the right correlative to each and every duty be God. There could still be some duties, e.g., duties that properly concern the magistrate, such that the bearers of the rights correlative to them are one's fellow human beings.

¹⁵ Indeed, in the brief discussion that first drew my attention to this passage from the *First Treatise*, Jeremy Waldron concludes that, the "prohibitions on suicide and voluntary enslavement concern the [property – actually the lack of property – that persons' have in themselves *vis-á-vis* God] while the claims about self-ownership involves only a man's right against other men" (1988, p. 178).

¹⁶ Richard Tuck (1979) masterfully provides the background for the felt need of seventeenth century anti-authoritarian theorists to deny the rationality of the surrender of rights to life and liberty. A passage from Locke's friend James Tyrrell's *Patriacha non Monarcha* (1981) nicely exemplifies the idea that only incompetents would (try to) waive their fundamental rights: "no man can be supposed so void of common sense (unless an absolute Fool, and then he is not capable of making any Bargain) to yield himself so absolutely up to another's disposal ... (p. 155). Some of the numerous passages supplied by Tuck, e.g., the passage from Anthony Ascham's *Of the Confusions and Revolutions of Government* (1649) seem to express the bolder claim that something about the nature of the right or some other fundamental moral fact about persons makes the waiving of these rights "one of our moral impossibilities" (p. 153).

¹⁷ This is how Simmons reads the passage (1983, p. 201).

¹⁸ However, Simmons demurs from describing Locke's theory as "*seriously* utilitarian" on the grounds that "it rests on a divine will foundation and is concerned to promote preservation (over happiness or pleasure)" (1983, p. 199, n. 55). Note that Simmons' claim is that Locke is committed to a (rule) utilitarianism *of preservation*, not a utilitarianism *of rights*. The former position defines persons' rights in terms of claims to the enforcement of those rules respect for which, compared with any alternative set of rules, would maximize human preservation (see, for example, 1983, n. 56, p. 199). The latter specifies persons' rights independently of considerations of what rules would maximize preservation and then favors the maximum satisfaction of those rights (or adherence to that set of rules which, compared to any alternative set of rules, maximizes the satisfaction of those rights).

¹⁹ Although Locke often talks about moral rules and about best preserving mankind, the distinctive feature of a *rule* utilitarian position seems absent. This would be some argument for preferring a principled adherence to a set of rules which might yield a lesser total preservation of mankind than would be generated by an enlightened case by case application of the master rule that mankind is to be maximally preserved. Perhaps the most plausible candidate for a *rule* adopted by Locke on utilitarian grounds would be, "prefer the innocent to the aggressor." But see the discussion below, including note 23).

²⁰ In Locke's *Essay* it is happiness rather than self-preservation that is the goal of rational action. According to Locke, " ... the highest perfection of intellectual nature lies in a careful and constant pursuit of true and solid happiness." But, the " ... pursuit of happiness in general, which is our greatest good, and which, as such, our desires always follow ... " consists, more precisely, in each person's pursuit of his own happiness (1959, p. 348). "[A]ll good, even seen and confessed to be so, does not necessarily move every particular man's desire; but only that part, or so much of it as is considered and taken to make a necessary part of *his* happiness. All other good, however great in reality or appearance, excites not a man's desire who looks not on it to make a part of that happiness wherewith he, in his present thoughts, can satisfy himself" (1959, p. 341). Thus, Locke's claims in the *Essay* that each ought to pursue human happiness must be read distributively, not aggregatively. This parallel supports the present reading of the *Second Treatise*.

²¹ Thus, I am arguing that, within the body of section 6, Locke is not concerned with the *duty* of self-preservation. The claim that we are the property of God is not used to support this duty and even the assertion that each is *"bound to preserve himself"* serves to mark the rationality of each person's pursuit of his preservation rather than any obligation (to God) to do so.

The claim that man, despite his "uncontrollable liberty to dispose of his person or possessions, yet ... has not liberty to destroy himself ... ", does indeed occur at the very beginning of section 6. In this opening sentence, Locke seems to me to shift temporarily from his account of man's natural freedom as reason, i.e., as the law of nature, requires us to understand it, to a separate and distinct refinement of his conception of our *"state of liberty,"* *viz.*, that *"it is not a state of license."* The prohibition on "license" – which includes a requirement that man not arbitrarily destroy "any other creature in his possession" – could not be defended and is not defended on the ground of our equal and independent status *vis-á-vis* one another. It seems *not* to be included by Locke among the dictates of the law of nature – dictates that regulate individuals' interaction with one another – to which topic Locke returns following this brief declaration against suicide.

This reading coheres with Locke's position in *A Letter on Toleration*. The limits on our natural freedom marked by others' rights (= the law of nature?) are a matter of our equal status as beings each of whom rationally pursues his own preservation. The limits on our natural liberty such that it is not a state of license are requirements not to sin, not to violate certain duties to God, even though such sinning would not trespass upon the rights of other men.

²² But the non-spoilage and the as good and enough conditions may be traces of a positive function of mankind's original common ownership of the earth (Locke, 1967, *Second Treatise*, pp. 32,33).

²³ It may well be true that following the rule "prefer the innocent" better preserves mankind as a whole than would adherence to any other rule. But if this were Locke's reason for endorsing the rule, he would not have to argue that, through their actions, the aggressors had forfeited their rights. The classification of some as innocents and others as aggressors who have forfeited their rights is independent of any calculation of consequences.

²⁴ The wife and children may retain less than they had once expected, less than the father could have legitimately bequeathed to them had he not forfeited his unencumbered estate. But they must receive all of what they have title to in virtue of their (innocent) relation to the aggressing husband and father.

²⁵ Although, according to Simmons, the basis for this right is the presence of a rule against killing (non-threatening) others in that set of rules adoption of which will most preserve human life at large.

²⁶ The arguments considered are the first two presented by Barnett. The third appeals to the derivation of rights from duties. "Rights ... may arise from duties owed to oneself. Suppose that it could be shown that one has a moral duty to live a good life or pursue happiness. Such a duty may imply that it would be wrong for others to interfere with such actions, which would mean that we would have rights to be free from such interference. Would not such a claim also imply that some of these rights may not be transferred to another by consent" (p. 191)? This is the argument for "mandatory claim rights" that I criticize earlier.

The fourth argument offered by Barnett is an argument from the high administrative costs of legally assuring that apparently voluntary and competent waivers of rights have been genuinely voluntary and competent to the conclusion that there should be "a blanket prohibition on the transfer of certain kinds of rights" (p. 193). However, it seems that, at most, such considerations would justify requiring those who pursue such transfers to bear the costs of proving their voluntariness and competence. Barnett also cites the distinct consideration of "the high cost [to the slave-to-be] of [our] erroneously deciding that a slavery contract was truly voluntary" (p. 193). But surely against this has to be weighed the cost to the agent who voluntarily and competently seeks slavery and has this option summarily forbidden.

[27] See, for example, Epstein (1985, p. 973): "One assumption implicit in the case for free alienation is that the buyer's use of the property after alienation [in the example here, B's use of A] does not violate a rule of tort or criminal law.

[28] Note that the alienability of "A's right to control his own person" is simply not at issue.

[29] Barnett, quite correctly, notes that this first argument leaves untouched all agreements to obey rights-respecting orders (p. 188). Thus, even if accepted, the argument would block no agreements not *already* subject to prohibition.

[30] This argument nicely captures the idea that the possession of rights is so intimately connected with one's status as an agent that, for one to alienate one's rights, one would have to undo one's agency. See, for example, that wonderful passage from Richard Overton's *An Arrow Against all Tyrants* (1646): "To every Individual in nature is given an individual property by nature, not to be invaded or usurped by any: for every one as he is himselfe, so he hath a selfe propriety, else could he not be himselfe ... " (quoted in MacPherson, 1962, p. 140).

[31] But it is not clear how much of a barrier there is, in fact, to A's transferring or, at least, waiving his right of agent-type control. There are, as Barnett notes (1986, p. 188, n. 34), numerous ways in which A's capacity to control himself can be undercut. Might not A, in fact, agree to and be subject to relevant drug treatment or neurosurgery? Barnett asserts that "... a promise to undergo a dependency-inducing procedure would be an unenforceable attempt to transfer an inalienable right: the right to control whether or not to submit to the operation" (1986, p. 188, n. 35). But why is *this* right inalienable? It cannot be that it is impossible for one to be subject to this operation!

[32] For a sample of recent defenses of this type of natural rights position, see Lomasky, 1987, Mack, 1989b, and 1992, and DenUyl and Rasmussen, 1991.

[33] For an important statement of anti-redistributive Lockeanism, see Nozick, 1974. Towards a statement of the appropriate property rights theory, see Mack, 1990.

[34] For a philosophically powerful presentation of this morally minimalist understanding, see H.T. Engelhardt, Jr, (in this volume; 1986, especially chapters 1-3; and 1996). On the need to understand value as "personal" or "agent-relative" in order to provide room and purpose for individual rights, see Lomasky (1987, especially chapters 2-4) and Mack (1989b).

BIBLIOGRAPHY

Barnett, R.: 1986, 'Contract remedies and inalienable rights,' *Social Philosophy and Policy*, 4, no. 2, 179-202.

DenUyl, D. and Rasmussen, D.: 1991, *Liberty and Nature*, Open Court Publishing, Peru, IL.

Boyle, J.: 1999, 'Personal responsibility and freedom in health matters: A contemporary natural law perspective,' in this volume, pp. 111-141.

Engelhardt, H.T.: 1999, 'The body for fun, beneficence, and profit: A variation on a post-modern theme,' in this volume, pp. 277-301.

Engelhardt, H.T.: 1996, *The Foundations of Bioethics*, second edition, Oxford University Press, New York.

Engelhardt, H.T.: 1986, *The Foundations of Bioethics*, Oxford University Press, New York.

Epstein, R.: 1985, 'Why restrain alienation?' *Columbia Law Review* 85, no. 5, 970-990.

Feinberg, J.: 1978, 'Voluntary euthanasia and the inalienable right to life,' *Philosophy and Public Affairs* 7, no. 2, 93-123.

Locke, J.: 1959, *An Essay Concerning Human Understanding*, Dover Publications, New York.

Locke, J.: 1967, *Two Treatises on Government*, second edition, P. Laslett (ed.), Cambridge University Press, Cambridge.

Locke, J.: 1983, *A Letter on Toleration*, J.H. Tully (ed.), Hackett Publishing, Indianapolis.

Lomasky, L.: 1987, *Persons, Rights and the Moral Community*, Oxford University Press, Oxford.

Mack, E.: 1980, 'Locke's arguments for natural rights,' *Southwestern Journal of Philosophy* 9, no. 1, 51-60.

Mack, E.: 1988, 'Liberalism, neutralism, and rights,' in *Religion, Morality, and the Law*, J. Chapman and R. Pennock (eds.), New York University Press, New York.

Mack, E.: 1989a, 'Dominos and the fear of commodification,' in *Markets and Justice*, J. Chapman and R. Pennock (eds.), New York University Press, New York.

Mack, E.: 1989b, 'Moral individualism: Agent-relativity and deontic restraints,' *Social Philosophy and Policy* 7, no. 1, 81-111.

Mack, E.: 1990, 'Self-ownership and the right of property,' *The Monist* 73, no. 4, 519-543.

Mack, E.: 1991, 'Ethical limits of economic liberty,' *Encyclopedia of Ethics*, Garland Publishers, New York.

Mack, E.: 1992, 'Agent-relativity of value, deontic restraints, and self-ownership,' in *Value, Welfare, and Morality*, R. Frey and C. Morris (eds.), Cambridge University Press, Cambridge.

McConnell, T.: 1984, 'The nature and basis of inalienable rights,' *Law and Philosophy* 3, no. 1, 25-59.

MacPherson, C.B.: 1962, *The Political Theory of Possessive Individualism*, Oxford University Press, Oxford.

Nozick, R.: 1976, *Anarchy, State, and Utopia*, Basic Books, New York.

Radin, M.: 1987, 'Market-inalienability,' *Harvard Law Review* 100, no. 8, 1851-1937.

Simmons, A.J.: 1983, 'Inalienable rights and Locke's *Treatises*,' *Philosophy and Public Affairs* 12, no. 3, 175-204.

Tuck, R.: 1979, *Natural Rights Theories*, Cambridge University Press, Cambridge.

Waldron, J.: 1988, *The Right to Private Property*, Clarendon Press, Oxford.

GEORGE KHUSHF

INALIENABLE RIGHTS IN THE MORAL AND
POLITICAL PHILOSOPHY OF JOHN LOCKE:
A REAPPRAISAL*

There is a long and venerable tradition of political theory that gives considerable credence to the notion that there are inalienable rights, rights such as those to life, liberty, and the pursuit of happiness (or property). Such rights provide the sufficient condition for resisting certain institutions like slavery, unlimited and arbitrary government, dueling, and euthanasia. However, recently inalienability has been broadly criticized (Kendall, 1941; Nozick, 1974; Stell, 1979; van de Veer, 1980; Simmons, 1983; Simmons, 1993; Mack, 1998). Several philosophers have considered and rejected any justification that has been provided for these rights, and, surprisingly, some of the most significant criticisms have come from staunch defenders of those types of rights traditionally associated with the thesis of inalienability.

To these critics, an assertion of inalienability amounts to unwarranted paternalism (Nozick, 1974, p. 58; Feinberg, 1978; Simmons, 1983, p. 186; for a criticism of that contention, see McConnell, 1984). Why should the holder of a right to life be prohibited from waiving that right in a duel or war game, or, even more significantly, be prohibited from voluntary active euthanasia, should suffering become too extreme? And while there may be only a small minority actively lobbying for the right to duel (Stell, 1979), within current bioethical debates euthanasia has a strong following. These debates, of course, are not put in terms of the alienability of the right to life, but the implications for such a position would seem to be clear.

When one goes to the literature searching for a defence of inalienability, it turns out that there is relatively little by way of philosophical justification.[1] More often than not, it is simply assumed; inalienability is the rock bottom, self-evident basis from which argumentation proceeds (as in the U.S. Declaration of Independence). And when one does find arguments for inalienability, as one does in authors such as Brown, Hart, and Frankena, for example, it is for something like hypothetical (as opposed to categorical) inalienable rights (Hart) or an inalienable right to *institutions* that protect moral interests,

M.J. Cherry (ed.), Persons and their Bodies: Rights, Responsibilities, Relationships, 177–206.
© 1999 *Kluwer Academic Publishers. Printed in Great Britain.*

persons, or estates (Brown) – in each case something very different from what was traditionally asserted (for additional attempts at justification, see Meyers, 1981; McConnell, 1984; Schiller, 1969; each considers the "myth value" of the inalienability claim). Thus, if one wants to evaluate the philosophical merit of inalienable rights, the most appropriate place to begin would be with the political theorist who is most often taken as the champion of such rights, John Locke.

In this essay I will consider two recent interpretations of Locke, which, in different ways and coming from very different directions, both argue that Locke did not advocate the inalienability of rights. I will attempt to show that these interpretations miss certain essential moves in Locke's moral and political theory, and that they fail to appreciate the relation between Locke's argument against unlimited and arbitrary government and his argument against an individual's unlimited and arbitrary control over his or her life. Inalienability will prove to be an important element in this linkage, and it will rest in a natural theological premise that enables Locke to move beyond a Hobbesian account of natural liberties to claim rights. After reconstructing this argument for inalienability in Locke, I will appraise its implications and viability for the current debate on inalienability. Before moving into this discussion, however, it will be helpful to start with a brief account of what is meant by inalienability, and how this can be distinguished from other attributes of rights such as imprescriptibility, indefeasibility, and nonforfeiture.

I. THE MEANING OF INALIENABILITY

There is actually some ambiguity in the meaning of an inalienable right, and it is important to specify exactly what I mean in the discussion. In order to somewhat narrow the topic, I focus on the inalienability of the right to life, although I often consider other rights, especially that to liberty. I regard the inalienable right to life as a negative, nonabsolute, forfeitable moral right, which cannot be morally waived or transferred by the individual that possesses the right. It may or may not be prescriptible and it may or may not be a mandatory claim right.

This definition can be unpacked as follows:

(1) The right to life is a moral right, not a legal right (for an account of this difference, see Khushf, 1992). In Lockean terms, it characterizes people in the state of nature, and provides a basis independent of any

government upon which the limits of government can be determined. It is a human right, accrues to all by virtue of their humanity, and thus does not depend on any human institution for its existence (Meyers, 1981, pp. 127-128; Feinberg, 1978, p. 97).

(2) The right to life is forfeitable through an act of violence against another. In Lockean terms, a violation of another's rights puts one into a state of war with that person.

> I should have a right to destroy that which threatens me with destruction: for, by the fundamental law of nature, man being to be preserved as much as possible, when all cannot be preserved, the safety of the innocent is to be preferred: and one may destroy a man who makes war upon him ... for the same reason that he may kill a wolf or a lion; because such men are not under the ties of the commonlaw of reason, have no other rule, but that of force and violence, and so may be treated as beasts of prey, those dangerous and noxious creatures, that will be sure to destroy him whenever he falls into their power (Locke, 1980, sec. 16).

(I quote this at some length because, as we will see, the tie between innocence, natural law, reason, and the state of war will become important later.) The inalienability of the right to life thus does not imply that capital punishment is illicit, as some have assumed.

(3) The right to life is not absolute, or, in other words, it is not indefeasible in the strong sense (McConnell, 1984, pp. 29-30; Feinberg, 1978, 97-104; for a discussion of the difference between the weak and strong senses of "indefeasible" see Schiller, 1969, p. 313). This follows from being forfeitable. It may also be overridden in certain circumstances; for example, when all are threatened by an outside aggressor, a subject may be drafted into armed service. (This may be an area where, contra Locke [Simmons, 1983, pp. 178-179, 184-185], a right to life and liberty is even prescriptible, as in the case of conscription; for an alternative interpretation of Locke on prescriptibility, however, see Kendall, 1941.) Some inalienable rights may be overridden in certain extreme circumstances, such as a limit on the right to liberty in the case of civil unrest. It is thus important not to confuse inalienability with the absoluteness of a right.

(4) The right to life is a negative right (for a discussion of the difference between positive and negative rights, see Khushf, 1992). This is probably the most contested part of the definition. However, even for

those who argue for positive rights, it is important to see that the right to
life would be a complex or cluster right, which is both positive and
negative, and one would need to distinguish between dimensions when
considering inalienability.[2] It may be that a claim on food or housing
needed to preserve life would be alienable (much as property), while the
negative right still would not be. Thus, my emphasis on the negative right
can be viewed as a minimalist option, on which one could, in principle,
add an additional positive right.[3] (However, one would need to also
recognize the inverse proportionality between positive and negative
rights; Khushf, 1992.)

Thus, the right to life will be regarded as more than a mere liberty
(contra-Hobbes), entailing a correlative duty on the part of others not to
inappropriately interfere with or take one's life; it will therefore be a
claim right, rather than a mere liberty (Feinberg, 1978, p. 95). On the
other hand, the right to life will be less than a positive right, since it will
involve no claim on others to provide that which is needed to preserve
life. The right to life will mark off a domain or jurisdiction where the
individual may pursue diverse individual and communal ends.

(5) That the right to life is inalienable means that it cannot be morally
waived or transferred. This does not mean that it cannot be immorally
waived or transferred. Joel Feinberg (1978, p. 112) has pointed out the
curious correlate of the doctrine of an inalienable, forfeitable right:
someone wanting to alienate this right could do so by violating the rights
of others, and thus forfeiting his or her right to life. This has led some to
conclude that a right that can be forfeited by a willful act is necessarily
alienable (van de Veer, 1980, p. 168; for critique see McConnell, pp. 27-
28). However, the act that would lead to its forfeiture is an immoral act,
and thus one could not "legitimately" alienate the right. The alienation
may be effective, but it is not a true transaction, since it takes place by
reducing the offender's status to that of an animal, and, in the ensuing
state of war, allows the offended individual to take possession of that
"beast." One way to put this is to say that one cannot alienate certain
human rights without the alienator being transformed into a subhuman,
and the fact that such a transformation would take place in any alienation
is one of the reasons why the right is morally inalienable.[4] Thus, as
Schiller points out, "the dispute about whether there are any inalienable
natural rights [is] a normative dispute" (p. 312). The dispute does not
address what could in fact be done (i.e., can you get another to kill you?),

but what could morally be done. It is thus important to distinguish the logical and moral senses of possibility.[5]

Here it should also be noted that Feinberg (1978, pp. 114-123) wants to separate waiving from transferring, arguing that it may be appropriate to alienate the right to life in the sense of waiving (thus voluntary active euthanasia is allowable), while it is inappropriate to alienate it in the sense of transferring (thus one cannot give a third party the right to decide whether or not one will be killed).[6] Whether such a distinction could be sustained will depend on the argument that justifies inalienability – a point that Feinberg did not sufficiently address. As we shall see, Locke's argument will not provide a grounding that could sustain such a distinction. I will thus use inalienable in the strong sense, entailing that the right can be neither waived nor transferred.

(6) Finally, it should be noted that this right may or may not be a mandatory claim right (Feinberg, 1978, pp. 104-110). A mandatory claim right has been viewed as contrary (at least in spirit) to the very notion of a negative or jurisdiction right. It says that one has a right to do that which one has a duty to do. It is a way of saying "ought implies can," and is thought of as a rather uninteresting right. However, the way one understands a mandatory claim right depends on how one interprets duty. If, for example, one argues that one's duty is freely to pursue happiness or, as another option, creatively to administer or perfect a certain domain, then a mandatory claim right may actually be a good account of a negative right, incorporating into the notion of that right a sense of its justification. Although Eric Mack is quite put off by the very notion of a mandatory claim right (Mack, 1999), he will actually, in another context, provide a justification of deontic constraints and with them rights that depend on an agent relative sense of values that implicitly entails a sense of duty to determine the scope of the right (Mack, 1989). He could thus, in a certain way, be viewed as an advocate of a mandatory claim right.[7] The importance of this possibility will become apparent later in this essay.

II. LOCKE ON INALIENABILITY: TWO INTERPRETATIONS

The basic structure of Locke's argument in the *Second Treatise of Government* is familiar, as is the role in that argument that has been traditionally assigned to inalienability. The argument runs like this: In the state of nature an individual has two powers:

> The first is to do whatsoever he thinks fit for the preservation of himself, and others within the permission of the law of nature; by which law, common to them all, he and all the rest of mankind are one community, make up one society, distinct from all other creatures. And were it not for the corruption and viciousness of degenerate men, there would be no need of any other; no necessity that men should separate from this great and natural community, and by positive agreements combine into smaller and divided associations.
>
> The other power a man has in the state of nature, is the power to punish the crimes committed against that law (Locke, 1980, sec. 128).

Because of the corruption of some, and because of the difficulty each individual has in punishing the crimes of those who offend, people transfer these powers to an association or civil order that will protect the interests that each individual has in the state of nature (Locke, 1980, secs. 129-131).

The problem is: does this transfer of powers provide the commonwealth with an unlimited and arbitrary power over the individual? This was the Hobbesian solution. But Locke will answer with an emphatic "no". This is where the doctrine of inalienability has been thought to play a role (Simmons, 1983, pp. 177, 183-184). People cannot alienate certain rights such as those to life, liberty, and the pursuit of property, and thus the commonwealth is limited in its scope and power. If a government claimed more than one could morally transfer, then one would be justified in revolting against that government. In Locke's words:

> shaking off a power, which force, and not right, hath set over any one, though it hath the name of rebellion, yet is no offence before God, but is that which he allows and countenances, though even promises and covenants, when obtained by force, have intervened (Locke, 1980, sec. 196).

The inalienability of the rights thus is the condition for establishing the limits on government and avoiding the Hobbesian state.

The question is whether this traditional interpretation of Locke on inalienability can be sustained.

1. Locke as a Rule Utilitarian: A. John Simmons

A. John Simmons (1983; 1993, pp. 108-119) has recently argued that Locke does not explicitly affirm the inalienability of rights. Simmons develops a middle position between the individualist interpretations of Locke, which see him as an egoist of a Hobbesian sort, and the theological communitarian interpretations, which focus on Locke's account of property in common.[8] Simmons argues that Locke is a rule utilitarian, who derives all rights and duties from the fundamental "law of nature," which, according to Simmons, is the preservation of humanity (1992, ch. 1; 1983, pp. 190-204). He quotes (1983, p. 190) from Locke's discussion of the prerogative of government, and, more specifically, the prerogative of the executive power:

> It is fit that the laws themselves should in some cases give way to the executive power, or rather the fundamental law of nature and government, viz. That as much as may be, all the members of the society are to be preserved. ... for the end of government being the preservation of all, as much as may be, even the guilty are to be spared, where it can prove no prejudice to the innocent (Locke, 1980, sec. 159; here I quote more than Simmons included).

This and similar passages are combined with Locke's statement that the state of nature is a "state of perfect freedom to order their actions, and dispose of their possessions and persons, as they think fit, within the bounds of the law of nature" (Locke, 1980, sec. 4). From the combination, the law of the preservation of humanity is understood as that which grounds and justifies individual rights. Simmons argues that this implies positive rights, although, contra the communitarian interpretations of Locke, he will defend fairly robust jurisdictional liberties, based on their role in preserving society (Simmons, 1983, pp. 192-200; on the priority of duty over rights, see 1992, pp. 68f.).

Simmons makes an important and valid point: the limits on rights, and thus the limits on the capacity to transfer power to the commonwealth, are fixed by the *duty* one has to the law of nature (Simmons, 1983, p. 191; 1992, pp. 68-79). This (wrongly) leads Simmons to conclude that there are no inalienable rights, i.e., no rights that cannot be transferred, because one never really had the power in the first place.

> When Locke denies that we can enslave ourselves by voluntary compact, he is not affirming that we possess certain rights which are

inalienable. We cannot enslave ourselves because we don't have the appropriate rights to transfer. Even if we transferred all our rights to another, we would not become slaves; to become our masters others would need rights over us which we do not ourselves possess. No claims of inalienability are required, and Locke quite clearly does not make any (Simmons, 1983, 191-192).

In Locke's words, "[n]obody can give more power than he has himself, and he that cannot take away his own life cannot give another power over it" (1980, sec. 23). Thus, for Simmons, the power of individuals over themselves and of government over individuals are equally limited by the law of nature, to which all are bound in duty. It is not the inalienability of rights but the presence of duty that leads to the famous Lockean restraints on unlimited, arbitrary government (Simmons, 1983, p193).

Simmons also suggests that Locke may be committed to certain types of inalienable rights, even though Locke does not explicitly argue for them. Drawing on the correlativity of duties and rights, he suggests that Locke's position entails inalienable mandatory claim rights, which maintain for individuals the domain for performing that which it is their duty to perform (1983, pp. 195-198). Thus, beginning with duty to the law of nature rather than jurisdiction rights, Locke's position leads to an inalienability thesis.

Simmons asks whether the grounds Locke gives for the duty, and thus for the mandatory claim right, are very plausible, and he is quite pessimistic in that regard (1983, p. 198). Locke's argument begins with God's existence, moves to an affirmation that God wills the preservation of humanity, and concludes that the fundamental law of nature is that of preservation. In his book, *The Lockean Theory of Rights*, Simmons further elaborates upon this grounding and seeks to eliminate the theological premises and reconstruct a Kantian-type argument that builds the claim right on freedom, equality, and the independence of individuals in the state of nature (Simmons, 1992, pp. 41-46; also see Tully, 1995, p. 107).

While Simmons undoubtedly addresses some important concerns in Locke, there are some problems with his rule utilitarian interpretation. Locke's *Questions on the Law of Nature* shows that his interpretation of natural law is extremely complex, and it cannot be reduced to a law about preservation.[9] In fact, the idea of power and obligation, rather than a particular rule, actually plays the more significant role in Locke's "natural law", and such an approach to power better serves the traditional interpretation of Lockean rights (Zuckert, 1994b, pp. 211-212, 218-219; note especially Zuckert's account of the law of preservation as a summary of the prohibition of harm to others and self coupled with the Lockean proviso). Simmons should have considered Locke's diverse accounts of the law of nature, rather than simply selecting one formulation.

Simmons also moves too quickly from the contention that duty constrains right to the claim that the same rule of preservation constrains both the individual in the state of nature and the commonwealth, and thus that there are no inalienable rights in the traditional sense. If, for example, individuals and natural society were bound by a law of preservation that entailed duties of charity, that would not necessarily imply that the law of preservation that directs the commonwealth involves a duty of charity. In other words, the law that directs government may be a more truncated version of the law that directs humanity in the state of nature. The reason for this truncation may be that individuals are barred from transferring certain aspects of their natural jurisdiction (and thus natural responsibility); namely, certain rights may be inalienable. The nature and scope of the law that directs the commonwealth is determined by the purpose and scope of the transfer that led to its formation, while the laws governing the transfer will be more foundational, and thus prescribe matters such as inalienability, which do not follow directly from the law of government.

To appreciate the limits of the law of preservation associated with government it is imprtant to remember that for Locke the transfer of power that comes with the formation of the commonwealth is a concession to the fallen state of humanity. Possibilities for human community are altered as a result. For Locke the state of nature need not consist of an atomistic individualism. Community and society do not first form when power is transferred, nor is the potential for such free community fully lost when the commonwealth is present. As noted in the previous quote regarding the way humans are transformed into beasts as a result of violence to the law of nature, the commonwealth is a response to the state of war, and the notion of preservation associated with it is primarily that of protecting against and punishing those acts of violence. "And were it not for the corruption and viciousness of degenerate men," then, as Locke explicitly says, "there would be ... no necessity that men should separate from [the] great and natural community, and by positive agreements combine into smaller and divided associations" (Locke, 1980, sec. 128). It is for the purpose of restraining evil that people transfer their powers, not to positively realize the ideal of the natural community. In other words, by means of government one seeks to guard against unnatural violence, thus allowing the natural possibilities of humanity to flourish. One does not, through the unnatural means of the state, seek to realize the ideal. Rapaczynski (1987, p. 118) rightly notes that:

> Political participation ... is not an activity in which men realize their very humanity (as Aristotle believed), nor is the state for Locke (as it was for the medievals) a divinely ordained paternalistic order. Instead ... it is a pragmatic arrangement designed to assure that men can freely and without interference from each other pursue those goals they already have *before* they enter the state for the first time.

We will find that the notion of inalienability is central to sustaining this space between free community and the state.

2. Locke as an Ethical and Psychological Egoist: Eric Mack

Eric Mack (1999) accepts much of Simmons' characterization of the limits that Locke put on what one could do with one's self. Those limits, he argues, are based on a duty that is religiously grounded. However, for Mack these duties are a separable strand in Locke from the jurisdiction rights, and Simmons' account of the derivation of these rights is rejected, along with his claim that Locke is a rule utilitarian of preservation (1999). By contrast, Mack contends that Locke is a psychological and ethical egoist, and he uses Locke's *Letter on Toleration* to argue that the restrictions on rights that Locke bases on individual, religious duties should not in any way be legislated by the commonwealth (1999). Rejecting the duties on which Simmons bases Locke's commitment to mandatory claim rights allows Mack to jettison completely the inalienability claim. All rights will be regarded as alienable. Summarizing, Mack claims that

> Locke's theory of human rights [as opposed to his theory of duty, which is separable and religiously based] does not encompass, as the rule utilitarian interpretation implies, even a *prima facie* duty to preserve oneself and, therefore, does not mandate particular ways in which persons must exercise their authority over themselves (1999, p. 145).

Mack brings some telling quotes from Locke against the interpretation provided by Simmons. The following, from Locke's *Letter on Toleration,* provides a good example:

> Will the Magistrate provide by an express Law, that such an one shall not become poor or sick? Laws provide, as much as is possible, that the Goods and health of Subjects be not injured by the Fraud or Violence of others; they do not guard them from the Negligence or Ill-husbandry of the Possessors (Locke, 1991, p. 28; see also p. 43).

This and other passages imply that, on the basis of the rights theory, one need only refrain from taking away or hindering what is needed by others for their own preservation, which leads to the conclusion that Locke's claim rights only function negatively.

There are, however, some significant problems with Mack's interpretation. First, the limitations on an individual's rights that result from the duty to God and from God's property rights in all of humanity do not depend, for Locke, on the type of particular religious commitments that he was addressing in his *Letter on Toleration*. Locke distinguishes between matters of belief and special revelation related to individual salvation and matters of natural revelation (discernable by reason) that are foundational for natural law and civil government. In the *Letter on Toleration*, Locke made clear that belief in God was a prerequisite for civil matters: "Those are not at all to be tolerated who deny the being of God. Promises, covenants, oaths, which are the bonds of human society, can have no hold upon an atheist. The taking away of God, though but in thought, dissolves all" (Locke, 1991, p. 47). In his *Questions Concerning the Law of Nature* Locke explicitly grounds any possible natural law on the existence of God. Zuckert (1994b, p. 188) nicely summarizes when he states "[f]or Locke, no God, no natural law." In Locke's philosophy, knowledge of God's existence and God's claim right to individuals (Maker's and Creator's rights) is dependent on reason alone, and cannot be separated from the broader theory of rights. One thus cannot lump everything pertaining to God together and put it on the side of those beliefs that Locke separates from the jurisdiction of the state.

In fact, as Mack in another place seems to recognize (1980, pp. 57-59), the movement from a Hobbesian type liberty to a claim right depends on the natural theological grounding that Locke gives to his moral theory. Mack may want to provide a secular argument that replaces this natural theological one, but he should then be explicit that he is in an important way deviating from Locke; it is not simply a matter of how Locke caries out his argument in the *Letter on Toleration*.

By jettisoning this aspect of Locke's argument, Mack undermines his ability to sustain the claim rights for which he wants to argue. The problem is as follows: if Locke is an ethical and psychological egoist, as Mack contends, then how can one generate an obligation not to interfere or violate another's life, liberty, and pursuit of property? As Leo Strauss (1953) argued, his state of nature reduces to that of Hobbes, where there are no claims against others that attend one's liberties, and thus no duty on the part of another to respect one's "rights". In an essay that addresses just this problem, Mack (1980) looks at three secular arguments that could be used to generate claim rights out of Locke's ethical egoism, but he has to concede that none of them succeeds. The only argument that Mack identifies that is sufficient for grounding Lockean protections against others and, in turn, against the unlimited and arbitrary government is the theological one that Mack seeks to privatize in his discussion of inalienability (pp. 57-59).

This brings me to a third criticism: Mack has not provided an account of why alienable rights do not raise the specter of Hobbes' Leviathan. If one can transfer any and all rights, then, as Mack admits, one can sell oneself into slavery, duel (an expression of the state of war), and, even further, transfer all one's rights to the state authority; a transference that some may make, for example, in order to obtain from the state nurture and protection. When this is coupled with the inability to generate a claim right, the result is a strong justificatory framework for a benevolent (or not so benevolent) dictatorship. Although Mack seems to appreciate the role that inalienability traditionally has played in justifying the type of limited government he favors (1999), he does not take this into account when he tries to extend the rights further in the libertarian direction. It may well be that the attempt to take a little more can lead to much less.

Fourth, although Locke's notion of happiness may be individualistic, Mack does not appreciate the degree to which even individual happiness (especially eternal happiness; Locke, 1991, p. 42) depends on certain social ends, including that of the common good. Locke directly contrasts those who simply follow the given desires for self-preservation – people who are no better than animals – and those who pursue happiness in a higher sense, which involves God's end for humanity, an end that includes a communal dimension (Locke, 1980, sec. 77; Zuckert, 1994b, pp. 202-203, 213).

Mack profoundly summarizes Locke's moral and political philosophy, when he states:

> Each person [for Locke] is teleologically required to maintain his own life, liberty, and property. And each person is deontically required not to trespass on the life, liberty or property of any other morally similar being (1980, pp. 58-59).

But Mack's failure to appreciate the overlap between the teleological end of each person and that of others makes him speak as if obligations to others are exhausted by their legal obligations (consider, e.g., Locke on charity; 1991, p. 28). Further, it makes him unable to account for those areas where the commonwealth can, for Locke, move beyond certain minimalist functions when the preservation of all are at stake. For example, in the *Letter on Toleration* Locke states that the commonwealth may not keep someone from sacrificing an animal, if the sole reason for doing so was its being religiously abhorrent. However, if there was famine, then the state could prohibit this in the name of the common good (Locke, 1991, pp. 36-37; note also p. 17). Mack's account provides no explanation for why Locke could say this.

Finally, it should be noted that Mack, in some important ways, misunderstands what the inalienability of a right would entail. He seems to think that there is a conceptual contradiction in the very idea of an inalienable right:

> to ascribe to William the right to life, the right over his life, while at the same time denying to William the moral power to decide what is to happen to his life – in particular and most dramatically, to decide whether that life will continue or not – seems to retract with one breath what has been asserted with another (1999, p. 145).

However, this is not the case, since one does not retract (or better, withhold) everything, but only certain areas of control. It is not contradictory to speak of a limited property right, for example. One may have jurisdiction over some land, but nevertheless not be allowed to put a trailer or commercial business on it, or not be allowed to transfer that land to another. In the same way, one may have a right over one's life, which entails a claim against others, while at the same time not having the capacity to alienate that right; i.e., one is, as it were, bound to that life, and thus has obligations (in the case of Locke, obligations to God and other people) not to transgress those boundaries.

III. RECONSTRUCTING THE INALIENABILITY CLAIM
IN LOCKE

The dilemma confronting any traditional interpretation of Locke is either: (1) Locke's argument is based on an egoism that is unable to move beyond Hobbesian liberties to claim rights, and further is unable to resist the transfer of right to an unlimited and arbitrary state; or (2) one calls on theological arguments to justify this limit, but such argumentation is also tied to the property in common tradition and to fairly robust duties that seem to make rights subordinate to some other principle (in the case of Simmons, to a utilitarianism of preservation). Either way, one does not end up with the jurisdiction claim rights and limited government that many want to see defended in Locke.

I believe the dilemma is a false one, and that the traditional interpretation of Locke can be sustained. To make this case, it will be helpful to consider briefly how Hobbes' position differs from Locke's. Then I shall provide the key natural theological argument in Locke that grounds this difference.

1. Hobbes and Locke on the State of Nature

Hobbes' state of nature does not entail claim rights, whereby each person has a claim against others to not interfere with the life, liberty, and pursuit of property, and whereby others have a duty to respect that claim (McConnel, 1984, pp. 47-52). Rather, the state of nature is a state of war of all against all (Hobbes, 1962, ch. 13). People have the liberty to pursue life and liberty, and thus no one has an obligation to subject oneself to any other. Even further, Hobbes argues that this natural liberty, called a right, is inalienable, because an obligation can never arise to submit to another who would take one's life.[10] However, nobody else ever has an obligation to respect one's natural liberty. "The right of nature, which writers commonly call *jus naturale*, is the liberty each man hath, to use his own power, as he will himself, for the preservation of his own nature." This means "every man has a right to every thing; even to one another's body" (Hobbes, 1962, p. 103). It is the unlimited scope of natural liberties, entailing overlapping jurisdictions (e.g., all have an equal claim to your body), which leads to the state of war. Thus, it is by violence and not by reason that one must resolve dispute.

For Hobbes, no one has security in such a state of nature. People agree, therefore, to give up their power – their natural liberty – for protection and self-preservation (Hobbes, 1962, ch. 18). However, since there are no limits in the state of nature on what can be done with others, there will be no limit drawn from nature or natural law on what the sovereign (or government) can do with its subjects. (While subjects may not have an obligation to submit to the sovereign who would execute that subject, the sovereign has no duty not to perform the execution). Thus there will be no natural constraint on unlimited and arbitrary government.

Locke's state of nature differs in important ways from that of Hobbes (Andrew, 1988, ch. 4; for a detailed discussion, see von Leyden, 1982). For Locke, individuals do have natural obligations not to interfere with others (Locke, 1980, ch. 2). If all people abided by natural law, there would not be a resultant state of war. Rather, war results when some individuals do not sufficiently regard natural law, thus transgressing their own natural boundaries (ch. 3). In order to guard themselves against this unnatural state of war, people transfer their powers to the government, which then enforces their natural rights (chs. 8-9). This government will be limited by that same law of nature that grounds natural claim rights (ch. 11).

The question, however, is how Locke justifies his contention that people have natural claim rights and not simply Hobbesian liberties. Some have attempted to justify such rights by appealing to the natural equality of all. However, as Eric Mack notes,

> Lockean premises about natural equality and lack of natural subordination and jurisdiction are premises that are also present in Hobbes. ... To defend the view that Locke has some philosophic warrant for asserting natural rights of a stronger sort than the liberties proclaimed by Hobbes, we must isolate either some line of philosophical reasoning which proceeds from the premises already noted but which Hobbes did not pursue or we must find further premises in Locke that are relevant support for his announced natural rights views (Mack, 1980, pp. 51-51).

Although many today are attempting to derive claim rights from Hobbesian premises alone, that is not the most fruitful path to pursue. Hobbes rightly developed their implications. Instead, our task must be to look for the additional premises in Locke.

However, to justify limited government in Locke, the derivation of a claim right will not be enough. If one could transfer such rights to a sovereign in the process of forming a government, then the protections against others that one has in the state of nature would no longer apply as protections against the commonwealth. This is where inalienability plays an important role. And it is important to note that for Locke the limits on the state come indirectly by way of the inalienability of the natural claim rights, rather than by a direct duty of the state not to interfere with the lives of individuals. This is because civil government is not a natural entity; it is derivative (Rapaczynski, 1987, pp. 117-118; Zuckert, 1994b, p. 229). Its laws and limits are generated by contract and depend on the consent of the governed. One thus cannot speak of natural duties that such government has to individuals. This means that any limit or duty of the state must be attributed to either (1) the duties and limits placed upon the state by those entering the contract, or (2) the natural duties and limits that are present among individuals in the state of nature. A corollary of this account is that the inalienability claim primarily says something about how people can or cannot relate to one another in the state of nature, and only secondarily does it say something about how subjects can relate to civil government (Locke, 1980, ch. 11). The limits on civil government depend on and arise from these natural duties and limits.

Thus, in evaluating Locke's political philosophy, we must consider how he generates (1) the natural duty to not interfere that gets attached to Hobbesian liberties to generate a claim right (I call this the claim right justification), and (2) the natural duty of the rightsholder to not transfer the resultant claim right (I call this the inalienability claim justification). I will show that the key to understanding Locke's political theory is to recognize that the same foundational argument justifies both the claim right and the inalienability claim; these two stand or fall together.

2. Grounding Inalienable Rights

Locke's basic moral argument can be put simply. God has absolute jurisdiction (Locke, 1980, secs. 6, 23). To put this in a more philosophical idiom: there is a transcendent ground of obligation, which has absolute jurisdiction. Part of this jurisdiction is transferred to humanity (secs. 4, 7, 25-26). It is transferred in common to all (property in common; secs. 25-26) and individually to all (liberty, equality, and the natural executive power; secs. 4, 7). To the degree it is transferred in common, it can be individually appropriated through labor (sec. 28). Locke's state of nature is then the result of the initial divine transfer coupled with individual appropriation and legitimate individual transfers of property.

For Locke each person is a ruler of his or her domain (sec. 44), and an important analogy is drawn between the way in which the self rules over itself and the way in which those in government rule over their subjects. It is an analogy that is structurally similar to the one that Plato develops between the virtues of the soul and those of the state (Plato, 1974, 368d-369b). By looking at the way in which the elements of the state are constituted, Plato seeks to show how those of an individual are similarly constituted. For Locke, the analogy can be used to show how claim rights and inalienability are generated. It runs like this: as the powers and limits of the state are generated by the way in which individuals transfer power to the state, so the powers and limits of individuals are generated by the way in which God transfers power to the individual.[11] These latter "natural" powers are inalienable claim rights.

The central concern that led Locke to write his two treatises on government related to the scope and limits of state power. At least in part, he was responding to Sir Robert Filmer's justification of the absolute, divine right of kings and the illicit character of any rebellion. To counter Filmer's claims, Locke asks: what is the limit and scope of state power and authority?[12] To answer this, he considers the conditions under which such power is generated. His argument is that the state's power is not given by God directly (Locke, 1980, ch. 1). Rather, the state and its rulers obtain their power from the people (ch. 8). In order to determine the limits and duty of the state, one need simply ask: for what purpose did people initially transfer their powers? The conditions of the transfer then determine the scope of the state's use of that power (chs. 9, 11; Zuckert, 1994a, p. 76).

Now one can move one step backward and ask: where did people get their powers? This question has the same form and structure as the initial question about state power. Locke answers it in a similar way: they obtained it by a transfer (sec. 135 with secs. 6, 26-27). In order to determine the scope and limits of the power that the people initially had, one can thus ask: for what purpose did God initially transfer these powers?[13] The answer to this question – namely God's intention in the creation and empowering of humanity – is called by Locke "natural law" (secs. 6, 32, 35). As the intent of people in forming a commonwealth serves to determine the scope and limits of the state's power, so the intent of God in creation (the telos of the natural order) determines the scope and limits of the powers that individuals have.

We are now in a position to see why both Simmons and Mack are wrong in the way they characterize the relation between rights and duties. Simmons makes a natural law of preservation primary, and then seeks to derive rights from this law (1993, p. 122). In the case of inalienable rights, Simmons attempts to derive the rights directly from the duty (his mandatory claim rights). In the case of other rights (e.g., liberties), he will allow for an independence of rights, but make their scope dependent upon the calculus of the utility of preservation (Simmons, 1992, ch. 2); those rights are simply what is left over by the law of nature. However, for Locke, there is a priority given to the powers and jurisdictions that are transferred. Duty, prescribed by natural law, only follows the transfer as a determination of the limits on the power. As Mack notes, there is an independence and priority given to rights understood as powers and jurisdictions.[14]

However, Mack (1999) is wrong when he seeks to separate the notion of right from that of duty, and relegate the latter to a separate religious domain. If that is done, one simply has Hobbesian liberties, where "every man has a right to every thing; even to one another's body" (Hobbes, 1962, p. 103). For Locke, the *limits* on rights are simply another word for the duties associated with claim rights and the inalienability claim. Simmons is thus correct when he says that "[o]ur rights end where natural law imposes duties" (1983, p. 192). Generally, people do not have a right to other's bodies. The duties that attend rights simply specify that the rights are not absolute in scope.

When the duties are viewed in this way, the duty that attaches to one's right is not the duty that others have not to harm the rightsholder. Rather, it is the duty that each has to respect the jurisdictions, as determined by God, who provides all with their rights. In other words, the duties that circumscribe my right (analogous to the duties that circumscribe the right of the commonwealth, once powers are transferred) is *my duty* to God, manifest in a respect of the rights of others, since they have been equally provided by God (Locke, 1980, sec. 6).

What has been called a claim right is actually a combination of an individual's jurisdiction coupled with the duties of all others, which follows from and constrains their rights. It is only the fact that all others have rights also, with concomitant duties, that leads to my having a claim right.

In this way Locke brings together two notions that were thought to be incommensurable, that of law and right (Hobbes, 1962, p. 103; for detailed discussion of this theme, see von Leyden, 1982). Natural right (*jus naturale*) is concerned with "the free use of something," while natural law (*lex naturale*) is concerned with "that which either commands or forbids some action" (from Locke's *Questions on the Law*, quoted in Zuckert, p. 192). Locke takes as his point of departure the right, understood as a gift, the result of a divine transfer. Then he introduces law, and with it duty and obligatoriness, as that which specifies the domain of the right. It is through law and duty that the particular configuration and extension of right is specified. This movement from right to law is clear throughout Locke's *Second Treatise*, and it is quite surprising that so many astute interpreters of Locke do not appreciate the way in which the two operate together. Locke does not derive right from duty; rather, the law of nature (and with it duty) comes in as a second step, specifying the limits and end for which the presupposed right is designated. The right is the result of the transfer.

A good analogy for this can be found in the relation between a feudal lord, his fiefdoms, and the vassals to whom the fiefdoms are given. (This analogy is more appropriate than the one suggested by Zuckert [1994b, pp. 220-221] between a slaveholder and slave, and it can answer Zuckert's problem regarding how one can simultaneously assert divine-ownership and self-ownership.) The lord may transfer land to a vassal, who then cultivates that land for the lord. In this case, one may speak of the land and rule transferred by the lord as the provision of the right. One does not begin with a duty of a subject properly to cultivate the land, etc., and then provide the land and rule as a correlative of the duty. (This is Simmons' interpretation of a mandatory claim right in Locke.) Rather, one begins with the transfer of the jurisdiction. This is the privilege and right. The scope of the right is then circumscribed by the duty one has to the lord and the lord's subjects; duty provides the *fact* (or *that*) of the transfer with its determinate content (or *what*).

In Locke's account, God is the Lord, and he transfers part of his jurisdiction and power to humanity. The fiefdom transferred is, first and foremost, that property of each person in his or her self, their right to life, liberty, and the pursuit of property (Locke, 1980, secs. 27, 44). Secondarily, humanity (in common and individually) is given dominion over the rest of creation (secs. 25-26). For Locke, God's creation involves the establishment and transference of dominion to humanity. The problem is then: how does one know the boundaries that God has placed upon the various fiefdoms and on the vassals that have dominion over them?[15] This is the central question, since the boundaries are exactly the duties given in the inalienability claim and the claim right.

Returning to the analogy, the question can be put as follows. Imagine that a lord has established fiefdoms, and has also given each subject sufficient power to protect the fiefdom and guard against threat to its rule (this is Locke's executive power; secs. 7, 130). Now all subjects gather together and, on the basis of their knowledge of the lord's intent in transferring the jurisdictions, must discover the extent of their rights and duties. How would they go about this?

Locke's answer to this is complex, and it entails several steps, including his well known and often debated account of property. A detailed evaluation of this would be helpful, especially since there are important parallels between the way in which labor gives a certain quality to the products of nature (resulting in property) and the way in which God's creation gives a certain status to the created, including humanity (God's property rights). However, such a detailed overview cannot be provided here, nor need it. It will be sufficient to show how Locke will move toward the two duties that we are considering, namely those not to harm others (generating claim rights) and those not to alienate certain aspects of one's jurisdiction (the inalienability claim). These elements of one's self-property precede and condition Locke's more detailed account of secondary property (thus ch. 2 on the state of nature comes before ch. 5 on property in Locke's *Second Treatise*, just as sec. 27 on self-property comes before sec. 28 on secondary property).

A central premise in Locke's account is that the jurisdictional boundaries are to be determined peaceably, without recourse to force, and on the basis of natural reason (Locke, 1991, p. 45; 1980, sec. 8). This premise follows from Locke's natural theological conclusions regarding God's intent in creating humanity, for it is by reason, which is the law of human nature, that humanity is distinguished from the animals (secs. 6, 16; see Zuckert, 1994b, p. 208). As a result of the link between reason and natural law, Locke avoids a heteronomous account of obligation, and he links Thomistic and Grotian natural law arguments regarding human self-realization and sociality with a Hobbesian account of power and law (for an overview of these, see Zuckert, 1994b, ch. 7; however, Zuckert's account of Locke on sociality [p. 233] is very problematic). When people use force to resolve conflict, the resultant state of war involves relating to people as one does to animals, bringing the "beast" into subjection (sec. 16). This only takes place when the laws of nature have been transgressed. Thus, in seeking to discover the laws of nature, one will seek for the sufficient condition of resolving disputes peaceably, i.e., humanly, and thereby avoiding a state of war. At the heart of the derivation of natural law, one looks for the sufficient condition to avoid a Hobbesian state of nature qua state of war.

This key link in Locke's argument leads full circle, back to the process by which I developed the two conditions needed to avoid a Hobbesian account of the state of nature and the unlimited sovereign. Once one establishes the transcendent grounding for the obligatoriness of natural law, and one adds the premise regarding God's intent in creating humanity as distinct from the animals, then the task of discovering natural law is simply that of deriving the (sole) sufficient condition for avoiding violent (= nonrational) resolution of jurisdictional boundaries. This means that, with one small variation, I have already provided the justification of the claim right and inalienability claim in the previous account that was given of the conditions needed for avoiding a Hobbesian outcome.

Minimally, each person (all of humanity) must be granted by others protection against violent intrusion upon their jurisdiction, and the pure possibility of such a jurisdiction and its protection is given in the right to life, liberty, and the pursuit of property. Without this self-ownership *vis-à-vis* others, there could be no fiefdom. The claim right is thus the expression of the pure possibility of peaceably resolving conflict.

However, the claim right is not enough. It captures the limits on what one may do to others, but it does not sufficiently embody the reason why one must respect those limits (note Locke on the obligatoriness of Law; Zuckert, 1994b, pp. 189-191). Alone, it can only have a hypothetical status: *if* one wants to resolve conflicting claims peaceably, by reason, then one must not violently interfere with another's jurisdiction. However, the claim right by itself does not sufficiently embody the obligation to resolve conflicting claims peaceably.[16] For Locke, the ground of the obligation is transcendent, and the status of the imperative is categorical (Zuckert, 1994b, p. 191). The duty not to harm another will thus follow from, and be an expression of, the more foundational and absolute obligatoriness. This more primary duty will be captured first in the inalienability claim, and then, by extension, in the duty not to harm others.

The priority of the inalienability claim *vis-à-vis* the duty not to harm another is clear in the pivotal section 6 of Locke's *Second Treatise*. After arguing that natural liberty is not license, and that it is subject to the law of nature, which follows from humanity being the "workmanship of one omnipotent and infinitely wise maker," Locke states:

> Every one, as he is bound to preserve himself, and not to quit his station wilfully, so by the like reason ... may [he] not, unless it be to do justice on an offender, take away, or impair the life, or what tends to the preservation of the life, the liberty, health, limb, or goods of another (Locke, 1980, sec. 6).

Here, inalienability, i.e., the duty "not to quit [one's] station wilfully,"[17] follows from one's obligation to God, the creator of humanity and source of obligation. In terms of our analogy, the vassal is bound to the lord, and cannot relinquish responsibility for the fiefdom. The duty to not harm others then follows "by like reason" from this same ground of obligation; it is an expression of respect for the jurisdictions that the lord has provided to others.

From Locke's argumentation, we can now summarize at least three reasons why certain central rights are inalienable:

(1) Alienation of those rights would involve a rejection of the transcendent source of obligation, and thus of the basis of morality. It would reduce all moral imperatives to hypothetical status.[18]

(2) Since the rights and their jurisdiction-defining duties are central to the constitution of human nature; or, in a more theological idiom, since God's command, which is natural law, is none other than the true nature of humanity, reason; if one alienates the rights, then one makes oneself less than a person. One reduces oneself to an animal, and as such becomes a threat to the rest of society.

(3) Since the duty to not harm others follows from and circumscribes the right, the alienation of the right involves an alienation of the basis for the duty. Again, one becomes a threat to society. This last point follows from the fact that in Locke a claim right is generated by one's right coupled with others' duties. If one is allowed to alienate the right, then others lose the protection that goes with a claim right. A claim right only follows from all people having rights and with them the attendant duties; it does not follow from one's right alone.

The necessary conditions of peaceable community thus involve everyone having rights, not just some.

Taken together, these three arguments show how central inalienability is for Locke's moral theory. Thus grounded, the inalienability claim will also provide the sufficient condition for resisting the unlimited, arbitrary state. It will therefore also be central for his political theory. One cannot jettison the inalienability claim without simultaneously undermining the whole of the Lockean moral and political project, since even claim rights will be intimately intertwined with inalienability.

IV. THE VIABILITY OF THE LOCKEAN THEORY

At the heart of a philosophical grounding of any ethical theory, whether it be a utilitarianism, deontology, or an explicitly communitarian or theological ethic, is an interpretive movement whereby one attempts to articulate the basic meaning of morality. People all have an experience of obligation, the good, etc. and in this primordial moment of interpretation, they seek to bring the experience to thought and language, so that it can serve as a basis for the development of moral norms and the resolution of concrete questions regarding practice. Many of the disputes between the great moral theories of our day rest upon fundamentally different interpretations of what is at the heart of moral experience.

Inalienable claim rights have a status for Lockean theory that is akin to the status of the principle of utility for utilitarianism, although they have a somewhat narrower scope, since they address the moral use of force rather than all morality. They articulate Locke's basic interpretation of moral obligation, and draw out some of its implications, thus providing the foundational notion from which classical liberal theory is developed. This is why inalienable claim rights are so often made the starting point of moral/political discourse, as they are, for example, in the U.S. Declaration of Independence.

The moral vision embodied in inalienable claim rights sees the mechanisms for coercive resolution of conflict – including those of the state – as a concession to forms of sociality that already involve a violation of moral obligation. Such mechanisms are thus excluded from the ideal; their function is restricted to restraining action that is already outside that ideal. Realms of individual jurisdiction and communities of free association are understood as the locus of human flourishing

(Rapaczynski, 1987, p. 118). This does not mean, however, that human community is understood only in atomistic terms, as so often contended. Nor does it mean that communities are viewed as constructed artifacts, the form of which is not essential to the identity of the isolated individual. Such an interpretation of Locke, advanced by libertarians and criticized by communitarians, fails to appreciate the degree to which the natural community, with its ends, is a different thing from the commonwealth, which is indeed "unnatural" and a human artifact. As long as they are not coercively legislated, communitarian ideals are thus fully reconcilable with a liberalism of a Lockean sort.

Inalienable claim rights provide the conditions of sociality and peaceable resolution of conflict, and they thus establish a bulwark against those forms of individual and communal life, which would threaten the peace and liberty of others. They mark off deontic constraints, leaving broad sweep for the realization of diverse individual and communal ends, as long as the means for the realization of those ends do not violate the basic conditions of sociality embodied in the inalienable claim rights. It does not follow from this, however, that all the ends that would be pursued within the latitude of the protected jurisdictions are appropriate. Nor does it follow that morality is exhausted by those norms that address the moral use of force. The only thing asserted is that the methods that can legitimately be used to lead one to reject inappropriate or insufficient pursuits that do not violate inalienable claim rights must be noncoercive. Thus, within the space of free association, there is ample room for argument, negotiation, and even nonviolent social sanction. But violence can only be used against violence; it cannot be used against the stubborn or ignorant, who will not willingly advance notions of communal flourishing that others think embody the ideal.

In excluding violence, inalienable claim rights also exclude certain forms of what, on first blush, may be regarded as "free association." Included among such illicit social forms are voluntary slavery, dueling, and euthanasia. These can only be pursued by eliminating the humanity of some, and by violating the transcendent ground of obligation. This contention regarding certain types of "free consent" is what is captured in the inalienability claim. It says that certain capacities such as those endowed in the right to life, liberty, and the pursuit of happiness are essential to human nature. One cannot relinquish them without abdicating the moral responsibility that marks humanity off from animals.

Behind Locke's interpretation of morality, one thus finds a notion of obligation that entails a transcendent source and ground. Theologically, one can speak of God, the mystery and source of human life, and of humanity's duty to respect limits upon what can be done with others and oneself. Today, in much of the literature on topics such as euthanasia, this interpretive matrix for the elucidation of inalienability is captured in the phrase "the sanctity of human life," a phrase that in many ways serves the same function as inalienable claim rights do in Lockean moral and political theory.[19]

There are many who would reject the inalienability claim (and with it the sanctity of life), arguing that it is based upon "religious" premises that are not viable in our current, secular, pluralistic context.[20] Eric Mack's (1999) attempt to isolate the inalienability claim from claim rights, rejecting the former, can be viewed as an example of such argumentation. However, such argumentation fails to appreciate the degree to which inalienability is at the heart of classical (as opposed to modern) liberal theory. Even the notions of religious tolerance, on which this criticism is based, depend on the very inalienable claim rights that they seek to criticize.[21] Inalienable rights do not entail a sectarian notion of religiosity, but rather a sense of openness and responsibility to the Mystery of Life, which, for Locke, is in principle (though often not in practice) accessible to all, and for which all are accountable, should they lack the notion of obligation that follows therefrom.

Liberal theory is all of a piece; limited government, tolerance, and peaceable resolution of conflict all arise out of that basic interpretation of moral obligation, which is given in inalienable claim rights. Without their root, these different parts of liberalism come apart, and they then get reconstructed in a weakened form on the basis of new principles, such as that of utility. This is what has taken place in modern liberalism. Simmons' reconstruction of Locke as a rule utilitarian provides an example of this new liberalism, where rights follow from duties that arise from a rule of human preservation. In such a reconstruction, instrumental reason takes center stage, and all of the protections associated with inalienable claim rights are put on unstable footing, subject to the vicissitudes of changing social forms of rationality. And the principle of utility is no less contingent or sectarian than that of inalienable claim rights (Engelhardt, 1986, pp. 34-35; Quinn, 1995, p. 40). The key question is thus whether inalienable claim rights express a notion of moral obligation that unites and accounts for the broad moral and political commitments we have. Even if one does not accept the natural theological matrix that led to the elucidation of the inalienable rights, one may still accept the rights as the point of departure, finding in them an embodiment of obligation that is reflective of one's own moral experience. In current society, and in reconstructed liberal theories, these diverse commitments have become fragmented, and a great, syncretistic attempt is made to hold them together. However, at their root we find an interpretation of moral obligation that gives them coherence and sheds considerable light on what we consider important in ethical discourse. Such coherence and light should not be under emphasized when considering the viability of inalienable rights.

Department of Philosophy and Center for Bioethics
University of South Carolina
Columbia, South Carolina

NOTES

* I would like to thank Mark Cherry, Dan Sabia, and Edwin Wallace, IV for helpful comments on earlier drafts of this essay. I also gratefully acknowledge the research assistance of Caroline Wheless.

[1] Thus van de Veer (1980, p. 165) notes that the justification of inalienability "has received scanty attention by philosophers."

[2] In a more detailed study, it would be helpful to consider different aspects of the cluster right. For example, Feinberg (1978, p. 94) adds "the right to be rescued from impending death" to "the right not to be killed," but excludes "the right to live decently" from the right to life. It is not clear, however, whether inalienability applies equally to the two aspects he includes. For example, one could reasonably distinguish between killing and letting die, and argue that one may alienate the right to be rescued. Thus lifesaving treatments can be legitimately withheld and withdrawn (others may let you die). But this need not imply that the right not to be killed can be waived. For further discussion of positive and negative rights, and of the rule of rescue, see Stell, 1979, pp. 10-13. Stell (p. 22) also does not sufficiently distinguish between killing and letting die in the discussion of inalienability.

[3] In addition to the need to evaluate the rule of rescue, a sufficient account of Locke on the right to life would require an evaluation of the Lockean proviso. Recently, Lustig (1991) has strongly argued in favor of including positive rights in the right to life. However, he has not sufficiently addressed MacPherson's account (1962, pp. 199-221) of the way in which Locke sought to get around this proviso, nor has he sufficiently addressed the degree to which Locke might have placed the moral requirements for charity within the domain of free association, and thus excluded them from the prerogative of the state. To address these concerns, we would need to consider in some detail Locke's account of property, and we will not be able to accomplish that in this essay.

4   This is an important point to consider in evaluating many of the arguments for euthanasia that begin by talking about what we all do with animals: we would not allow them to suffer. Then one says: how can we do less with humans? But this argument begs the important questions. The point of the debate is: would we reduce humans to animals, if we treat them as you are willing to treat your dog?

5   There is a long history of confusion between the moral and logical senses of inalienability, which dates back to writings of Grotius and Hobbes (Andrew, 1988, pp. 82-83). For an example of a current writer who distinguishes them, but then goes on to argue as if there is no distinction, see Schiller, 1969, p. 309.

6   When Meyers argues that "an inalienable right must be a right that a person cannot obligate himself to allow others to infringe" (1981, pp. 130, 138), she only captures what Feinberg means by saying that such a right cannot be transferred; it could be waived, however. Her account (and Feinberg's) works well for Hobbes, where the right in question is only a liberty and inalienability means no more than the absence of an obligation to submit to another. Since the other never had an obligation to respect the right in question, one does not need to consider the stronger question: can a person allow the other to infringe upon the object of the right? Before one can address this question, one must first consider the justification that moves one from Hobbesian liberties to Lockean claim rights.

7   The problem with Mack's account is like the problem Locke identifies with a certain strand of natural law that demands that individuals realize themselves (Zuckert, 1994b, ch. 7): such law can at best have the status of a hypothetical imperative, and one can always ask: why am I obligated to realize my nature?

8   Thus Tully (1995, p. 106) notes that "Simmons' interpretive strategy is to sort the prevailing interpretations into two opposed positions and to argue that Locke's theory lies somewhere in between."

9   Simmons himself discusses Locke's argumentation in the *Questions on the Law of Nature*, and it would take us too far afield to respond to the details of his argumentation. However, it should be noted that Locke has provided many different formulations of the law of nature, and the one Simmons has chosen can be regarded as *a* not *the* law of nature. Closer to the essence of the law is the "command of the divine will, knowable by the light of nature, indicating what is and what is not consonant with a rational nature, and by that very fact commanding or prohibiting" (fols. 11-12; quoted in Zuckert, 1994b, p. 188, also p. 210). For Locke this will of God is in principle knowable by all, but in practice it is "'hidden' and 'secret', mostly if not entirely unknown by other human beings" (Zuckert, 1994b, p. 209). In fact, Locke's very complex account of this law calls into question any simple rule or principle, since any such principle would not pose the difficulties that Locke's law of nature poses for understanding. Simmons does not sufficiently appreciate the transcendental character of this Law and of Locke's notion of obligatoriness, thus concluding (contra-Locke) that for Locke it is clear and plain to all. It would have to be, if the law is simply Simmons' rule of human preservation. To support that the law is plain to all, Simmons (1992, p. 55, esp. note 99) provides two quotes from Locke: (1) the law of nature is "as intelligible and plain to a rational creature, and a studier of that law, as the positive law of commonwealths, nay possibly plainer" (Locke, 1980, sec. 12); and (2) "reason teaches all mankind, who will but consult it" what the law of nature demands (sec. 6). However, it should be noted that what reason teaches in sec. 6 is not simply Simmons' rule consequentialism (we will consider this passage in some detail later in this essay, pp. 196f.), and the "who will but consult it" should not be under emphasized, since for Locke many will

not expend the effort involved in consultation (a very rigorous task). Note the same qualification in sec. 12: it is only to a studier of the law of nature that that law will be clear. However, Zuckert (1994b, p. 240) will go too far in the other direction when he argues that the right of preservation is simply a pragmatic result, given that no one knows the transcendental law of nature.

10 Even if one agrees to yield one's life, such an agreement would never be binding (Hobbes, 1962, p. 110).

11 An appreciation of this analogy can lead to a reappraisal of the relation between the conditions of the transfer of a right and the conditions of its genesis. This is a topic that Simmons directly addresses, when discussing Locke's one use of inalienability (1983, pp. 186-187), and Simmons can only see there a confusion.

12 One could say that Locke's *First Treatise* sought directly to counter Filmer (see Locke's title page), while the *Second Treatise* sought indirectly to counter Filmer by providing a concrete alternative. To do that he has to provide an account of the limits of state power and authority.

13 Or, in a more philosophical idiom, one can ask: how is the absolute jurisdictional right of the transcendent ground of morality transferred to humanity?

14 Another way to put this is to say that rights are basic, and thus they "are not supported by yet more fundamental principles that make no mention of rights" (L.W. Summer, quoted in McConnell, 1984, p. 26). However, this assertion need not imply that no moral justification can be given for the rights (as McConnell contends). Such a contention too quickly aligns the theory of rights with a foundationalist empistemological account of that which is properly basic. But Stell (1979, p. 9) goes too far in the other direction when he argues "[t]o predicate a right presupposes the existence of a legal or moral rule (or principle) which confers it (or from which it may be derived). ... [T]o assert that someone has a right and simultaneously to deny that there is a rule (or principle) which confers it is to speak nonsense." In both cases important metaethical and epistemological issues are insufficiently addressed.

15 Note the parallel between this question and that posed by Locke in his *Questions Concerning the Law of Nature*, especially his account of the promulgation of natural law (Zuckert, 1994b, pp. 195-204).

16 For a current example of a similar problematic, one can consider the work of H. Tristram Engelhardt (1986). Like Locke, he is committed to the peaceable resolution of conflict, and argues for negative claim rights on that basis. But like Grotius (Zuckert, 1994b, p. 191), Engelhardt does not sufficiently capture the obligation to resolve conflicting claims peaceably. Thus his whole account has a hypothetical status: *if* one wants to resolve claims peaceably, *then* one must ... In order to provide the categorical status to his claim rights, Engelhardt needs to incorporate the inalienability claim. However, this would radically alter many of his conclusions; for example, his account of euthanasia and assisted suicide (Engelhardt, 1986, pp. 301-317).

17 It is important to note that those passages where Locke speaks of the "duty not to quit one's station" should be taken as directly addressing inalienability. In fact, the tradition on inalienability that precedes Grotius and Hobbes on that topic, actually focused on the inalienability of sovereignty; namely, on a ruler's inability to quit his or her station (Riesenberg, 1956). This important context for interpreting Locke on inalienability is, to my knowledge, completely unaddressed in Locke scholarship. When we take such passages in Locke into account, then we must modify Simmons contention that there is only one place where Locke explicitly takes up the issue of inalienability (1983, p. 186).

[18] Another way to put this is to say that inalienable claim rights provide a necessary condition of the possibility of moral interaction. Meyer's (1981, p. 140) rightly notes that "the prospect of moral interaction collapses without them." However, she does not sufficiently appreciate the role of the natural theological grounding in the justification of categorical obligation, and thus wrongly criticizes the role given to "divine authority" in accounts of inalienable rights.

[19] For an overview of the literature on the sanctity of life, see Khushf, 1995.

[20] This approach to the natural theological argument in Locke is found in most interpreters of Locke, and it is exemplified in McConnell (1984, p. 47), Stell (1979, p. 16), and Simmons (1983; 1992; 1993).

[21] One can distinguish between two notions of tolerance, one involving the restraint of force, the other involving a suspension of judgment (Khushf, 1993). The former is tied to the Lockean tradition, and depends on inalienable claim rights.

BIBLIOGRAPHY

Andrew, E.: 1988, *Shylock's Rights*, University of Toronto Press, Toronto.

Engelhardt, H.T.: 1986, *The Foundations of Bioethics*, Oxford University Press, Oxford.

Feinberg, J.: 1978, 'Voluntary euthanasia and the inalienable right to life,' *Philosophy and Public Affairs* 7(2), 93-123.

Frankena, W.: 1955, 'Natural and inalienable rights,' *The Philosophical Review* 64, 212-232.

Hobbes, T.: 1962, *Leviathan: or the Matter, Forme and Power of a Commonwealth Ecclesiastical and Civil*, M. Oakeshoot (ed.), Macmillan Publishing Co., Inc., New York.

Kendall, W.: 1941, *John Locke and the Doctrine of Majority-Rule*, the University of Illinois Press, Urbana.

Khushf, G.: 1992, 'Rights, public policy and the state,' in T. Bole and W. Bondeson (eds.), *Rights to Health Care*, Kluwer Academic Publishers, Dordrecht, pp. 355-374.

Khushf, G.: 1994, 'Intolerant tolerance,' *Journal of Medicine and Philosophy* 19, 161-181.

Khushf, G.: 1995, 'The sanctity of life: A literature review,' in K. Bayertz (ed.), *Sanctity of Life and Human Dignity*, Kluwer Academic Publishers, Dordrecht, pp. 293-310.

Locke, J.: 1980, *Second Treatise of Government*, C.B. Macpherson (ed.), Hackett Publishing Co., Inc., Indianapolis, Indiana.

Locke, J.: 1991, 'A letter concerning toleration,' in J. Horton and S. Mendus (eds.), *John Locke, A Letter Concerning Toleration in Focus*, Routledge, New York, pp. 12-56.

Lustig, B.A.: 1991, 'Natural law, property, and justice: The general justification of property in John Locke,' *Journal of Religious Ethics* 19, 119-148.

Mack, E.: 1980, 'Locke's arguments for natural rights,' *Southwestern Journal of Philosophy* 11, 51-60.

Mack, E.: 1989, 'Moral individualism: Agent-relativity and deontic restraints,' *Social Philosophy and Policy* 7, 81-111.

Mack, E.: 1996, 'The alienability of Lockean natural rights,' in M. Cherry (ed.), *Persons and Their Bodies: Rights, Responsibilities, and Relationships*, Kluwer Academic Publishers, Dordrecht, pp. 143-176.

MacPherson, C.B.: 1962, *The Political Theory of Possessive Individualism*, Oxford University Press, Oxford.

McConnell, T.: 1984, 'The nature and basis of inalienable rights,' *Law and Philosophy* 3, 25-59.

Meyers, D.: 1981, 'The rationale for inalienable rights in moral systems,' *Social Theory and Practice* 7, 127-143.

Nozick, R.: 1974, *Anarchy, State and Utopia*, Basic Books.

Plato: 1974, *The Republic*, D. Lee (trans.), Penguin Books, New York.

Quinn, P.: 1995, 'Political liberalisms and their exclusions of the religious,' *Proceedings and Addresses of the American Philosophical Association* 69(2), 35-56.

Rapaczynski, A.: 1987, *Nature and Politics: Liberalism in the Philosophies of Hobbes, Locke, and Rousseau*, Cornell University Press, Ithaca.

Risenberg, P.: 1956, *Inalienability of Sovereignty in Medieval Political Thought*, Columbia University Press, New York.

Riley, P.: 1982, *Will and Political Legitimacy: A Critical Exposition of Social Contract Theory in Hobbes, Locke, Rousseau, Kant and Hegel*, Harvard University Press, Cambridge, Massachusetts.

Schiller, M.: 1969, 'Are there any inalienable rights?' *Ethics* 79, 309-315.

Simmons, A.J.: 1983, 'Inalienable rights and Locke's *Treatises*,' *Philosophy and Public Affairs* 12(3), 175-204.

Simmons, A.J.: 1992, *The Lockean Theory of Rights*, Princeton University Press, Princeton, New Jersey.

Simmons, A.J.: 1993, *On the Edge of Anarchy: Locke, Consent, and the Limits of Society*, Princeton University Press, Princeton, New Jersey.

Stell, L.: 1979, 'Dueling and the right to life,' *Ethics* 90, 7-26.

Strauss, L.: 1953, *Natural Right and History*, University of Chicago Press, Chicago.

Tully, J.: 1995, 'Property, self-government and consent,' *Canadian Journal of Political Science* 28, 105-132.

van de Veer, D.: 1980, 'Are human rights alienable,' *Philosophical Studies* 37, 165-176.

von Leyden, W.: 1982, *Hobbes and Locke: The Politics of Freedom and Obligation*, The Macmillan Press, London.

Zuckert, M.: 1994a, 'Hobbes, Locke, and the problem of the rule of law,' in I. Shapiro (ed.), *The Rule of Law*, New York University Press, New York.

Zuckert, M.: 1994b, *Natural Rights and the New Republicanism*, Princeton University Press, Princeton, New Jersey.

SECTION THREE

METAPHYSICAL QUANDARIES
AND MORAL QUESTIONS

THOMAS M. POWERS

THE INTEGRITY OF BODY:
KANTIAN MORAL CONSTRAINTS ON THE PHYSICAL
SELF

I. INTRODUCTION

The moral permissibility of organ transplantation is taken for granted by
most biomedical ethicists and practitioners. Of contemporary concern is
not whether, but by what arrangements, we ought to allow organ
transplantation. Should we institute markets for organs, thereby
increasing their availability and saving many lives? Should organs be sold
to the highest bidder? Should we allow the *post mortem* taking of organs
without prior consent?[1] Among moral theorists, the Kantians are
suspected of being the least enthusiastic with respect to these and similar
questions. I will show the elements in Kant's theory that account for this
lack of enthusiasm and how contemporary Kantians might answer
questions about the permissibility of various arrangements for organ
transplantation. I will argue that Kant would have had a permissive
position on organ transplantation *post mortem* and a restrictive position
on organ markets. These results will be based on a broader Kantian view
of obligations to the body which stem from a non-formal theory of value.
This non-formal theory, I will argue, is at the foundation of Kantian
ethics.

II. THEORY AND APPLICATION

The two main pillars of Kant's ethical theory offer a predictably
ambiguous treatment of duties to the self. The formal pillar, seen in the
command to act only on universalizable maxims, is thought to generate
moral rules for types of actions. Any particular case – say, of Jones'
prospective organ transplantation – is supposed to fall under the general
rule for the type: "One cannot universally will organ transplantation."
There are no personal exemptions, then, to universally-applicable rules. It

M.J. Cherry (ed.), Persons and their Bodies: Rights, Responsibilities, Relationships, 209–232.
© 1999 *Kluwer Academic Publishers. Printed in Great Britain.*

is this part of Kant's ethics that is infamous for its austerity and adulation of rules and which is generally seen to fail when it comes to the specification of content-full duties to the body.[2] The non-formal pillar, on the other hand, operates through the command to treat persons as ends-in-themselves and not merely as means. The content of this position is further spelled out in the Kantian claim that persons are alleged to have a certain status – dignity or intrinsic value. Since the self is also a person, the agent is proscribed from treating herself as a means or from doing anything to eviscerate dignity. On the surface, this non-formal part of Kant's ethics avoids cold rule-worship, but perhaps at the expense of poetizing the notion of the person through possibly false (and certainly dated) enlightenment rhetoric. Even before the fashionable post-modern complaints against enlightenment ideals, Kant's critics were chiding him for his rather maudlin evocation of the dignity of humankind and our moral transcendence of "brutalizing nature."[3]

In several writings, Kant tried to apply this non-formal theory of value to questions of social morality. Whatever one thinks about Kant's theory, his applications of the theory to "real-life" problems, in *Lectures on Ethics* and the *Metaphysics of Morals,* are thought to be particularly unsuccessful. In these applications, Kant is presumed to have made some serious errors, especially in the area of self-regarding duties towards one's body. Kant's prudishness about sex, his frequent harangues against suicide, and his apparent denial of the right to sell our own hair all seem to show that he is a reactionary. One can make a strong *prima facie* case that Kant would have opposed all organ transplantation, since he writes that even the selling of our teeth violates duty. If this is right, we are tempted to disavow Kant on humanitarian grounds due to the contemporary need for organ transplantation.

A thorough reading of Kant's works would show the presumption of Kantian errors in his applications to be hasty if not itself erroneous. After all, Kant claims that we must be sure to *stimulate* part of our animality,[4] that suicide with honor is better than a life of disgrace,[5] and that selling one's hair is no *moral* transgression, though it may be in poor taste.[6] However, of the two ways to defend Kant – on the reasoning in his applications or on the fundaments of his theory – the former way is inferior as a means to get clear on contemporary issues about bodies.

In particular, consideration of Kant's applications alone will not answer questions about the Kantian position on markets for organ transplantation and the conditions under which donation might be permissible. Kant could not have reasoned about these applications through the lens of our contemporary medical science. Since I want to try to give Kantian answers to the moral questions surrounding organ transplantation, my defense of the Kantian view will focus on the fundaments of his non-formal theory of value: the doctrines of dignity and humanity in the person and of bodily integrity.

The necessary propaedeutic for a Kantian position on organ transplantation is a clearing away of the prevailing and rather blunt consequentialist sentiments behind the call for increased, compelled, or financially-rewarded organ donation. This is easy enough, since "pure" consequentialism adopts the unlikely position that consequences are *all* that should matter in moral decisions. So while it is of course true that more people would live if more organs were donated, it is also true that more people would live if we sacrificed one healthy two-kidney "donor" for every two patients in need of kidneys, or if we cleared trees from highways so as to reduce traffic fatalities, or even if we *encouraged* motorcyclists to drive without helmets, assuming a greater than 1:1 ratio of organ recipients saved to motorcyclists killed.[7] If the desired consequence is an increase in the number of living human beings, then any of these proposals would be morally required. Clearly they are not. All that a non-consequentialist position must maintain, Kant's included, is that *something else* matters besides the number of people to be saved by a more vigorous program of organ transplantation.[8] The task for the Kantian is to say what this "something else" might be.

III. KANT IN HIS SETTING

The problem in Kant's theory regarding obligations to one's body has its roots in both Cartesian dualism and Humean phenomenalism. For Descartes, our bodies lie across the epistemic curtain erected by the clarity-and-distinctness criterion of truth, since ideas about bodies – even our own bodies – are not clearly and distinctly apprehended by the mind. The wax discussion of Meditation II prepares the reader for this strange result. By Meditation VI, when Descartes insists that he can exist independently of his body, it is obvious that the mind is the center of the Cartesian self, and that body is somehow incidental (Descartes, pp. 21-23, 50-62). Little changes as modern philosophy takes on an empiricist direction. On Hume's account, impressions that purport to be from and about the body cannot be recovered from the chaos of other perceptions in which they are lost. Impressions of solidity, redness, and pain are all "on the same footing" (Hume, pp. 190-93); no impression brings with it an obvious mark by which we can tell that it is a *bodily* impression.

Kant finds himself stuck between these two unpalatable doctrines. Unlike Descartes, Kant wants to allow that the body can effect the mind, and especially that the body can have a corrupting influence on moral reasoning. For Kant, the body must be disciplined so that it "does not exercise any compulsion" on our moral and intellectual endeavors (LOE, p. 157). Bodies have a positive role to play too; physical talents on Kant's view are instrumental for moral and intellectual enterprises (MM, p. 445). So Kant's epistemology cannot be skeptical of bodily perceptions. Yet Kant cannot go too far in trusting mental events about bodies. He cannot let the body dictate moral reasoning, as the feelings of approval and disapproval do when taken, as Hume does, for a "moral sense" with which to derive moral distinctions. An ambitious ethical theory – and Kant's doctrine is certainly that – must integrate the moral, intellectual, and physical selves without sacrificing the control of practical reason on moral matters.

What is impressive in Kant's project to integrate the three selves is the breadth of scientific knowledge that he brings to it. Kant seems to have been deeply interested in developments in the biology and medicine of his time. In his last unpublished work, a contribution to the philosophy of science which survives as the *Opus Postumum,* Kant shows his preoccupation with the leading medical works of his time: John Brown's influential *Elementa Medicinae* [1780] and Albrecht von Haller's *De partibus corporis human sensilibus et irritabilibus* [1753] (OP, pp. 103-04), William Cullen's *Nosology* [1785] and Friedrich Hildebrandt's *Lehrbuch der Physiologie* [1799] (OP, p. 122).

Beyond showing Kant's generic concern with medical science, the *Opus Postumum* contains his doctrine of integrity as it applies to the life sciences. Kant thinks that, while plant parts can be integrated into another system within the species, animal parts cannot be removed from their proper organic system (OP, pp. 182-83). Kant seems not to be exactly sure of the integrating principle of organic beings, but he is sure that the search for such a principle is crucial to the unity of the life sciences.

Connected to this search is Kant's interest in the notion of the "vital force" (*Lebenskraft*). During Kant's lifetime the posit of the vital force – a non-chemical and non-physical force inhering in and sustaining living beings – was one of the central disputes in biology. Kant sides with the vitalist doctrines of Georg Ernst Stahl and commits to the existence of the vital force.[9] He seems to conceive of it in a way similar to the elastic medium of the pervasive ether. What ether allegedly does for space, the vital force does for living beings. The vital force on Kant's view infuses living organisms and explains the continuity of the processes which sustains them. This doctrine sounds mysterious because it is; nonetheless, it is Kant's rejoinder to the mechanistic understanding of body which comes on the heels of the Newtonian revolution. For Kant, the vital force is the "something else" about living entities which the mechanists cannot capture.

IV. THE FUNDAMENTS OF KANT'S MORAL THEORY

What seems mysterious for the science of life is much less so for the science of morals. One need believe neither in souls nor in auras – indeed, one need not bring in "the divine" at all – to believe that there are some sustaining elements to a person's moral life which are not captured by chemistry, biology, or physics. This belief in a "something else" for moral life is Kant's motivation in distinguishing body, life, and person, and in ordering them in a moral hierarchy.

Person occupies the highest position in Kant's hierarchy and is an explicitly normative concept. "A person," he tells us, "is a being who has rights of which he can become conscious." God is also a person, though perfect, unlike any human being (OP, pp. 210-11). Our imperfections afford us duties, of which we *should* also become conscious (OP, p. 203). Kant's famous categorical imperative – in all of its various formulations – is a heuristic device to aid in our becoming conscious of our duties. Since such moral consciousness is a function of reason, any action which impedes reason in us also impedes consciousness of rights and duties and hence "stunts" our moral personhood. Reason's contribution to moral consciousness is vulnerable, however, since reason is ultimately dependent on our physical condition. In this way reason is held hostage to the body. To his credit, Kant recognizes that many bodily circumstances, e.g., addictions, habits, and even laziness, harm reason's project for moral consciousness.[10] If Kant sounds offensively preachy on these subjects, it is because he is strongly convinced that reason's project for moral personhood is the most important undertaking of the moral life.

When reason is successful in its project it creates *dignity* in the person, a mark of having transcended the brute animality of nature. This success connects a category of being with a way of acting. Namely, an agent's "dignity (prerogative) of being above all the mere things of nature implies that his maxims must be taken from the viewpoint that regards himself, as well as every other rational being, as being legislative beings (and hence are they called persons)" (Gr, p. 438).

Dignity, then, is a mark of moral success in acting only on maxims which are universalizable. It is in this sense that persons are legislative beings, for their maxims could serve as universal laws of action. Beings without the opportunity for dignity are just those beings that cannot make their own laws of action but must follow slavishly the laws set for them by nature. Sensuous impulse is their master.

We must pause here to consider two misguided views concerning Kant's notion of dignity. The first view has its origin in natural law and claims that the dignity of a person is unconditional and constitutes the "absolute value" of persons. Were this the case, nothing would follow from the fact that persons have dignity, since there is neither content to nor conditions on the attainment of dignity. Specifically, no rights and duties would follow from such an automatic and unconditional notion of dignity. Only if there is a special (and difficult) way to attain dignity will there follow from the conditions on dignity a meaningful set of rights and duties which sustain the practice of *becoming* a person with dignity.[11] In the language of Kant's transcendental argumentation, the conditions for the possibility of dignity yield *a priori* the structure of the concept. Nonetheless, the manifestation of the concept, just like the manifestation of a cause, awaits an empirical exemplar in the world of phenomena.[12] Put plainly, our actions must manifest dignity in order to keep it from being a chimerical concept.

The second mistake is to see the ground of human dignity in the fact that we are not, for Kant, *merely* animals. Since the ascription of dignity is honorific, this view would have to rest on the position that humans are intrinsically better than animals and hence are "rewarded" with dignity by some metaphysical judge. This is not Kant's view; he is interested in one particular way in which we distinguish ourselves from animals. There are many ways in which humans differ from the rest of the animal kingdom. We build cities, travel into space, write books, translate languages, and have knowledge of our own mortality. But these differences in themselves are morally irrelevant. The only distinction which makes a *moral* difference is that we have the potential for practical rationality free from the determination of sensuous nature (a synthetic *a priori* claim) and that we actually exercise that potential (a defeasible empirical claim) by acting on a conception of moral law. Mere potentiality without actuality would not be enough for Kant, since it is the determination of the will by other-than sensuous motives which introduces the Ought into a world of Is. For Kant, actions manifesting dignity are the *sine qua non* of a world containing value. There is nothing bad about the world of sensuous animality; it merely lacks the Ought, and that is our special contribution.

Now to revisit the person. Insofar as dignity connects a way of being (person) with a way of acting (morality), it also connects the first formulation of the categorical imperative with the second formulation – the duty that we treat persons as ends-in-themselves and never merely as means. The status of a person as an end-in-himself simply reiterates the fact of his moral success, for as Kant says "just this very fitness of his maxims for the legislation of universal law distinguishes him as an end-in-himself."[13] The dignity of the person explains why the second formulation serves, in Kant's terms, as the "supreme limiting condition" for self- and other-regarding action. Objects or things can be used by anyone as means for various sensuously-conditioned ends. Only persons enjoy a status, that of being ends-in-themselves, which separates them from the world of everyday objects and things. The very notion of an end-in-itself, then, denies that something external to the self could condition or generate ends for the subject. The ends (goals or projects) of every end-in-itself are found *internal* to the self, in reason. Action motivated by reason alone, on Kant's view, is the only action that escapes the slavery of the sensuous world. Rationally motivated moral action is the only way we *become* persons.

We can summarize this nexus of concepts in the following manner. A person's dignity provides the condition for her to transcend sensuous animality, and her humanity, her status as an end-in-herself, delivers the person from the instrumentality of objects and things. Dignity and humanity, then, describe the moral status owing to persons. They are not emotive terms. The point of the dignity claim is a requirement that one not harm one's conscious of duties. The humanity claim, on the other hand, provides a "supreme limiting condition" on the use of persons; we may not use them as tools or instruments. As we shall see, this claim includes the restriction on using one's own body parts, under certain conditions, as tools to prolong the lives of others.

The notion of the person occupies such a large space atop Kant's moral hierarchy that it threatens to occlude the notions of life and body. Nonetheless, these notions do have importance for his ethics, even though they are not as thoroughly normative as the notion of the person.

Life and body can be treated together if we mean by 'life' merely the physical body over time. The animating force of life, which we saw above is the vital force for Kant, both affects and is affected by the circumstances of the body. Kant is genuinely concerned with this interrelationship and seems to ask the questions that biomedical ethicists want answered: "What are our powers of disposal over our life? Have we any authority of disposal over it in any shape or form? How far is it incumbent upon us to take care of it?" (LOE, p. 147).

Kant answers these questions by focusing on the place of body in life. If bodies were interchangeable or exchangeable without risk to life, i.e., if bodies were like organic furniture, then he sees that we could dispose of our bodies however we will. But since life is entirely conditioned by the body, we are not absolutely free. One of the supervenient characteristics of life is will, so to will the destruction of the very body on which life is conditioned is to will to negate the will. In this Kant finds a self-contradiction.

Apart from will- and life-destroying acts, what moral constraints does Kant put on us with respect to our bodies? At first, it might seem that we are only minimally constrained. He says that "we may treat our body as we please, provided our motives are those of self-preservation" (LOE, p. 149). But Kant's notion of self-preservation masks a complex teleology which is grounded in his view that all organic bodies contain the concept of purpose (OP, p. 211). Since there is a division of labor in the organs of a system and preservation depends on the cooperative functioning of the individual parts, self-preservation turns out to be quite a tall order. Individual parts have purposes, and they are each necessary for the successful completion of the overall "organic" purpose, a full life. We may indeed amputate parts for the purpose of self-preservation, but Kant seems to think that any reason less dire than self-preservation is not good enough. Since we will consider below the specific conditions for such medical procedures, let us postpone for now the application of this theory to the removal of body parts.

We have been considering the notion of physical life and its position slightly ahead of body in Kant's moral hierarchy. Surprisingly, Kant holds that "life is not to be highly regarded for its own sake" (LOE, p.

150). In discussing the common practice in warfare of sending soldiers to the slaughter, Kant says that the physical life is not as important as courage, devotion, and the other generic virtues of soldiering. For Kant, how and why one acts is much more important than the result of one's acts. Hence the soldier who throws himself into a hopeless battle dies a noble and moral death, while the soldier who instead commits suicide is regarded (with disgust) as "carrion" (LOE, p. 151). Such examples allow Kant to introduce the notion of the "moral life" which, unlike the physical life, can end through the violation of the person's dignity (LOE, p. 156). Moreover, Kant thinks that the physical life *should* end if the moral life cannot be sustained. He claims that "there is much in the world far more important that life. To observe morality is far more important. It is better to sacrifice one's life than one's morality. To live is not a necessity; but to live honorably while life lasts is a necessity" (LOE, p. 152).

The notion of a moral life, or the character or spirit of one's actions over time, is captured in Kant's conception of *Gesinnung*. In English translations of Kant's work the sense of this term is unclear, but is perhaps best rendered 'attitude'.[14] In a moral context, *Gesinnung* can mean 'character' in the sense of a "fundamental attitude" (*Einstellung*). One says in German "*Er ist ein gut gesinnter Mensch*," or "He is a decent person." To be *gesinnungslos* is to be unprincipled, usually with the implication that this is an enduring personal trait, though not necessarily one which cannot be corrected. For Kant, *Gesinnung* plays an important role in that it connects the agent's attitude and her action based on the moral law. Harold Köhl calls Kant's ethic on the whole a *Gesinnungsethik*,[15] which I would translate "Ethics of Principle." The contrary of a *Gesinnungsethik* is what Köhl sees as a *Folgenethik* (Ethics of Results or Consequences) and Max Scheler calls an "Ethics of Success" (Scheler, 1973).

In the *Grounding*, Kant uses the term *Gesinnung* in its narrow sense to indicate the disposition of the will in any given choice of maxims. *Gesinnung* is the spirit or moral frame-of-mind behind the choice of maxims, as opposed to the tangible effects of maxims through actions (Gr, pp. 406, 422, 435-436). It is precisely in this attitude or spirit behind maxim formation that Kant locates moral value.[16] The difference between acting in accordance with duty and acting from duty is a difference in *Gesinnung*. Kant insists that the agent herself can never be sure whether there is also an inclination to do the act which lurks somewhere behind the putatively rational choice of the maxim. The existence of such an opaque inclination reduces the maxim to one which merely accords with duty and thus is not done from a sense of duty. The hidden inclination robs the maxim of moral worth. Because of the opacity of moral motivation, we are never really sure that our motivation is purely moral.[17] Hence Kant's narrow account of *Gesinnung* in the *Grounding* remains epistemically problematic.

In the *Lectures* and *Metaphysics of Morals* Kant adopts a broader view of *Gesinnung* according to which "ethics is the philosophy of *Gesinnung*" (LOE, p. 71). Here *Gesinnung* is more than the specific disposition of the will in choosing a particular maxim, as in the narrow view. The broad sense of *Gesinnung* is the character of one's moral life over time. Conveniently, the broad notion is constructed out of the narrow one. The character of one's moral life is a function of how one chooses the many particular maxims over a lifetime. Since *Gesinnung* in this broad sense is constructed by individual moral choices, it can also be taken apart by the wrong choices. This explains why in the passages above Kant seems so dismissive of physical life. Making the wrong choices, morally speaking, begins to chip away at the character of the moral life. At some point, the damage is irredeemable and the duty to preserve one's physical life no longer has a purpose, since the project of morality in such a being has been abandoned.

V. DUTIES TO THE PHYSICAL SELF

The reader familiar with contemporary accounts of Kant's ethics will no doubt feel a certain uneasiness at this point due to the scarcity of the categorical imperative in my account. The categorical imperative has remained in the background because we were concerned with the "why" of the Kantian moral project and not the "how" of Kantian pure practical reason. Now it is time to turn to this latter question to address the duties to the physical self. By piecing together what Kant says in the applications of his moral theory and what I have portrayed as the fundaments of his moral theory, we will be able to answer our original questions about organ transplantation.

Kant is concerned in the *Lectures* and *Metaphysics of Morals*, his only works in "applied" ethics, with roughly the same set of duties towards one's physical self or "animal being." Duties against suicide, drunkenness, gluttony, self-mutilation, and carnal crimes are the foci of

Kant's applications. In general, violations of one's duty to the physical self can be total or partial. Total violation of one's duty is suicide, while partial violations are ones in which the agent "deprives himself of certain integral parts (organs) by dismembering or by mutilation," or "deprives himself ... of the physical (and hence indirectly also the moral) use of his powers ..." (MM, p. 421). Let us focus on the partial violations of duty.

Kant's main prescriptions are to maintain the integrity of the body as a purposive organism and to respect the humanity of the person – the non-thing-like character which denies its use as an instrument. Concerning the issue of integrity, Kant considers it partial self-murder "to deprive oneself of an integral part or organ (to mutilate oneself), e.g., to give away or sell a tooth so that it can be planted in the jawbone of another person, or to submit oneself to castration in order to gain an easier livelihood as a singer ..." (MM, p. 423). Kant's reasoning here does not depend on a strictly "moral" notion of integrity, as we see in his later comment that "integrity is not the opposite of depravity (perversity) but of loss (as of a limb)" (OP, p. 239). Rather, his reasoning turns on a teleological conception of biology. Kant's teleological view of the integrated body parts and organs rules out certain modifications to the organism because they are irreversible and may thwart the purposive ends of the person as a whole. Self-mutilation was not uncommon in his time, as his mention of the phenomena of the *poltron* or *pollex truncatus* indicates.[18] This term describes the army recruit who severs his right thumb in order to avoid military service. Having achieved that end, the man no doubt discovers that the right thumb serves many other purposes of which his body is no longer capable.

Kant's doctrine of the humanity of the person insists that the body is not at the disposal of the person, since the body is not a mere thing. Hence the agent "is not entitled to sell a limb, not even one of his teeth" (LOE, p. 165). Kant's thematic arrangement in his applied works aligns self-mutilation, prostitution, and carnal crimes. Thus he can make the argument that, when the agent sells a tooth, he engages in "irreversible" prostitution; and in the event that he does not profit from the mutilation, his act is tantamount to a carnal crime. For Kant, selling body parts and leasing body parts are in the same category.

Integrity and humanity, as we saw in section IV, are central to the Kantian moral project for a type of freedom conditioned by reason alone. Kant holds that "the principle of all duties is that the use of freedom must be in keeping with the essential ends of humanity" (LOE, p. 124). This principle is violated both when the agent's use of her body is conditioned by money, as with the selling or leasing of body parts, and when her use is conditioned by the satisfaction of another's desires. To engage in such actions is "to dispose over oneself as over a thing and to make oneself a thing" (LOE, p. 165). The peril in such actions, on Kant's view, is that others may treat us as we treat ourselves. If the agent uses himself as a thing, others are then entitled to use him as a thing, since "he has made a thing of himself, and, having discarded his humanity, he cannot expect that others should respect humanity in him" (LOE, p. 151). Not only is the person imperiled as the recipient of another's actions, he is also undercut as an agent. The agent's own violation of his humanity makes him "incapable of doing his duty towards others," since the "prior condition of our duty to others is our duty to ourselves" (LOE, p. 118).

Many of us will be inclined to think that Kant's arguments go too fast here. The first difficulty is that it does not follow from the agent's disregarding of her own humanity that others may disregard it too. Similarly, the agent's use of her body as a means – to dig a garden, let us say – does not therefore permit others to use her as a means. The first objection is not entirely to the point, since Kant is not offering a social psychology of human interaction, but a moral psychology for practical reason. Kant's argument is that, since the dignity of the agent is dependent on the character of her self-regarding actions, other agents *can* deny her dignity since she has failed in the moral performance that is necessary for dignity. She has failed to create or sustain the dignity that would protect her, morally speaking, from others' disregard. To answer the second objection, Kant needs (and nowhere supplies) some further argumentation that distinguishes permissible from impermissible ends of actions. Then he can argue that the use of one's body as a means – but not simply as a means[19] – for permissible ends (digging gardens) does not violate humanity, while use of the body as a mere means for impermissible ends (earning money through prostitution) does violate humanity in the person. The permissible ends are what Kant calls the "ends of reason" (MM, p. 395).

A third objection would be that it does not follow that the agent becomes incapable of duty towards others when she fails in duties to herself. For Kant, however, the moral project which *results* in the agent's recognition of other-regarding duties *begins* in the dignity of the self. Dignity creates moral consciousness *simpliciter* and hence provides the condition for the extension of obligation outside the personal sphere.

Without a sense of her own dignity, the agent will have no basis in her own moral consciousness for extending the constraints of dignity – rights and duties – to persons generally.

So far I have described Kant's application of the second formulation of the categorical imperative to one's duties to the physical self. It should be clear that Kant's argument is not as simple as the mere application of the rule "treat others and oneself as ends and not as means merely." The second formulation cannot stand alone without the moral content provided by Kant's notions of the dignity and humanity of the person and the integrity of the body. Where Kant is more apt to offer a free-standing rule is in his use of the first formulation of the categorical imperative, the requirement that we act only on maxims that are universalizable. Advocates of the formalist interpretation of Kant's ethics – and sometimes Kant himself – see this rule as being sufficient for the generation of all duties in a straightforward way.[20]

While Kant tries to show in *Grounding* that both the first and second formulations of the categorical imperative will deliver the same set of duties, in the *Lectures* and *Metaphysics* he has trouble getting the duties to the physical self from the free-standing requirement to universalize one's maxims. This "supreme principle of morality" is supposed to be our sole means of discriminating right from wrong (LOE, p. 43; MM, p. 222). By making our maxims of action conform with universal rules – thus making them *regelmäßig* – we show the essential impartiality towards the self which combats sensuous inclination. However, in his actual examples Kant abandons the first formulation in duties to the self and uses the second formulation of the categorical imperative.[21] That is, the thrust of his arguments rely on the humanity and dignity of the person and not on the universalization procedure.

Kant's second formulation is better suited to the talk of duties to the body because bodies can be put to so many uses and that formulation explicitly addresses the notion of use. While there are many ways we might use our bodies, reason demands hegemony over the use of the body. The body is allowed to be instrumental for no other consideration – one's own or another's pleasure, money, or amusement – because it must be instrumental to reason's ends. Practical reason needs the body for moral enterprises (e.g., helping others with physical labor, earning one's keep, and acts of heroism) and theoretical reason needs the body for intellectual enterprises. These enterprises may depend on some *particular* body part, organ, or talent which is integrated into the whole. This is

Kant's doctrine of bodily integrity. So Kant must hold that it would be wrong to get rid of any body parts prematurely, i.e., before we know that we can no longer use them. That point comes only at the end of life. While we live, we are obligated to "the cultivation of all of our capacities in general in order to promote the ends set before us by reason."[22]

The reasoning against living organ donation would go as follows. Suppose the agent donates a kidney and then suffers serious damage to the remaining kidney at a later date. Did the agent not harm the functional integrity of his body in donating? Kant's example of the agent encountering a person in distress serves as a good analogy for the problem regarding live organ donation. Supposing that we meet someone in need,

> We should relieve his want and affliction, but only insofar as we can do it without detriment to ourselves .... I ought not to give the poor wretch more than I can spare, for if I were to give away what I cannot spare, I should myself be in want, I should myself have to seek the charity of others, and I should cease to be in a position to act morally (LOE, p. 19).

Just as we run the risk of falling into poverty by being "overly beneficent," so too can we risk harm to our bodies by donating organs before death. This issue may well turn on the probabilities of harm, but Kant would likely reject *any* donation for which there is a risk.[23]

The common objection to this Kantian position, and one voiced by Joseph Boyle (in this volume, pp. 111-141), points out the "obvious" moral permissibility of the sale of other body parts (blood, semen, hair, etc.) when they are not *necessary* to maintain the health of the subject. While it is true that Kant would have erred had he claimed that selling hair is morally wrong (as opposed to merely in poor taste, as I have argued), it is a fallacious parity of reasoning to ally modest donations of fluids and regenerable parts with the donation of vital organs. The fallacy originates in not seeing how Kant's humanity claim goes hand-in-hand with his integrity claim. In fact we would *not* allow someone to give or sell blood to the point of anemia, deliver sperm to the point of dysfunction, or even to expectorate until the onset of dehydration. Under normal and controlled conditions, the surrender of these parts of the body in no way threatens to harm its integrity. So while it may be obvious that donating small amounts of blood is morally permissible, it is likewise obvious that this act is fundamentally different from selling a kidney or, for that matter, from the practice of surrogacy. One could object that the law allows amounts of blood to be donated to the point where, statistically, the donor has the same mortality profile (runs the same risk) as the single-kidney donor, but this would merely beg the question. Likewise, pointing to the current practice of "renting" (and perhaps one day, of "selling") other body parts, such as wombs, begs the question of moral permission, *pace* Boyle.

Given that organ transplantation was not feasible in his time, Kant's restrictive position on procedures as simple as transplanting teeth is understandable. No doubt, this position is informed by the state of 18th-century Prussian medicine, with its general lack of antisepsis. It is also relevant to an understanding of Kant's writings that his Königsberg was rather backward when it came to medical science.[24] There was a scarcity of trained physicians, and the plague almost wiped out Königsberg. We must remember that even the removal of a tooth brought with it the certainty of a great amount of pain and the likelihood of infection, and Kant was skeptical about the smallpox vaccine for similar reasons (cf. MM, p. 424). On a scale much less serious than organ removal, the removal of a tooth in order to sell it would have been foolhardy and perversely greedy. Kant's point in repeatedly using this example is that anyone who would risk such pain and danger for monetary gain is far from having the priorities of reason in mind.

I believe that, far from explaining away Kant's position, such prudential considerations over determine his position on organ transplantation. When we focus just on moral considerations, we see that Kant's broader views about the duties to the self support his reasoning concerning bodily obligations. He says that "to have a healthy mind in a healthy body is a duty to oneself" (LOE, p. 142). But the two parts do not have equal status; our first duty towards the body is to create an "autocracy of mind" over it (LOE, p. 157). We must do this, Kant thinks, because the body can influence the mind to such an extent that reason no longer has mastery over the body. He claims that self-mastery "is the objective condition of morality,"(LOE, p. 138) since it is necessary for all self-regarding duties and, as mentioned above, the self-regarding duties are necessary for the other-regarding duties. The entire project of morality, then, seems to weigh foremost on reason's ability to control the body.

What Kant's rather puritanical rhetoric amounts to are proscriptions against overindulgence, ones which are not significantly different from

Aristotle's doctrine of the mean. For Kant, the body "must be made frugal in its needs and temperate in its pleasures;" and though he believes that we should strengthen and harden the body "in every useful way," he is against mortification of the flesh (LOE, p. 158). Kant is very clearly opposed to excessive drinking and drug addiction, and it was his servant Lampe's non-stop "boozing from morning until night" that contributed to his being dismissed after forty years of service.[25]

VI. THE ARGUMENT FOR ORGAN DONATION FROM BENEFICENCE

We have treated the moral question of organ transplantation up until now as an issue concerning self-regarding obligations only. This cannot be the entire picture, since the person to receive the organ – a person who would perhaps die without it – must weigh heavily in the evaluation. Especially for Kant, *other* persons must be taken seriously as ends-in-themselves. This means in part that their subjective ends – including their happiness – must become objects of my duties insofar as I can accommodate them (MM, p. 393). Certainly the prolongation of another's life is a legitimate object of duty. How then can Kant deny the duty to sustain another's life, when one can, through donating a vital organ?

Kant is aware of this tension in his own theory. He asks, "How can one require as a duty that everyone who has the means show beneficence to those who are needy"(MM, p. 452)? Kant's answer comes in an argument about reciprocity. He reasons as follows:

1) Everyone would wish for aid when in need.
2) When we have the means, not giving to others in need contradicts the universalized maxim in premise 1.

Hence:

3) Our selfish maxim when in need would have to become a maxim of beneficence when we have the means.

When we apply considerations of reciprocity to organ donation, we get the following argument:

1*) Everyone would want organs should they find themselves needing a transplant.
2*) When we have the means (and almost all of us do after death), not donating organs contradicts the maxim in premise 1*.

Hence:

3*) We ought to donate organs when we have the means (especially after death).

This reciprocity argument "from beneficence" is one that a Kantian can accept, since we seem not to harm the moral or intellectual ends of reason by giving up organs after death. This position is also consistent with the argument "from dignity" put forth above, that organ donation while one is alive is absolutely forbidden. Where the two positions clash is on the issue of markets for organs. If it is the case that creating markets for organs will save more lives, markets seem to further the duty of beneficence to others. However, markets for organs are a perfect manifestation of what Kant's principle of humanity in the person condemns: the conception of the "use value" of persons and parts of persons for monetary and utilitarian ends. So on the issue of organ markets, Kant's doctrine seems to be pulled in two directions.

If the self-regarding duty of dignity conflicts with the other-regarding duty of beneficence, then the thesis of *Kantian rigorism* must be false. The rigorist thesis holds that, in Kant's ethics, there can be no conflicting duties, and Kant supports the thesis (in word if not in deed) himself. He says that, in *apparent* conflicts of duties, the resolution lies in deciding which obligation has a stronger ground (MM, p. 224). We can then recast our question. Which is stronger: the obligation grounded in the dignity of the self or the one grounded in the happiness of others?

Kant believes that the duty to self prevails, since "to sacrifice one's own happiness, one's true needs, in order to promote the happiness of others, would be a self-contradictory maxim if made a universal law" (MM, p. 393). Here we must recall the argument I gave earlier, that the duties to others are all grounded in duties to the self. The case for markets, since it is grounded in a duty to others, looks to suffer if markets for organs conflict in any way with the duty to self.

Ironically, any system of organ distribution which compels organ donation or entices donation for monetary gain would rob the act of donation of its moral merit, since "action as such is immaterial; it is the source of it which matters" (LOE, p. 75). Of the three sources for beneficent action under consideration here – force, money, or the spirit of doing what is right – only the last one carries moral merit on the Kantian view. Leaving aside markets, the issue about organ donation is whether the good of others is enough to outweigh any squeamishness that the prospective donor might feel in knowing that her body will be carved up after death. I tend to think that, given the tremendous benefit served by donation to others, one ought simply to get over such feelings. For a Kantian, these feelings do not amount to much, morally speaking, and the lives of others do amount to a great deal. Organ markets, however, do more than produce squeamishness. (They could conceivably be arranged in such a way that they would just be in bad taste, but this is not necessarily so.) Even though markets may benefit recipients, they are wrong for other reasons.

From the standpoint of theory it may be easy to distinguish those body parts which are "attached" and hence integral to a living person from those which are "unattached" and can no longer serve the ends of reason for their original owners. Further, it is possible from this standpoint to distinguish the use of one's own "attached" parts for monetary gain (as in prostitution) from the use of "unattached" parts for monetary gain. Markets for transplant organs are committed to the latter type of use and are complicated by the issue of who stands to gain monetarily. Is it right that other family members, "middlemen," and transplantation surgeons might profit? It seems likely that middlemen and surgeons would profit, since there would be a large increase in the "raw materials" of their industry.

So markets for transplantation are committed to the use of body parts for *someone's* monetary gain. Let us distinguish two cases.

Case one. Where the persons who profit are other than the donor, their vocation is a form of trade in body parts.[26] One can imagine, perhaps, more macabre forms of such trading, but this form is troubling enough. The supporters of transplantation are simply wrong to believe that the psychic cost of transplantation is merely tied to the donor's "reluctance to consider their own mortality and the idea of being operated on after death" (AMA, p. 582.) Much more grievous is the idea of having body parts become yet another commodity in a capitalist scheme. For Kant, the argument from humanity rejects such practices because they would destroy the moral psychology necessary for the internalization of the intrinsic status of persons.

Case two. Now suppose the easier case, where only the donor profits from her future contribution. (This scenario seems unlikely insofar as any self-sustaining market creates jobs to support the trade itself, but we will disregard that problem.) Still, it is a form of profit that the donor would not be allowed, on Kantian grounds, were the parts still "attached." Kant's argument about the integrity of the body does not come into play here, as the "unattached" parts are no longer beholden to the purposes of reason. The donor's reason is extinguished with her life. Rather, Kant's principle of respect for humanity in the person comes into play, as it restricts the use of bodies as things or objects. What more is a body, though, after life has left it?

The social psychology of burial tells us a great deal here. Bodies are clearly not the persons they were after life has passed, but nor are they mere clumps of decaying tissue. From the scientific standpoint, of course, that is all they are, but we would find disposing of bodies as though they were merely tissue to be quite offensive. Hence we see even in wartime a strong aversion to mass graves. Granted, it is a philosophical confusion of the highest order to think that we can harm the person after his death, regardless of what we do to his body. There are no *post mortem* duties toward others or the self, on Kant's view, as those duties would stem from the moral life of the person. We can only have duties *towards* ourselves and other persons (MM, p. 442). Nonetheless, Kant would hold that we do have duties *regarding* dead bodies which are at the same time duties towards ourselves.

Kant's examples of duties regarding other things encompass all objects in the natural environment (MM, p. 442). For instance, we have obligations regarding animals not to treat them cruelly. Cruelty weakens "that feeling in man which is indeed not of itself already moral, but which still does much to promote a state of sensibility favorable to morals ... " (MM, p. 443). Kant is allowing here that psychology has a role to play in morality, if for no other reason than that the principle of discrimination of our duty does not carry with it a "principle of the performance of our obligation" (LOE, p. 36). The first is the measuring-rod, the second the mainspring of morality. This principle "of the moral incentive to action lies in the heart." My claim – and it depends entirely on psychology – is that markets for organs would undercut this proper feeling about bodies that Kant's doctrine of the humanity of the person seeks to instill. The psychological hurdle is not death or *post mortem* operation. Rather, it is the central principle that bodies and parts of bodies are not things that ought to be converted to monetary values.

There is however a compromise position that is consistent with Kant's doctrine: a system of barter for organs. Recall that Kant's duty of beneficence towards others invokes the reciprocity argument concerning giving and receiving aid. This argument is well suited for the issue of organ donation. The prospective donor's maxim would read: "I will accept organs only if I will donate organs *post mortem.*" Such a system of deferred barter would respect the humanity of the person, since parts are neither bought nor sold and profit is not taken. Deferred barter would approximate the desired effect of the organ market – to increase the number of organs available to those in need. There are rational (game-theoretic) motivations for donating, given the seriousness of the need should a transplant situation arise. To encourage early commitment to the barter agreement, administrators could create incentives tied to the length of commitment. Perhaps "tenured" parties to the agreement would have priority for available organs. Still, the deferred barter system would not be coercive. In emergency situations, patients could consent to future donations of some organs in exchange for the needed organ at the moment.

VII. CONCLUSION

My position has been to defend a moderate, content-full Kantian position on organ transplantation. The deferred barter system could increase the number of available organs without sacrificing the Kantian notions of dignity, humanity, and bodily integrity. In comparison, we can see how the organ market would be coercive, at least for those who are so poor as to be driven into selling the last things they have, their bodies. Defenders of the free-market laud the "opportunity" that the market offers for the desperately poor.[27] For Kantians, this "opportunity" is little different from the ones presented by prostitution rings, freak shows, and, in earlier times, the institution of indentured servitude. Further, the economic opportunity comes at too high a cost. It is not merely that, for the laissez-faire capitalist, nothing is sacred. For such a person, whom Marx calls "Free-trader Vulgaris" (Marx, p. 176), nothing is safe from being turned into a good or service. One need not be a Marxist to object to the commodification of at least some set of things, among them bodies. For Kant, the barrier to commodification is drawn around those things which are integral to the moral project as he sees it.

Department of Philosophy
Santa Clara University
Santa Clara, California

NOTES

[1]   These and similar issues are discussed in 'Financial incentives for organ procurement,' (AMA, pp. 581-89).

In this paper, Kant's individual works will be cited as follows: Gr = *Grounding for the Metaphysics of Morals*; KrV = *Critique of Pure Reason*; LOE = *Lectures On Ethics*; MM = *Metaphysics of Morals* (Part II, Tugendlehre); OP = *Opus Postumum*.

[2]   See H.T. Engelhardt's contribution in this volume (pp. 277-302). My interpretation of Kantian ethics will avoid Engelhardt's overly narrow elucidation of Kant's position.

[3]   Max Scheler (1973) for one, complains of Kant's defense of the dignity of humanity that it is a false "enlightenment pathos."

[4]   "The continual deliberate stimulation of the animal in man is a duty of man to himself" (MM, p. 445).

[5]   "If a man cannot preserve his life except by dishonoring his humanity, he ought rather to sacrifice it" (LOE, p. 156). The context of the discussion makes it clear that Kant was only rigidly against suicide out of self-love. This position is also suggested by MM, p. 422: "The deliberate killing of oneself can be called *self-murder* ... only when it can be shown that the killing is really a crime committed either against one's own person, or against another person through one's own suicide" (Cf. MM, pp. 423-25).

[6]   Kant believes that cutting off one's hair presents no moral problems, though "selling one's hair for gain is not entirely free from blame" (MM, p. 423). The relevant difference for Kant between selling hair and selling *extracted* organs and parts turns on the danger of further harm. We will take up this issue in section V.

[7]   This last point was suggested to me by Mark Cherry.

[8]   One could imagine a more sophisticated consequentialist position that chooses utility instead of life as the consequence to be maximized. Still, the counter-examples would go through under plausible assumptions about the uniformity of utility over lives. That is, the two kidney recipients would produce more utility (over their lifetimes) than the one "donor" would have produced in potential utility, adjusted for any actual disutility upon being divested of her kidneys.

[9]   Kant portrays revolutionary scientific achievements such as Stahl's as providing a model for his *Critique of Pure Reason* (KrV B xiii). Kant's commitment to the notion of vital force can be seen most clearly in OP (pp. 53, 66, 103, and 137); cf. OP, p. 275, where he shows that he was also familiar with the man who would ultimately defeat vitalism, Lavoisier. See also Eckart Förster's depiction of Kant's vitalism (in OP, p. 258, n.9).

As for Kant's fondness for mistaken scientific doctrines, we must remember that not until Einstein was the ether shown to be scientifically superfluous.

[10]   Most famously, Kant is concerned with various forms of sloth, addiction, and self-abuse in Gr (pp. 423, 430) and MM (pp. 421-447).

[11]   Another way to view this controversy is to ask whether *human beings* have dignity automatically, that is, regardless of successful moral performance or status as persons. There is considerable disagreement amongst Kant scholars. I am defending the view that they do not, that while individual agents should be *presumed* to have dignity on the basis that we generally have no evidence of their moral failures, dignity is defeasible and indeed *is* forfeited in acts of egregious immorality. I believe that, among other positions, Kant's supportive position on capital punishment in the *Rechtslehre* is evidence for my view. For the

view that dignity is unconditional, see Thomas E. Hill, Jr.'s *Dignity and Practical Reason in Kant's Moral Theory* (1992, pp. 50, 204-205).

[12] This is the same condition Kant requires in the first *Critique* in order to prove the objective reality of the categories. Kant says in Transcendental Deduction A:

> An a priori concept which did not relate to experience would be only the logical form of a concept, not the concept itself through which something is thought. Pure a priori concepts, if such exist, cannot indeed contain anything empirical; yet none the less, they can serve solely as a priori conditions of a possible experience. Upon this ground alone can their objective reality rest (KrV, A 95).

And similarly in Deduction B:

> the only intuition available to us is sensible; consequently, the thought of an object in general, by means of a pure concept of understanding, can become knowledge for us only in so far as the concept is related to objects of the senses (KrV, B 146).

[13] Gr, p. 438. See also Kant's long discussion of means and ends in the LOE, pp. 120 ff. Here he states quite precisely that "man is not free to dispose of his person as a means."

[14] Lewis White Beck uses 'intention', which can be thoroughly misleading (1949). Ellington's translation of *Grounding* uses 'disposition' (Gr, 1983) and Paton (1948) uses 'attitude' or 'attitude of the mind'.

[15] Hence the title of Köhl's superb book, *Kants Gesinnungsethik* (1990).

[16] Kant intends to locate moral value in the notion of duty, but more precisely, in the distinguishing characteristic of a will which acts from duty and not merely in accordance with duty. Such a "good will" manifests itself most conspicuously in the cold and indifferent person who nonetheless carries out duties to others for whom he has absolutely no fellow feeling. For Kant, "*gerade da hebt der Wert des Charakters an*," [just there does the value of the character begin] (Gr, p. 399).

[17] Kant begins *Grounding II* with a discussion of this very epistemological problem. He hints at the possibility that the *Gesinnung* of the maxim might be "laid open" by the action ("*Gesinnungen, ... die sich auf diese Art in Handlungen zu offenbaren bereit sind*"). He then proceeds by chastising those philosophers "who have absolutely denied the reality of this disposition (*Gesinnung*) in human actions and have ascribed everything to a more or less refined self-love" (406). Unfortunately, the rest of his argument supports their position! Kant admits that even the "most acute self-examination" cannot rule out 'some hidden impulse of self-love" as the actual determining ground of our allegedly good will (407). Kant unequivocally proclaims that the relationship between *Gesinnung* and action is opaque, even to the agent herself; *a fortiori* the attitude behind actions must remain hidden to the public and is hence not "*zu offenbaren*" – it cannot be laid open.

[18] See Eckart Förster's discussion in OP (p. 288).

[19] This is the important qualification that Kant makes to the second formulation, that we must treat humanity in our person or in the person of another "always at the *same time* as an end and never *simply* as a means" (Gr,429, my italics).

[20] Kant says in the *Grounding* that "I need no far-reaching acuteness to discern what I have to do in order that my will may be morally good ... . I only ask myself whether I can also will that my maxim should become a universal law" (403). The formalist interpretation of Kant's ethics is by and large dominant in the English-speaking analytic tradition. Its most sophisticated defenders are Christine Korsgaard, 'Kant's analysis of obligation: The argument of foundations I,' (1989), and 'Kant's formula of universal law,' (1985); Onora

O'Neill, *Acting on Principle* (1975) and *Constructions of Reason* (1989); and John R. Silber, 'Procedural formalism in Kant's ethics,' (1974).

[21] Kant's indecision is shown when he introduces the problem in LOE (p. 43):

> May a man, for instance, mutilate his body for profit? May he sell a tooth? May he surrender himself at a price to the highest bidder? ... I apply my understanding to investigate whether the intent of the action ... could be a universal rule.

He then answers his questions straightaway by using the second formulation and disregarding the issue of universality.

> What is the intent of these cases? It is to gain material advantage. It is obvious, therefore, that in so acting man reduces himself to a thing, to an instrument of animal amusements. We are, however, as human beings, not things but person, and by turning ourselves into things we dishonor human nature in our own persons.

[22] MM p. 393; cf. Gr, pp. 423, 430. The cultivation of capacities and talents is what Kant calls an imperfect duty, or one in which the "fulfillment is merit, [but] transgression is not forthwith an offense" (MM, p. 390).

[23] Perhaps a responsible Kantian could soften this line considerably by allowing "insignificant risks" to the functional integrity of the body in donation. While there is always a chance of death from any operation to remove an organ, it would seem that some risk might be worth taking. In principle, though, there seems to be no way to draw the line for "significant" risk.

[24] A fascinating account of Königsberg in this regard is offered in Oskar Ehrhardt's 'Dr. Laurentius Wilde und die Anfänge der medizinischen Wissenschaft in Preussen,' in *Abhandlungen zur Geschichte der Medizin*, vol. 14 (1905). Ehrhardt reports that no trained physician could be found in Königsberg until 1513; only barbers, monks, and mystics were available, and for a hefty fee. In 1549 the plague killed 14,000 residents of the town.

[25] A note to this effect appears in the margin of the *Opus Postumum*. See Förster's account in OP, pp. 279 n.114, and 287 n.162.

[26] I do not want to suggest, however, that the transplant surgeon who would profit from a market would be merely engaged in a "parts trade." Surely, the fact that the surgeon saves lives as a primary part of his or her vocation distinguishes that practice.

[27] See for instance Tom Bole's essay in this volume, (pp. 331-351).

BIBLIOGRAPHY

American Medical Association, Council on Ethical and Judicial Affairs: 1995, 'Financial aspects of future contracts for cadaveric donors,' in *Archives of Internal Medicine* 155, 851-859.

Beck, L.W. (ed. & trans.): 1949, *Kant's Critique of Practical Reason and other Writings in Moral Philosophy*, University of Chicago Press, Chicago, Illinois.

Bole, T.: 1999, 'The sale of organs and obligations to one's body: Inferences from the history of ethics,' in *Persons and Their Bodies*, M. J. Cherry (ed.), Kluwer Academic Publishers, Dordrecht.

Boyle, J.: 1999, 'Personal responsibility and freedom in health care: A contemporary natural law perspective,' in *Persons and Their Bodies*, M. J. Cherry (ed.), Kluwer Academic Publishers, Dordrecht.

Descartes, R.: 1984 [1641], *The Philosophical Writings of Descartes*. J. Cottingham, *et al.* (ed.), vol. 2, Blackwell Publishers, Cambridge, England.

Ehrhardt, O.: 1905, 'Dr. Laurentius Wilde und die Anfänge der medizinischen Wissenschaft in Preussen,' in *Abhandlungen zur Geschichte der Medizin*, vol. 14, Breslau.

Engelhardt, H.T., Jr.: 1999, 'The body for fun, beneficence and profit: Variations on a post modern theme,' in *Persons and Their Bodies*, M. J. Cherry (ed.), Kluwer Academic Publishers, Dordrecht.

Hill, T., Jr.: 1992, *Dignity and Practical Reason in Kant's Moral Theory*, Cornell University Press, Ithaca.

Hume, D.: 1896 [1739] *Treatise of Human Nature*, vol.2 'Of understanding,' L.A.Selby-Bigge (ed.), Oxford University Press, Oxford, England.

Kant, I.: (1900-) *Gesammelte Schriften*, edition Royal Prussian (German) Academy of Sciences, Berlin.

Kant, I.: 1983 [1785], *Grounding for the Metaphysics of Morals,* James W. Ellington (trans.), Hackett, Indianapolis, Indiana, Academy pagination.

Kant, I.: 1965 [1781], *Critique of Pure Reason,* N. Kemp Smith (trans.), St. Martins Press, New York, with the pagination of the Royal Prussian Academy edition of Kant's work.

Kant, I.: 1963, *Lectures On Ethics,* Louis Infield (trans.), Hackett, Indianapolis, Indiana.

Kant, I.: 1983 [1797], *Metaphysics of Morals* (Part II, *Tugendlehre*), James W. Ellington (trans.), Hackett, Indianapolis, Indiana, Academy pagination.

Kant, I.: 1994, *Opus Postumum*, E. Förster (ed. with an introduction), Cambridge University Press, Cambridge, England.

Köhl, H.: 1990, *Kants Gesinnungsethik*, De Gruyter, Berlin, Germany.

Korsgaard, C.: 1985, 'Kant's formula of universal law,' *Pacific Philosophical Quarterly*, 66, no. 1, pp. 24-47.

Korsgaard, C.: 1989, 'Kant's analysis of obligation: The argument of Foundations I,' *The Monist* 72, 311-340.

O'Neill, O.: 1975, *Acting on Principle*, Columbia University Press, New York.

O'Neill, O.: 1989, *Constructions of Reason*, Cambridge University Press, Cambridge.

Paton, H.J.: 1948, *The Moral Law: Kant's Groundwork of the Metaphysics of Morals*, HarperCollins, New York, New York.

Scheler, M.: 1973, *Formalism in Ethics and Non-Formal Ethics of Value,* 5th ed., M. Frings and R. Funk (trans.), Northwestern University Press, Evanston, Illinois.

Silber, J. R.: 1974, 'Procedural formalism in Kant's ethics,' *Review of Metaphysics*, 28, 197-236.

DREW LEDER

WHOSE BODY? WHAT BODY?
THE METAPHYSICS OF ORGAN TRANSPLANTATION

I. INTRODUCTION

What rights and responsibilities do we have in relation to our own bodies? The question arises in many areas of medical ethics and practice. Are we obliged to take good care of our physical condition or can we legitimately engage in whatever form of self-abuse we desire? Can we will our own bodily degradation, mutilation, even death, without transgressing some moral boundary? Can we rent our wombs for child-bearing purposes, or sell our organs on an open market? Are there any limits to such practices in the name of human dignity, divine sanction, or state sovereignty, or is my body, after all, mine to do with as I wish?

It is tempting for the bioethicist to survey this terrain from a birds-eye view provided by ethical theory. For example, one could take the Kantian categorical imperative, with its test of universalizeability and respect for persons, and seek to apply it here. The hope is that the answers thereby uncovered have a transcultural validity sanctioned by reason itself. Kant dreamed of such a reason that would overleap psychological vagaries, human inclinations, and mere social prescriptions, to arrive at objective criteria of morality.

However, Kant's own critical enterprise, highlighting the importance of the subject in constituting the experienced world, helped open the floodgates for the modern and "post-modern" questioning of pure reason and its claim to objectivity. In the twentieth century, philosophers working within fields such as hermeneutics, feminism, and virtue ethics have heightened our awareness of the dangers of this pretense to moral truth. There is a latent parochialism in our very attempt to escape parochialism. That is, if we take our modes of thought and moral intuitions as universal truths, we fail to see how much they are shaped by our own culture with its particular history, technologies, socioeconomic structures, etc. As thinkers such as Gadamer (1984) and MacIntyre (1981) have explored, we are the products of specific traditions. From such

M.J. Cherry (ed.), Persons and their Bodies: Rights, Responsibilities, Relationships, 233–264.
© 1999 *Kluwer Academic Publishers. Printed in Great Britain.*

traditions, which we inhabit as a fish does water, we derive our ontologies, and our canons of morality and rationality. Thus, in asking what rights and responsibilities we have toward our own body, we must recognize that our notion of rights and responsibilities is itself predicated on concepts, embedded in Western culture, of the individual self qua moral agent. Similarly, our sense of the "body" – what it is and how we use and experience it – has been profoundly shaped by cultural factors. Is there then any way out of a self-enclosed circle where our questions already imply their own answers and we merely arrive at the intuitions we had to begin with?

Thinkers such as Gadamer (1984) and Bernstein (1983) have stressed that the move away from objectivism need not bring us to a dead-end relativism. Though we can never entirely escape our cultural and historical situatedness, in the very recognition of limitation we stretch our boundaries. We can reflect upon our sedimented ways of thought and their restrictions. We can explore alternatives from other cultures and times. We can critique our paradigms both in terms of their internal logic and their external consequences *vis-à-vis* our shared world. From this communal dialogue new intuitions and possibilities can come to light. This article seeks to further such a process of reflection.

Rather than approach the issue raised at the beginning of this paper – one's rights and responsibilities *vis-à-vis* one's own body – in a global and hence diffuse way, I will focus specifically on the topic of organ transplantation and whether individuals should be allowed to sell their body parts on an open market. There are two reasons I choose to concentrate on this question.

First, its medical import. Organ transplantation is a rapidly growing medical industry. Already we transplant kidneys, hearts, livers, lungs, corneas, bone marrow, spleens, pancreases, along with a multitude of soft tissues. Increasingly, transplantation has moved from the status of experimental procedure to routine therapy (Fox and Swazey, pp. 8-13), with both the demand and the numbers performed ever increasing. Novel methods for keeping the organs of the newly brain-dead suitable for transplantation, new immunosuppressive drugs which help prepare the host to receive the donor organ, advances in surgical technique, all are making transplantation a feasible option in a wider range of situations. Whereas in 1978 there were less than thirty transplant centers in the United States, ten years later there were 226. Moreover, the possibilities for future growth of this industry are almost unlimited. For example, there are about one million insulin-dependent diabetics in the United States who could in theory benefit from new pancreases (to the tune of 100 billion dollars). Then too, the morally controversial use of fetal tissue, which regenerates rapidly and is far less likely to provoke host rejection, could be the source of a multitude of clinical benefits if permitted. The dream of treatment by transplantation is a seductive one both for the medical community and for society at large.

At the same time, the increased demand for organ transplantation has come into conflict with limitations of supply. Certain organs or organ parts, such as kidneys and pieces of liver or bone marrow can be given while the donor is still alive. Others can only be taken from cadavers. In both cases there are chronic shortages. The Uniform Anatomical Gift Act, drafted in 1968 and adopted in some form in all states by 1973, permitted individuals to will their organs for scientific and medical usage after death, or for their relatives to donate the cadaver. The hope that demand for cadaver organs would be met by such voluntary contributions has proved a disappointment. Even recent "required request" legislation on the federal and state level requiring hospitals to ask families for permission to donate organs from their recently deceased family members has done little to alleviate the problem: health care workers are often loath to make such a request hard on the heels of a loved one's death. The waiting list for organs continues to grow, for example from 8,000 in 1987 to 18,000 in 1989, without any sign of abatement. Meanwhile thousands of patients die while waiting for a needed organ.

It is in this context that a cash market for organs has been proposed as one method for alleviating the shortage. The National Transplant Act of 1984 made it a crime to buy, sell or otherwise benefit from commerce in organs in the United States. However, this is an accepted practice in some other countries, and there has been a continued debate about whether it should be allowed here. If individuals (or their family members) could sell organs during life or after death, the mechanisms of the marketplace might bring a life-saving commodity to tens of thousands in need.

The issues surrounding organ transplantation are clearly of great medical significance. However, I choose to focus on this area for another reason as well: because I believe it to be highly revealing of the sense of embodiment characteristic of modern medicine and of the culture at large. Mark Dowie writes:

> Sometimes it is insufficient to examine a technology by itself, it being merely a symptom of a larger paradigm. Such may be the case with

organ transplanting, which is surfacing as the clearest metaphor we
have for contemporary Western healing. Transplantation must be
regarded, therefore, as more than a stand-alone technology. It should,
in fact, be observed as a window on modern medicine (Dowie, 1988, p.
131).

A window on modern *culture*, I would add. Transplantation and the sense
of embodiment upon which it relies is emblematic of a more general
trajectory whose arc has done much to define modernity. Our very sense
of the body, and with it of the human self and the natural landscape, has
been shaped by a collection of philosophical, technological, and
socioeconomic forces which have in turn contributed to the development
of organ transplantation. I will here examine two such forces whose
influence has been pervasive: the Cartesian sense of the body-machine,
and the structuring within capitalism of the body qua producer, product,
and consumer. Again, I do not presume that we can entirely free
ourselves from our tradition through such critical reflection, so much as
stretch its boundaries and open up its latent possibilities.

II. CARTESIANISM AND THE BODY-MACHINE

As is well known, Descartes defined the human being as a compound of
two substances whose natures were essentially opposed. The true self was
identified most closely with the mind – "I can infer correctly that my
essence consists solely in the fact that I am a thinking thing" (Descartes,
Vol II, pp. 54). The body, on the other hand, is something I *have*, a
material object with which I am intimately conjoined, but without which I
could continue to exist. Though Descartes himself at times recognizes a
kind of mind-body union which seems to challenge his own dualist
metaphysics and psychology (Zaner, pp. 106-129), it is this dualism that
has had the most profound effects in the culture and which I will hereafter
refer to as "Cartesian."

Within this Cartesian paradigm, the human body is thought of as a part
of *res extensa*, a machine operating according to material principles.
Impressed by the automata of his day, Descartes suggested that the body
itself could be thought of as a complex automaton run by thermal and
hydraulic forces. Material nature, and thus the body as a part of nature,
was seen as devoid of intrinsic telos, consciousness or subjectivity.

This "death of nature," to use Merchant's phrase (1980), played a crucial role in subserving the modernist project of mastery; the natural world was reconceptualized as passive, pliant, available to human analysis and control. Descartes writes in the *Discourse on Method* of a "practical philosophy" through which

> We could know the power and action of fire, water, air, the stars, the heavens and all the other bodies in our environment, as distinctly as we know the various crafts of our artisans; and we could use this knowledge – as the artisans use theirs – for all the purposes for which it is appropriate, and thus make ourselves, as it were, the lords and masters of nature (Descartes, Vol I, pp. 142-143).

If we but understand the forces which guide the material world, we can turn them to our ends, harnessing, imitating or altering nature at will.

This dream of better living through science and technology has been pervasive in the modern era, transforming diverse aspects of our world. But Descartes' dream also had a *medical* specificity. He continues:

> This is desirable not only for the invention of innumerable devices which would facilitate our enjoyment of the fruits of the earth and all the goods we find there, but also, and most importantly, for the maintenance of health, which is undoubtedly the chief good and the foundation of all the other goods in this life ... .we might free ourselves from innumerable diseases, both of the body and of the mind, and perhaps even from the infirmity of old age, if we had sufficient knowledge of their causes and of all the remedies that nature has provided (Descartes, Vol I, p. 143).

Descartes regarded this pursuit of medical knowledge and power as a cornerstone of his life's work. His goals were not exclusively altruistic, but also rooted in a personal concern with aging and dying. For example, in October, 1637 he wrote to Huygens,

> The fact that my hair is turning gray warns me that I should spend all my time trying to set back the process. That is what I am working on now, and I hope my efforts will succeed even though I lack sufficient experimentation (Vrooman, p. 141).

Two months later, in another letter to Huygens, Descartes writes:

> I have never taken such pains to protect my health as now, and whereas I used to think that death might rob me of thirty or forty years at most,

it could not now surprise me unless it threatened my hope of living for more than a hundred years (Vrooman, p. 142).

I present this biographical material because I take it to be revelatory concerning the modern era and its project of mastery. With the loss of unproblematic religious faith, death has become an increasingly fearsome thing, a specter of the unknown and uncontrolled. Descartes' battle to regain mastery over the mortal body continues to be waged in modern-day America with the development of ever-new medical technologies and the ever-rising proportion of the GNP devoted to health-care. We fight desperately to stave off death, no longer so sure that our destiny and immortal souls rest in God's loving hands. Absent this security, we rush to take over the divine office. As the protocol for an early computer conference proclaimed, "We are as gods and might as well get good at it" (Stone, p. 90).

Organ transplantation is a potent result and symbol of this cultural trajectory. We see this not only in the goal of transplantation, its battle against illness, aging, and death, but also in the methodology it employs. The Cartesian mechanization of the material world helped overturn all prohibitions against tampering with nature. When the natural world was thought of as sacred and ensouled, certain invasive interventions, such as mining, were condemned. Similarly, there were often taboos against cutting up the live and dead human body. These become mere foolishness if nature and body are but mindless matter. Thus, the extraction and transplanting of organs would be permitted, even encouraged within a Cartesian framework, since we are meant to be the "lords and masters of nature."

Moreover, transplantation exemplifies the paradigmatic bent of Cartesian science: to understand and control the workings of natural bodies by *analyzing* them into their component parts. Descartes rejected the substantial forms of Aristotelianism and their essential holism, in favor of a reductionist strategy. A thing is viewed as the sum of its parts and forces in interaction; we come to understand an object by "taking it apart." This analytic knowledge then yields power, for as Jonas writes of modern science, "To know a thing means to know how it is or can be made and therefore means being able to repeat or vary or anticipate the process of making" (Jonas, pp. 203-204). We see this scientific/technological power reflected in the medicine of transplantation. The body is analyzed into its component organs which can be replaced by machine analogues or parts from live human donors, cadavers or animals. We stretch toward the Cartesian dream of remaking the body at will.

We cannot dismiss this dream lightly for it has spared us many a nightmare. Modern medicine has arrived at many powerful and healing modes of treatment through its application of analytic science. However, the Cartesian world-view and its medicine has given rise to nightmares of its own from which we are perhaps just beginning to awake.

For the modernist attempt to overcome death proceeded from the start through its own strategics of death: the reconceptualization of the physical world as lifeless matter. This has indeed brought about a kind of "death of nature": nature conceived of as without intrinsic claims, holism, intelligence, moral status, or sacred dimension. A path has thus been cleared for unlimited technological manipulation with sometimes catastrophic outcomes. Pollution of the air and water, disruption of ecological systems, depletion of the ozone layer, global warming, are but a few phenomena that speak to us of the very real death of nature that imminently threatens.

Then too, one could speak of the "death of animals", Cartesianism effects, both conceptually and practically. Animals, as a part of *res extensa*, were thought of by Descartes as mere machines devoid of consciousness. This paved the way not only for the brutal, mechanized treatment of animals on our modern-day "factory farms," but for the widespread abuse of animals in medical experimentation. An eyewitness account of seventeenth century Port-Royal experimenters provides an early example:

> They administered beatings to dogs with perfect indifference, and made fun of those who pitied the creatures as if they felt pain. They said the animals were clocks; that the cries they emitted when struck were only the noise of a little spring that had been touched, but that the whole body was without feeling. They nailed poor animals up on boards by their four paws to vivisect them and see the circulation of the blood which was a great subject of conversation (quoted in Singer, pp. 201-202).

While modern medicine is often (though not always) more humane, the use and abuse of animals continues unabated, and has played a prominent part in the development of organ transplantation. As Dowie notes, "An incalculable number of animals – mostly dogs, cats, pigs, horses, rabbits, calves, goats, rats, mice, chimpanzees, baboons or other simians – have died for the advancement of transplantation" (Dowie, 1988, p. 217).

As we can speak of the modernist "death of nature" and "death of animals," so a sort of "death of the body" has been brought about in the name of perpetuating bodily life. In Andrew Kimbrell's words:

> The body is not a machine. That is the "pathetic fallacy" in reverse. The original pathetic fallacy had the unruly passions of the human spirit inhabiting stones, trees, and rivers. Now we seem to believe that nothing has soul: We are all *inanimata*, analogous to machines or factories, and can be treated as such (Harpers, p. 49).

The modern doctor and hospital have often been criticized for delivering cold and impersonal care, but the roots of such treatment lie in this metaphysics. The body is medically understood according to the model of a machine. The rest follows logically: need the mechanic speak lovingly to a car in the shop? Test it and fix it and give its owner the bill. Though there are many humane practitioners, the disease categories, technologies, treatment protocols, and institutions of medicine, all built upon a Cartesian foundation, exert a powerful pull toward depersonalized care (Leder, 1984).

Thus, it is not suprising to find that organ transplantation, as "the clearest metaphor we have for contemporary Western healing ... a window on modern medicine," is a window that opens onto some troublesome vistas. Clearly, transplantation has provided the gift of life and healing to thousands of individuals, and this is an incalculable gift. But the technology comes bearing with it a slew of costs: not only the human costs associated with the chronic rejection syndrome, which causes the rejection of over half of transplanted kidneys within ten years, and 1/3 to 1/2 of other organs, such as the heart, within five years (Fox and Swazey, p. 10); not only the enormous economic cost of transplants and their sequelae – e.g., for a heart-lung transplant, an estimated $240,000 for initial treatment, $47,000 a year for follow-up medications and care (Fox and Swazey, p. 84) – but an additional series of social costs that are not as immediately apparent.

For one thing, the ever-growing popularity of transplantation, and its metaphorical power within the public consciousness, can serve to perpetuate many of the diseases that transplantation treats. We have seen that transplantation exemplifies the modern sense of body-as-machine, and of disease as residing within a specific organ. This tends to discourage "holism" not only in medical treatment, but in understanding and addressing the etiology of disease. Thus, transplantation does little to encourage preventive approaches, wherein indviduals take responsibility for a healthy life-style and for proper self-care. On the contrary, if the body is a possession whose diseased parts can be replaced, why worry? Nor need the social causes of disease be tackled, such as air, food and water pollutants, occupational toxins, poverty, malnutrition, and stress-filled environments. Disease is conceptualized not as something "out there," a product of the social field, so much as confined within a replaceable organ. Meanwhile, the public monies alloted to transplantation represent resources that might otherwise have been used to address these social problems or foster preventive health care and education.

In addition to this problematic side of organ transplantation, there is something disquieting about its implicit reductionism *vis-à-vis* the human being. At first, this seems paradoxical. The Cartesian project appeared to be the opposite of reductionist, seeking to unlock the Godlike potentials of human reason. Yet when we attempt to be lord and master over the natural world, we find that we ourselves are part of it, and thus discover ourselves in a position of subjugation. We are so intimately one with our bodies, that if the body, as part of physical nature, is treated in a reductionist way, the self as well is often diminished. Hence the paradoxes of modern medical treatment – the patient reaching out for all the powers of medicine, but feeling powerless and degraded under the doctor's gaze; our elderly promised the dream of immortality, but suffering needlessly-prolonged deaths through high-tech care. The sense of powerlessness and suffering envelops even physicians, whose rates of suicide and substance abuse are inordinately high. Called on to do the impossible, driven by self-imposed and projected expectations, doctors daily meet the spectre of failure, patient-rage and malpractice litigation.

In the specific area of transplantation we find the reduction of embodiment and self most pronounced in regard to those whose selfhood is least established. For example, there has been a call to utilize the organs of newborn anencephalic babies who will shortly die, or fetuses who are being electively aborted. Least personlike by standards of developed consciousness and sociality, these beings can be most easily conceptualized as *just* a body, a collection of organs representing a valuable resource. At the other end of the life-span, we can do the same with the corpse. Though almost all cultures recognize the inherent significance, and often dignity or sacredness, of the corpse, so strongly identified with the once-living person, from the dualist perspective it is just meat. Why not utilize it for our medical purposes?

This reductionism, while most evident in relation to the anencephalic, fetus, and neomort can be extended to others whose personhood is socially marginalized. There are reports from the People's Republic of China that the government has turned the execution of dissidents into hard currency by selling the organs of dead prisoners. In third-world countries, such as India, the sale of kidneys from the poor to those seeking transplants has developed into a multi-million dollar industry (Fox and Swazey, pp. 68). One can imagine further extensions (as science fiction authors often do) wherein large groups or entire societies are converted into organ farms to be harvested by the privileged. Whatever possibilities actually play out, they are grounded in a logic intrinsic to Cartesianism: in viewing nature and thereby the body as mere resource, our quest for human lordship can lead to human degradation.

III. THE BODY WITHIN CAPITALISM

So far I have dwelled on what might be called the "ontology of transplantation" – the view of nature, self, and body that helps to ground the practice of organ replacement. However, the ethical issue at the focus of this paper is not simply transplantation per se, but whether individuals should be allowed to sell their organs on an open market. This draws our attention to the capitalist structure of production and exchange prevalent in countries where transplantation technology has been developed and utilized to the greatest extent. I will not attempt anything like a detailed politico-economic analysis of market structures; this is beyond the bounds of both my expertise and the goals of this article. I will only suggest ways in which the social system loosely known as "capitalist," (albeit with many of the social-welfare hedges characteristic of Western democracies) has profoundly shaped our sense of embodiment, both reinforcing and supplementing the lessons of Cartesianism. It has done this not through the medium of an articulated philosophy, such as that of Descartes, so much as through a set of lived practices which help determine how we use, experience, and understand our bodies in the modern world.

First, the body takes its place as an agent of *production*. The body of the laborer, using the power of muscles, the dexterity of hands, produces a variety of goods and services for the consumer. The more efficiently and productively the body can be employed, the more profitable the results. Hence most of us receive training from early childhood on to be able to regard the body as an instrument of productivity, and this can become a pervasive style of inhabiting the world. Even after the workday ends, we talk on the phone with one hand and put away dishes with the other, striving for ways to utilize the body with task-oriented efficiency.

Undoubtedly, one reason the Cartesian machine-body has proved so popular in our culture is that it resonates with and supports the sort of body appropriate to capitalism. If maximum productivity is the goal, one wants one's employees to operate as machines – rapidly, dependably, efficiently, requiring little maintenance or training, and easily replaceable when necessary.

This paradigm no longer rules unchallenged on the contemporary scene; there is a move in certain American companies, influenced by Japanese and European examples, to grant the worker greater latitude to exercise creativity, autonomy, and managerial skills. But this still remains the exception rather than the rule. As Garson argues in *The Electronic Sweatshop*, new technologies, employed for the monitoring and regulation of activities, have often served to make the workplace more rather than less dehumanized. The intelligence involved in running a McDonald's restaurant is located in sophisticated protocols, computer programs, and electronics – but what of the people? As an ex-worker describes:

> You follow the beepers, you follow the buzzers and you turn your meat as fast as you can. It's like I told you, to work at McDonald's you don't need a face, you don't need a brain. You need to have two hands and two legs and move 'em as fast as you can (Garson, p. 20).

We are reminded of the Cartesian body-automaton, devoid of intrinsic subjectivity and telos.

Within the capitalist system, the body is taken up not only as this producer-machine, but also as *property* for possible exchange. In the modern era the body has sometimes been considered the paradigmatic example of private property, such that other forms of ownership are derivative, based upon the body's labor (Engelhardt, 1986, pp. 127-130). As property, the body can itself take on the role of a commodity exchangeable on the open market. In a sense, this lies at the foundation of the wage-labor system for the worker sells his/her bodily powers to the employer for a period of time. It might be denied that this is selling the body per se, for labor is but one of the body's renewable expressions. However, the distinction is tenuous. The McDonald's worker quoted above is selling (or we might say "renting") his "two hands and two legs" to the corporation for it to use as it wishes. If he experientially retained full ownership of his body, he would undoubtedly choose to do something with it other than flipping burgers. This body for a time is no longer his, but a commodity exhanged in return for a paycheck.

Again, this style of regarding the body can be extended into other realms. For example, in pornography and prostitution even the intimate domain of sexuality gets commodified through the sale of body images or services. In fact, the detachment and commodification of the sexualized body-image has become a pervasive principle of modern-day commerce. Products of all sorts are posed next to an alluring model, usually female and often seductively presented. There is a hidden promise to such advertising: buy our product and with it you'll acquire the beautiful model and her/his embodied delights. This visual and voyeuristic mode of presentation seems to dominate twentieth-century sexuality, placing its stamp on our private imaginings and public practices. No doubt, this is because the visual image of the sexual body is much more easy to *detach and sell* then, for example, the tactile or odiferous. But one need not stop with the visual image. The new craze for telephone sex shows the possibilities latent in the human voice for detachment and commodification.

In all cases, whether selling bodily labor, images, powers, or parts, the ability to commodify the body depends upon being able to regard it as other, alienable from the essential self. As Marx writes, "A commodity is, in the first place, an object outside us" (Marx, p. 35). Again, we see why the Cartesian model is so congenial to capitalist economics. Descartes notes that I "have" or "possess" a body which is essentially distinct from the thinking thing that I am. We have little problem regarding our bodies as alienable commodities for we have been well-prepared by dualistic habits of thought.

In addition to the body as producer and commodity, the body takes its place within capitalist markets as the *consumer* of goods. That is, the body is the locus of needs and desires which provoke purchase. We not only have physical requirements which must be satisfied – for food, clothing, shelter, etc. – but a virtually unlimited set of potential cravings which it is the business of advertising to provoke or create. We are called to ornament the body, give it pleasures and comforts, exercise it, change it, heal or intoxicate it. As part of this call to corporeal completion we are informed that our body is essentially deficient. Our bad breath may drive away potential consorts; our cellulite is intolerably unattractive. This strategics of bodily insecurity serves to provoke the purchase of that which will make us "whole."

Just as organ transplantation is an exemplar of the Cartesian project and sense of embodiment taken to the limit, so it is in relation to capitalism. That is, one can say that transplantation represents a kind of ideal case of the body taken up as producer, commodity and consumer.

As discussed above, the productive-body is ideally something of an automaton, fashioning its product cheaply, reliably, mindlessly. One wants an indefinite supply of workers who require little training and are easily replaced. But what if the product made is itself a body part, a liver for example? Ideal conditions are fulfilled. Anyone and everyone can make one without training or special talent. The supply of workers is unlimited, the job itself requiring no intervention of the conscious mind. The producer is simply the body-machine fashioning its tissues and organs automatically. There is just one catch: given the limited nature of production, each worker producing only one or two of each key organ, he or she must in many cases die appropriately (e.g., through a youthful motorcycle accident) before the product can be "harvested."

Just as the production mode fulfills a certain ideal, so is the body organ a kind of ideal commodity and the body-purchaser the stuff of entrepeneurial dreams. In terms of the commodity, one no longer needs to sell a detachable physical image or service, or to use the body to create a separate product; body parts themselves become the product for exchange. Moreover, this commodity promises to satisfy the greatest need a consumer-body can have, that for continued life or health. In many cases the commodity thus all but sells itself. The deathly ill person need not be reminded of some corporeal deficiency to be remedied. She or he lives with it on a daily basis, and the requirement for the product seems undeniable.

Moreover, the potential market is almost indefinitely extensible. After all, everyone has a body that ultimately runs down. Most people would, if they could, choose to prolong their life if this were compatible with a decent standard of functioning. The promise of transplantation to fulfill this desire helps provoke an ever-growing demand which expands with each new technological breakthrough, or story advertising the miracle of a life-saving transplant. In the United States we currently face a situation where demand for organs well outruns the supply. But this demand, we must remember is not simply a natural fact, but a cultural construction, built upon the Cartesian dream of immortality, and the capitalist profit-motive which has encouraged the technology.

In fact, transplantation has proven to be immensely profitable to the health-care industry and no doubt will become more so in the future. Far more than the low-cost, disbursed techniques of preventive medicine, organ transplantation is capital-intensive, can be centralized in the hands of an elite, and lends itself to the accumulation of large profits. Thus, one might expect to see this technology receive extensive business and government support, just as nuclear energy has historically received more support than solar energy development. There are capitalist pressures toward high technology, high-profit modes of production, and the ever-growing push toward organ transplantation is but one example.

IV. A MARKET IN ORGAN PARTS?

Up to this point, I have suggested how organ transplantation illuminates and exemplifies our cultural sense of embodiment, shaped both by Cartesian and capitalist forces. But I have yet to even attempt an answer to the question with which this paper began: should individuals be allowed to sell their body organs? On the contrary, I have sought to show how the question itself would first come into being – that is, how individuals could see the body as a collection of organs, how technologies for organ replacement could be conceptualized and developed, how these organs could then be schematized as potential commodities whose exchange is motivated by dreams of immortality and profit. In certain ways reflection on how our culture could have arrived at this question is of greater interest than the specific answer we give. Nonetheless, this answer will have significant repercussions, and even be a life and death matter for some awaiting transplantation.

What then if we said yes, and permitted the sale of organs from living donors and cadavers? The practical scenario has many alluring features. We can surmise that the supply of organs might increase dramatically, though this is by no means a sure thing (Hansmann). There might also develop an improved system for matching up those in need with those in possession of the organs; market structures have often been a spur to efficiency and ingenuity. In such an exchange, the purchaser receives a life-saving or life-enhancing item. The seller, or seller's descendants, may also be given the ticket to a better life in return for an item for which they have little or no need. After all, most of us can get by quite nicely on one kidney while alive, and the dead have exhausted their demand for viscera. The market exchange thus seems ideal for all parties.

Moreover, one can construct a powerful defense of the practice on theoretical grounds. H. Tristram Engelhardt, Jr.'s article in this volume is a good example. Engelhardt argues, along with MacIntyre and others, that we live in a culture that no longer exhibits a moral consensus based on traditional, metaphysical assumptions. While many of us hold onto fragments of the old morality – for example, the notion of the human body as having a sacred dignity – the Christian world-view that once undergirded such presumptions is no longer universally shared, or even necessarily dominant within our secularized social context. In this milieu it is highly problematic to restrain individuals from exercising the freedom to contract with one another on matters of mutual agreement. If one person wants to sell an organ and another purchase it, why not let them? The coercive prohibition of this exchange is usually based on a morally contentful stance concerning human dignity, proper use of the body, and the like. But as Engelhardt suggests, our secular society no longer has any canonical basis for arriving at these moral notions. In fact, the market serves a crucial function in our pluralistic society, allowing for exchanges to take place between individuals simply on the basis of mutual consent. No particular world-view or set of moral commitments is presumed or imposed upon the people involved; they can contract with one another on the basis of their own values and wishes.

In this reading, the marketplace is largely value-neutral, and hence ideal for actualizing human freedom. However, I would question this presumption. My own discussion of the body in capitalism has suggested that the marketplace comes laden with an implicit world-view, set of roles and images, which shape our lives in a morally contentful way. The embodied self is constructed as machine-like producer of value, a commodity for sale on the open market, or the locus of desires and cash. Nature takes its place as a useable and potentially profitable resource. Then too, where the marketplace dominates it shapes relations between individuals in morally significant ways. Self-interest serves as the unspoken but assumed engine running capitalist exchange. Each party is out to make the best deal, and their interaction tends toward the common denominator of mutual use. You want my money, and I, your product. Hence, the rule of the market is not simply appropriate in a context where shared community and tradition have broken down, but can actively further this process of disintegration.

Moreover, where the market predominates, those with the most money and goods are in an elevated position. The result is differential standards of living, power inequalities, and often modes of dehumanization directed toward those with less. We have learned to accept these features through applying specific notions of "justice" and "freedom" built into the conceptual foundations of capitalism. It is *unjust* to encroach on basic rights of ownership and exchange. Meanwhile, the inequalities that may result from capitalism – unemployment, poverty, and the like – are read as justified, though highly unfortunate. Similarly, freedom tends to be understood in terms of the liberty to enter contractual relations. The poor individual who sells her kidney to help feed her child is exercising this basic freedom. But this is a thin sort of freedom indeed. Ignored are the coercive features of her position which leave a highly limited set of options – for example, a malnourished baby, a mindless and low-paying job, or the extraction and sale of her body organs. The latter may in fact be her best option, and yet still be coerced by circumstances.

For such reasons, it is incorrect to think of the market as embodying a kind of ideal neutrality *vis-à-vis* competing metaphysical and valuational systems. It bears within it its own implicit metaphysics and set of values – a sense of selfhood, the human body, nature, social relations, justice and freedom – which are highly contentful and highly problematic. Of course, the problems with the unfettered market have been widely recognized, and almost every society has ameliorative social programs, as well as domains that are conceptualized as outside the marketplace. We value realms of human relationship, community, interaction with nature, achievement, artistry, and privacy that can unfold according to a logic different than that laid out by commercial transactions.

The problem with permitting the sale of body organs, as with other practices such as surrogate mothering or the patenting and sale of life-forms, is that it sanctions the spread of market values and metaphysics into an ever wider domain. Everything becomes subjected to commodification – babies, species, body parts. This only furthers the modes of reductionism which I suggested were implicit within Cartesianism and reinforced by capitalist roles. The human body, and by extension the human being, becomes more firmly entrapped in an objectifying imagery and set of practices. If the sale of organs is permitted by law, it invites individuals to survey their own bodies in terms of its component organs and their market value. Even if one is uninterested in the transaction, its very possibility as constructed by our society necessarily shapes one's self-conception. Especially those driven by financial need will find it hard to avoid thinking of their body as a thing to be carved up and sold in pieces. And who, in such a situation, might not be tempted to sell off parts of a deceased parent or a dying child? We might well wish to avoid these new "liberties" and the trajectories of thought they would all but mandate. As usual, their reductionist and oppressive aspects would fall most heavily on the disadvantaged.

This raises a second reason to be suspicious of opening a market in body parts: it extends the notions of "freedom" and "justice" undergirding capitalist exchange to its most troublesome conclusion, with consequences that may strike us as highly coercive and unjust. On the side of the seller, we have the spectre of individuals being all but forced to sell off body parts to deal with financial hardship. On the side of the buyer, it is likely in our current system that only those with enough money or proper insurance will be able to acquire the life-saving goods (Caplan, pp. 158-177). Others, with scarce resources, will have to live without, or in many cases, to *die* without. The upshot is the possibility, already playing itself out in third-world countries, of an impoverished social class providing the organs and tissues to sustain the lives of the well-to-do. As May writes,

> A society that would exploit the penurious to sell a part of themselves demeans itself and its members and fails to solve its problems fittingly. The desperately ill ought not to solve their health care needs through the desperately poor (May, 1991, p. 180).

If saying yes to the sale of organs has these disturbing consequences, it seems logical to maintain the current legal ban on such exchanges within the United States and hope that this policy is universally adopted. However, upon examination this position itself is far from representing a "satisfactory solution."

On the practical level, the poor would still be poor. In fact, one avenue out of poverty would simply be disallowed by an international ban on organ sales. While India is considering making such sales illegal, and poor donors may be exploited by paid middlemen (Hazarika), individuals can sometimes make six times the average annual wage by selling a kidney, and use the funds to buy a small farm, shop or business (Bailey, p. 368). The disadvantaged, deprived of this option, would remain subject to the oppressive effects of poverty, and to other forms of dehumanizing labor and treatment, some no doubt worse than a kidney operation.

Moreover, it is questionable as to what extent legal bans ultimately work. Some predict it would be circumvented in third-world countries where sheer need and opportunity would drive a thriving black market (Hazarika). Even in America, one transplant physician estimates that in perhaps 15%-20% of organ donations between living relatives, some economic benefit is being exchanged under the table.

Whether such predicitions and suspicions are true, the fact is that a ban on organ sales would by itself do little to remove the entire business of organ transplantation from the logic and abuses of the marketplace. As May writes,

> The problem of commercializing transplants does not hinge solely on whether we secure organs by purchase or gift. The organ itself may be a gift, but the system by which we extract, preserve, transport, and implant organs may be so expensive and market-driven as to pose serious questions of justice, not in their acquisition, but in their distribution (May, pp. 197).

In this context, it can at times seem to heighten, rather than ameliorate, the injustices of the marketplace to deny the donor any recompense. In the words of Dr. John Dossetor, a nephrologist and bioethicist:

> I can't see why the only persons not to make a legitimate degree of financial advantage from transplantation are the people who give the organs. Everybody else is living by it, including myself (Bailey, 1990, p. 367).

Just as saying no to organ sales fails substantially to challenge the logic and injustices of the market, so it does little to overturn the large-scale paradigms of embodiment discussed above. The body is still reified into a productive machine and saleable commodity by contemporary business practices. Banning one form of commodification will leave a multitude of others intact. Likewise, the Cartesian vision of the mechanized body still rules, and will search out new avenues if limited in any given direction. For example, the current shortage in human organs, traceable partially to the forbidden nature of organ sales, merely heightens interest in alternatives such as animal xenografts or implanting artificial organs. But these exemplify, perhaps more evidently than human transplants, the reductionist nature of Cartesianism. As mere machine, devoid of consciousness, the human body is essentially no different from an animal (another natural machine), or the automatons fashioned by human hands. Parts of human bodies, animal organs, and artificial organs are freely exchangeable because they already share an ontological identity.

What then if we are uncomfortable with permitting organ sales, but unsatisfied that a prohibition would meaningfully address the sorts of problems discussed in this paper? A third option arises: a refusal to resolve the question on the terms set forth. Admittedly, this is a luxury reserved for theoreticians who can stand back from the immediacy of clinical decision-making and public policy debate. Clearly, we do have to decide this social issue, and bioethicists can help us arrive at the best answer possible given our shared intuitions concerning embodiment and selfhood. However, there is also a place for a different kind of bioethical inquiry: one that questions or challenges these very intuitions. The bioethical dilemma becomes not so much something to be resolved, as a provocation to hermeneutical questioning. How did we arrive at this dilemma? What does it tell us about our cultural understanding and its fractures? And finally, how might we think beyond the options presented? The yes/no's, do/don'ts of conventional bioethics sometimes have a hollow ring. The questions, the possible answers, have already been predelineated by a cultural understanding that seems to set the boundaries of thought, and to do so in such a way that little of substance will be challenged. Sometimes then the best response to a bioethical question might be to question the question itself, as I have tried to do in this paper, and to use this questioning to suggest responses that were nowhere on the menu. In this case, critiquing our cultural sense of embodiment can serve to reveal alternative paradigms which, among other consequences, might reframe our very understanding and use of transplants. In closing, I will attempt some remarks in that direction.

V. A PARADIGM OF INTERCONNECTION

In many of the problems addressed above, we see a recurring theme played out in different keys. This is the theme of disconnection. In our cultural framework, the self is understood as largely disconnected from its embodiment, which can be treated as mere machine or commodified. By much the same logic, the self is alienated from the natural world, for the latter is devoid of subjectivity, a mere resource to be exploited for human ends. Moreover, selves are disconnected from one another. In the Cartesian epiphany of the *cogito*, we find knowledge grounded in an enclosed consciousness which can, at the start, but infer the existence of other people by a complex chain of reasoning. This alienation finds practical expression in capitalism. Self-interest rules the marketplace, and relations between people tend toward the attenuated form of commodity exchange without the need for fuller modes of communion. Finally, the self, disconnected from body, nature, and other humans, often finds itself disconnected from any sense of sacred power or symbol of the All. This is not to deny a continued widespread belief in God and interest in spiritual pursuits. However, our public discourse and practices have become widely secularized, our culture of the marketplace rendering the sacred largely irrelevant as a category, or something itself to be commodified. Even in Descartes, the existence of the self comes first, God a subsequent and derivative theorem.

In this context, transplantation cannot but play itself out as a tactic in the battle for the survival of the separate self. The body becomes something to be controlled, its contingency overcome. Natural processes of transformation and decay are challenged and, hopefully, for a time defeated. Other people become valuable as repositories of spare parts and surgical techniques one can put to use. In the face of death, one turns not toward a sacred realm, but to human ingenuity for help. So the self battles on for self-preservation.

But one can imagine a shift in cultural understanding that would alter both the use and meaning of transplants. Where the Cartesian/capitalist model stresses separation, a different paradigm might highlight quite the opposite: the modes of continuity and connection amidst which we dwell. This might seem rather fanciful, utopian at best. But it need not be so dismissed. This model is suggested both by biological data and phenomenological reflection upon human experience.

The dualistic concept of the person has recently been challenged on many fronts. Within medicine, there is an increasing realization of the complex intertwinings of the so-called "mental" and "physical" realms, to the point where these categories begin to break down. A number of medical developments have highlighted this more integrated picture of self: research into the multisystem effects of stress; the study of "psychosomatic" diseases, and disease prone personality-types; attention to the import of the placebo effect in determining clinical and research outcomes; the burgeoning field of psychoneuroimmunology; the development of "behavioral medicine," including treatment techniques such as biofeedback, visualization, and meditation; research into the correlation between social stressors and subsequent disease or mortality; studies that evidence the import of the patient's attitudes, interpretations, and relationship with his or her caretaker, as predictive of treatment outcomes; etc. Such work, taken as a whole, tends to challenge the model of self as a thinking thing separate from the body, and of the body as separable into component organs. We are beckoned toward a different understanding: perhaps it is the *embodied self* as a unity that falls ill (Leder, 1992). Sickness must then be understood not simply as machine malfunction, but as something which befalls a conscious, embodied subject, arising out of and affecting the person's world of involvements.

In reconceptualizing this subject, the work of twentieth-century phenomenologists proves helpful. Philosophers such as Edmund Husserl (1989), Maurice Merleau-Ponty (1962), Erwin Straus (1963; 1966), and Richard Zaner (1981; 1988) sought to uncover and describe the structures of lived experience. What they found there was not the fractured self of Cartesianism, but a structure often termed the "lived body." The body, they suggest, is not simply a machine, but a way in which we constitute and inhabit an experienced world. The universe of landscapes, objects and other people is opened up to me through my bodily senses. My embodied needs and desires help charge this world with meaning and affective pulls: a chair invites my weary bones to rest, while a cup of tea calls to me from the kitchen. Such pragmatic engagements with the world are only possible because I have a body, or in some sense *am* a body. Husserl and Merleau-Ponty wrote of the bodily "I can" at the root of human competence: I *can* sit down in the chair, walk to the kitchen and fetch tea, or contemplate a hundred other options. Even in relation to "higher-level" cognition the body always plays its part. The intellectual work I am engaged in right now is dependent upon *words* which are embodied on a computer screen and, eventually, black marks on paper to be spoken or read. The physical capabilities of the computer on which I work subtly shapes my writing style. Phenomenological reflection thus reveals how the body and its technological extensions grounds human experience, from the most blatant physical cravings to the loftiest of intellectual pursuits (Leder, 1990).

Again, we find continuity primary within this perspective, not Cartesian disconnection. The mind is not an immaterial power of reason, nor the body an unconscious machine. Rather the self is an integrated whole whose subjectivity is embodied, and whose body is "mentalized" through and through.

Just as this perspective highlights the continuity of mind and body, so it suggests a continuity of the embodied self with its world. This is not to say that self and world completely merge. Clearly, there are forms of private interiority: I have thoughts that no one else hears, and feel my own pains, not yours. Biologically, my skin draws a limit between what is inside and outside me. However, both phenomenological and biological data challenge the ultimacy of a model which overemphasizes inner-outer separation.

Experientially, I am not locked inside my consciousness, as the *cogito* moment might suggest, but "out there" in the world with which I am involved. In this moment, I am gazing upon my computer screen and enjoying a needed snack. The screen, the food, fills up my embodied awareness such that "I" and the "it" are intertwined. My perceptions, emotions, and movements are directed toward an experienced world that inhabits me, just as I inhabit it.

We find much the same phenomenon from a biological perspective. While my skin does form a boundary, my very life depends upon its constant transgression. In breathing I am continually exchanging inner for outer. In eating, I take the substance of other living things and fashion from them my own bodily flesh. The gift will be returned after my death. Even the inner body is not exactly *mine*. Residing in me are ten quadrillion bacteria that do my physiological work while I provide hospitable environs. Even my "own" cells probably formed from prokaryotes enveloping each other in endosymbiotic relations, then reproducing and colonizing to form something like me. My carbon was given over by exploding stars. Such data begins to call into question the model of a self disconnected from the natural world. We are *of* this world, enveloped from within and without by a cosmos that is more than a manipulable thing.

This self reconceived as continuous with body and world, is also intimately related to other people. The Cartesian/capitalist model begins with the "I" constituting an island of self-awareness and self-interest. But is this so? Biologically, one first comes to be from within the body of another, sustained in the months of formation by her blood and breath.

Though we gain a measure of physical separation upon birth, our life continues for a time to depend upon embodied connection: we need to be fed, clothed, touched, and nurtured to survive and thrive physically/emotionally. Throughout our lives, we bear the genetic imprint of those who gave us birth, and inhabit communal settings constructed by our companions and the work of previous generations. We in turn play our part in creating new life, but only by intimate coupling with one another. The "separate self" is then something of an abstraction from this rich structure of social intertwinings.

But do I not find my isolate self within my consciousness and thought, mine and no one else's? Not necessarily. The bibliography at the end of this paper only begins to suggest the variegated sources from which "my" thought has been drawn. I do not wish to disclaim authorship of these ideas, but neither can I fully claim them as mine, for they have arisen from a communal conversation in which I but play a small part. Even when sub-vocalized "within my head," these words and ideas have also been given me by others. This shaping of consciousness by the social world plays itself out on multiple levels. Emotions flow from person to person through a rich membrane of empathic identification, such that another's tension, tears, or laughter, soon inhabits me from within. Then too, anthropologists inform us that our most basic perceptions of the world – of space, time, nature, objects, bodies – are profoundly shaped by culture and language.

I have been suggesting a paradigm that emphasizes continuity between the self and body, world, and other people. This stands in contrast with the Cartesian/capitalist predication of the multiply-disconnected "I." That is not to claim that separation is devoid of ontological or experiential significance. Clearly there are modes where the body surfaces as *other* to the essential self; where we withdraw from the world into privacy; where we discover our perspective, experiences and interests are discontinuous with those around us. I am, however, suggesting that a paradigm that overly stresses such discontinuities, and reifies them into a metaphysics and mode of life, is in significant need of correction. We are differentiated, but only within a rich web of connections revealed by biological and phenomenological data alike.

VI. INTERCONNECTION AND TRANSPLANTATION

In closing I will but briefly suggest how this recognition of embodied connection might change our use and understanding of organ transplantation.

In the first place, it might diminish the focus on transplantation as emblematic of the "medicine of the future." Clearly transplantation can be a powerfully life-enhancing and life-saving mode of treatment, sometimes in situations where no other good options exist. It is thus hard to imagine our society abandoning the procedure. However, organ transplants might not play as central a role, either practically or symbolically, in a medicine which emphasized interconnection. For example, what if we begin from the model of the integrated self, the subject embodied, the body "mentalized"? This suggests a close look at the emotions, lifestyle choices, and attitudes that help lead to disease. Intervention is often more appropriate at this level then at that of the end-stage diseased organ. Moreover, if the self is fully integrated with its society, the locus of sickness and treatment must be broadly conceived. Many of the diseases of the modern Western subject directly arise from interactions with its environs. We are surrounded by carcinogens in the air and water, offered unhealthy foods, frequently work at hazardous and debilitating jobs, and are exposed to a multitude of stressors and modes of anomie. Cancer, heart disease, strokes, psychiatric illnesses, etc. result. A medicine of interconnection would attend to the broader "social body" wherein disease finds its etiology, not simply an affected body part. Admittedly, such a shift of focus would be radical. Not only transplantation is challenged, but the whole model of health care as centered upon individual treatment. If the individual is embedded in a system of family, friends, and workplace, this system in turn part of a broader society, and this society in interaction with a natural world, different levels of "treatment" become appropriate. Can we deny that humanizing an abrasive and alienating work environment is health care for its employees? And so too is health care a matter of addressing poverty and malnutrition, or the depletion of the ozone layer. This is not wooly-minded liberalism, but hard-headed fact. A medicine always dealing with consequences rather than causes, and treating the problem organ by organ, may be dramatic and certainly lucrative, but is neither the most efficient nor humane. While there is clearly a place for "high-tech" interventions against end-stage disease, the costs must be counted as well.

Social resources poured into this level of care can often be more profitably used elsewhere – for example, in preventative care and education.

A paradigm of interconnection might thus diminish our focus on organ transplantation. At the same time, given that the procedure is unlikely to disappear, nor should it, from the medical bag of tricks, this paradigm also gives us a way to reconceptualize the meaning of transplantation. The question is not simply how many transplants are done, but also how they are performed, and what role they assume within the social iconography.

As discussed, from the standpoint of the Cartesian/capitalist model, transplantation plays itself out as a battle at any cost for the preservation of the separate self. But the procedure can be otherwise interpreted. I have already mentioned the myriad ways we flow into each other: thoughts, perceptions, emotions are constantly "transplanted" from one person to the next. In organ transplantation, we see this interpenetration actualized on the deepest visceral level. An organ from another's inner body enters into mine and sustains me from within. I receive not simply a gift of ideas or emotions, but of life itself. Dowie notes that this procedure can give "the human family" a whole new meaning: "By sharing organs and discovering ways to make them function in each others' bodies, we confirm our interdependence and expand our sense of community" (Dowie, 1989, p. 20).

The closest analogue in nature might be that given in gestation and birth. In organ transplantation we see a sort of metaphorics of birth: a person who is sick, disabled in certain ways, and quite possibly dying, is given a new life through the procedure. But whereas in natural gestation, one's body is formed from within the visceral organ (womb) of the other, here there is a curious reversal. One is "reborn" as it were, when the other's visceral organ takes root within one's own body. Moreover, when a cadaver is the organ donor, this new life paradoxically is born through the other's death.

If we reconceived of transplantation as a mode of deep interconnection between embodied selves, even between life and death, it might influence the ways in which we accomplish the procedure. For one thing, it would tend to lead us away from the metaphors and practices of the marketplace. We have seen that organs can be thought of as commodities to be bought and sold at will. But this notion is based on a series of externalities: the self as external to its bodily organs, which can thus be commodified, and external to other selves with whom exchanges are made on the basis of self-interest. A model of interconnection suggests a very different meaning to transplantation. Selves which are fundamentally embodied are in deep interdependence with one another such that each can sustain the other's life from within. To try to reduce this vital mutuality to the level of commodity transactions would seem bizarre, and such a method of organ procurement would probably be rejected.

Where then, and how, would organs be procured? One might begin with gifted organs. Murray (1987), drawing on Mauss (1967), discusses how certain societies clearly distinguish the ceremonial exchange of gifts from commercial trade. Whereas the marketplace forges limited relationships to accomplish specific purposes, gifting is widely understood to sustain deeper modes of human relatedness. Giving a gift is a way of both recognizing and creating forms of moral obligation, dependency, and affective bonding. In this sense, the gifting of organs seems far truer to the model of interconnection than market practices.

In response to the heretofore failure of voluntary gifting to provide all the organs needed, May suggests that our religious traditions and institutions can play an important part in encouraging the practice (May, pp. 187-191). For example, organ donation deeply resonates with the message of the Christian gospel. Individuals are called to self-donative love. Moreover, this is exemplified by Christ's gift of his very body that others may be saved, a gift that is repeated sacramentally through the Eucharistic ritual. We see here unfolding on a religious plane the same elements mundanely identified with organ transplants: from death, new life; one self revitalized from within by the gift of another; body entering body. May is doubtlessly right that our religious traditions help reveal the rich significance of organ donation as a mode of caring for one another and confirming our interdependence.

However, one could go further. Another way of securing organs, used for example in many European countries, is that of "presumed consent." Though an individual or their family can opt out of participating, unless they specifically do so it is presumed that cadaver organs can be used for transplantation. At first glance, this seems to undermine the rich moral and affective resonances of the gift discussed by Murray. Who can really give a gift if its taking is presumed? Moreover, "presumed consent" can be interpreted, as May does, as representing the final totalization of a coercive state, exerting its control over the person even post-mortem (May, pp. 182-187).

Within the Western paradigm of the separate self, "presumed consent" can easily be taken to signify pernicious state intrusion, and one that potentially abrogates the freedom and dignity of the individual. But might it not have a different meaning in a society that had shifted to a paradigm of interconnection? To illustrate this point, it may be helpful to step entirely, if briefly, outside our Western heritage for alternative world-views. Even elements of our tradition that profoundly emphasize mutuality, such as the Christian thematics discussed above, still retain a strong stress on the individual as the site of salvation, the soul-body distinction, and other elements we have been seeking to question.

In the non-Western world, Neo-Confucian philosophy provides one example of a metaphysics of interconnection. In this tradition, developing out of Confucian, Taoist and Buddhist strands, and flourishing in eleventh to seventeenth-century China, a central notion was repeated in the work of several philosophers: the idea that we "form one body" with the universe (Tu Wei-ming, pp. 35-50; Wang Yang-ming). This was taken to be an ontological fact. The human being was composed of "ch'i", a kind of (nondualistic) psychophysical power that also flowed through, and animated, other people, animals, plants, the earth, wind, water, and stars. However, it was not enough simply to know that we thus were part of a cosmic process conceived of as a single body; in order to lead a good life, one must morally/spiritually actualize this interconnection through modes of compassion and service to those around us (Leder, 1990, pp. 156-173).

We find a similar recognition of profound interconnection in the world-views of many indigenous peoples. The human being is seen as continuous with natural forces embodied in animals and elements of the natural world we would classify as inanimate – the wind, mountains, rivers, and the like. Abram writes of this ontology:

> The "body" – whether human or otherwise – is not yet a mechanical object in such cultures, but a magical entity, the mind's own sensuous aspect, and at death the body's decomposition into soil, worms, and dust can only signify the gradual reintegration of one's elders and ancestors into the living landscape, from which all, too, are born (Abram, 1991, p. 37).

Such cultures practice their own form of "organ transplantation," for example dismembering the body and leaving parts in locations where they will be consumed by condors, leopards, or wolves, thus facilitating the reincarnation of the individual into that animal realm (Abram, p. 37).

At death, the individual does not so much vanish from the natural world, as merge with it, giving back the elements from which new life will spring.

While the modern Western sense of interconnection could not mimic such systems, contemporary work in ecology and, as I have suggested, medicine, phenomenology, and a plethora of other disciplines, has surely brought us to a deeper appreciation of these paradigms. In them we discover ways of schematizing the world which are not primarily based on dualities of body-soul, self-other, or even that of the living-dead. Such a sensibility, should it take root in our culture, might do much to lessen the demand for organ transplantation, fueled as we have seen by a furious preoccupation with overcoming death. If the self is not primarily a separate entity, but embedded in and identified with a much larger cosmos, death loses some of its terrors. Moreover, when we do opt for transplantation, such models also suggest a different way of looking at "presumed consent" as a way of securing organs.

If we do, indeed, "form one body" with one another; if life is constantly regenerating itself from death; then the use of organs from a cadaver to give new life to another, is not *prima facie* a violation of individual sovereignty, so much as a confirmation of the way of things. All depends on what we are presuming with "presumed consent." If the presumption is in the unlimited power of a coercive state, then this manner of organ collection will tend to be felt as intrusive, and enacted in pernicious ways. However, in a society embracing interconnection what is presumed is something quite different: that we are so intimately interwoven in nature and society that the body is never simply one's own, but part of a larger circulation. The taking of organs need not then be schematized as the Cartesian extraction of a useable resource, so much as a ceremonial offering with resonances of humility, compassion, and affirmation of life.

If a society employs a system of presumed consent, the question arises, of whom can this consent be presumed? Again, to answer this, one must go beneath the surface method of organ procurement to examine what paradigm underlies it. A model of state sovereignty might answer "everyone, if the government so declares." But if presumed consent is based upon a recognition of deep mutuality, it would only be applicable in situations where individuals are fully part of the circulation, givers and receivers alike. We could not then presume consent from, for example, electively aborted fetuses. As long as abortion is sanctioned, it is a statement that the society has no commitment to provide these beings with embodied sustenance. There is a legally recognized right to withdraw from the fetus the bodily organs (e.g., the womb) it needs to survive. To then take from the aborted fetus tissue to enhance or sustain another's life is a one-sided bargain indeed. Mutality refused on one end cannot be rightfully asserted on the other. For the same reason, we clearly could not employ the logic of mutuality to justify the appropriation of animal organs. In our culture, animals are routinely subjected to coerced abuse and slaughter in service to meat production, medical experimentation, and product testing. Our religious traditions have taught us little of genuine mutuality between humans and other species, an idea further undercut by the Cartesian model of the animal-automaton. Moreover, in a human society filled with inequalities, presumed consent could not be applied to individuals who would not have had an equal chance to be organ recipients if the need arose. When the rich are predominately those who receive transplants, the principle of mutuality again is broken and the bodies of the poor become mere resources for exploitation. A notion that "we form one body" with one another, and a system of organ circulation that develops out of it, only makes sense if there is equal access to the transplantation technology that puts this circulation into effect. This would obviously imply important changes within our health-care delivery system. One can imagine a society deciding, based on communal resources, which modes of transplantation should be made available to its citizens, and which not – but then those that are available are made equally accessible to all.

Within a social context genuinely grounded in interconnection, presumed consent need not undermine the resonances of the *gift* that Murray, May, and Fox and Swazey address. Rather, it takes this sense of gift to a different level. To view the body primarily as a thing one may gift to another is still to remain somewhat within the Cartesian/capitalist sense of the body-as-possession owned by a separate self. From the perspective of interconnection, the body was from the start a gift of which one was but a recipient, and over which one can never claim full ownership. I did not fashion my limbs, nor weave together the intricate tapestry of vessels and nerves that sustains my life. This was accomplished by an anonymous wisdom coursing through me, and one which I neither fully understand nor control. My body and life were gift to me, and it is a gift renewed from moment to moment. My heart beats on, my kidney filters, my bone marrow produces blood cells, and thus I live. But it is not exactly correct to say that "I" do it, or that the whole process belongs to me. I am a site wherein mysterious powers unfold, whether they are conceived of as purely natural or transcendent in origin.

Murray writes of the obligations that flow out of the reception of a gift: to accept it with a measure of gratitude, to use it well in the mode of stewardship, and in the appropriate form and time to reciprocate the gift (Murray, p. 32). If the body itself is seen as a gift, never simply a possession, then the act of organ donation becomes itself a way of graciously fulfilling such obligations. The gift is, as it were, reciprocated in a way that benefits others. With the living donor, interdependence is confirmed and manifested in the most vital way through the sharing of what one received. In the case of cadaver donation, new life springs from death through a passing on of the gift. One participates in a larger circulation that both enfolds and outruns the individual self.

VII. CONCLUSION

I began this paper with the question of what rights and responsibilities one has toward one's own body. However, I have sought my responses through the indirect route of questioning the question. Implicit within its construction are a number of presumptions that prescribe and delimit our mode of understanding.

The very phrase, "one's own body" has a series of significances that I have explored – the vision of the separate self, the Cartesian body as a thing one possesses, and this manner of possession as tied to capitalist sensibilities. When one steps outside of this framework, for example into what I have termed the "paradigm of interconnection," the questionableness of the question comes into focus. What if "one" is not the primary term? What if the "body" never belongs to an individual, nor is, in fact, an "ownable" thing? That is, what if there never was "one's own body" to begin with?

Department of Philosophy
Loyola College in Maryland
Baltimore, Maryland

BIBLIOGRAPHY

Abram, D.: 1991, 'The ecology of magic,' *Orion*, Summer, 29-43.

Bailey, R.: 1990, 'Should I be allowed to buy your kidney?' *Forbes*, May 28, 365-372.

Bernstein, R. J.: 1983, *Beyond Objectivism and Relativism: Science, Hermeneutics, and Praxis*, University of Pennsylvania Press, Philadelphia.

Caplan, A. L.: 1992, *If I Were a Rich Man Could I Buy a Pancreas? And Other Essays on the Ethics of Health Care*, Indiana University Press, Bloomington and Indianapolis.

Descartes, R.: 1984, *The Philosophical Writings of Descartes*, Vol. I and II, J. Cottingham, R. Stoothoff, and D. Murdoch (trans), Cambridge University Press, Cambridge.

Dowie, M.: 1988, *We Have a Donor: The Bold New World of Organ Transplanting*, St. Martin's Press, New York.

Dowie, M.: 1989, 'Transplant fever,' *Mother Jones*, April, 19-20.

Engelhardt, H. T., Jr.: 1986, *The Foundations of Bioethics*, Oxford University Press, New York and Oxford.

Engelhardt, H. T., Jr.: 1999, 'The body for fun, beneficence, and profit: A variation on a post-modern theme, in this volume, pp. 277-301.

Engelhardt, H.T., Jr.: 1996, *The Foundations of Bioethics*, second edition, Oxford University Press, New York and Oxford.

Fox, R. C., and Swazey, J. P.: 1992, *Spare Parts: Organ Replacement in American Society*, Oxford University Press, New York and Oxford.

Gadamer, H-G.: 1984, *Truth and Method*, Crossroad, New York.

Garson, B.: 1988, *The Electronic Sweatshop*, Simon and Schuster, New York.

Hansmann, H.: 1989, 'The economics and ethics of markets for human organs,' in J. F. Blumstein and R. A. Sloan (eds.), *Organ Transplantation Policy: Issues and Prospects*, Duke University Press, Durham.

Harper's Magazine: 1990, 'Forum: Sacred or for sale?' October, 47-55.

Hazarika, S.: 1992, 'India debates ethics of buying transplant kidneys,' *New York Times*, August 17, A20.

Husserl, E.: 1989, *Ideas Pertaining to a Pure Phenomenological Philosophy*, Book II, R. Rojcewicz and A. Schuwer (trans.), Kluwer Academic Publishers, Dordrecht, Boston and London.

Leder, D.: 1984, 'Medicine and paradigms of embodiment,' *The Journal of Medicine and Philosophy*, 9, 29-43.

Leder, D.: 1990, *The Absent Body*, University of Chicago Press, Chicago.

Leder, D.: 1992, 'A tale of two bodies: The cartesian corpse and the lived body,' in D. Leder (ed.), *The Body in Medical Thought and Practice*, Kluwer Academic Publishers, Dordrecht, Boston and London.

MacIntyre, A.: 1981, *After Virtue: A Study in Moral Theory*, University of Notre Dame Press, Notre Dame, IN.

Marx, K.: 1967, *Capital*, Vol. 1, International Publishers, New York.

Mauss, M.: 1954, *The Gift: Forms and Functions of Exchange in Archaic Societies*, I. Cunnison (trans.), Free Press, Glencoe, IL.

May, W. F.: 1991, *The Patient's Ordeal*, Indiana University Press, Bloomington and Indianapolis.

Merleau-Ponty, M.: 1962, *Phenomenology of Perception*, C. Smith (trans.), Routledge and Kegan Paul, London.

Murray, T. H.: 1987, 'Gifts of the body and the needs of strangers,' *Hastings Center Report* 17, pp. 30-38.

Singer, P.: 1975, *Animal Liberation*, New York Review of Books, New York.

Stone, A. R.: 1991, 'Will the real body plese stand up? Boundary stories about virtual cultures,' in M. Benedikt (ed.), *Cyberspace: First Steps*, The MIT Press, Cambridge, MA.

Straus, E.: 1963, *The Primary World of Senses*, J. Needleman (trans.), The Free Press of Glencoe, Glencoe, NY.

Straus, E.: 1966, *Phenomenological Psychology,* Erling Eng (trans.), Basic Books, New York.

Tu Wei-ming: 1985, *Confucian Thought: Selfhood as Creative Transformation*, State University of New York Press, Albany.

Vrooman, J.R.: 1970, *Rene Descartes: A Biography*, G. P. Putnam's Sons, New York.

Wang Yang-ming: 1963, 'Inquiry on the great learning,' in Wing-tsit Chan (ed.), *A Sourcebook in Chinese Philosophy*, Princeton University Press, Princeton.

Zaner, R. M.: 1981, *The Context of Self: A Phenomenological Inquiry Using Medicine as a Clue*, Ohio University Press, Athens.

Zaner, R. M.: 1988, *Ethics and the Clinical Encounter*, Prentice Hall, Englewood Cliffs, NJ.

CHRISTIAN BYK

THE IMPACT OF BIOMEDICAL DEVELOPMENTS ON
THE LEGAL THEORY OF THE MIND-BODY
RELATIONSHIP

Spiritus promptus est,
caro autem infirma
(Holy Scriptures)

Can the difference of nature between our mind and our body lead to the
dis-aggregation of our being (Kauffmann,1993, p. 83)? While it is certain
that a full answer to this recurrent philosophical problem would assist in
the search for an appropriate resolution to the current questions
concerning the role played by the human body in our law, it is not my
intention in this short study to present a complete theory of mind and
body. However, it is my impression that the time has come for a careful
interrogation into this theme because recent biomedical interventions into
the human body reveal that it is possible not only to treat the body over
against the mind, but also to modify the body and to transform it.

In 1806 a professor of law could teach without hesitation that a "man is
both body and mind and each of the two needs the other" (Lombois,
1993). Yet today, as elements of the human body, including blood,
gametes, and organs, are now distributed outside of the body for
transformation or transplantation, we may be in danger of losing many of
our legal certainties regarding the human body (Kauffmann, 1993; UN
Covenant on the Rights of Children).

I. THE LOSS OF LEGAL CERTAINTIES

In modern French history, the human body, as such, has never been
regarded as *per se* a legal concept. Two reasons may explain this fact:
first, the body is the material which gives life to other recognized legal
concepts, such as the human person; and, second, the human body is a
living substance so special that the body benefits from legal qualifications
which hold it essentially to be sacred.

M.J. Cherry (ed.), Persons and their Bodies: Rights, Responsibilities, Relationships, 265–274.
© 1999 *Kluwer Academic Publishers. Printed in Great Britain.*

A. Definition and Nature of the Human Body

1. *Legal Definition of the Human Body.* Professor R. Dierkens, an eminent expert in medical law, considers the legal definition of the body to be as follows: "that each product of conception reaching a certain degree of development is a body. This body is neither a man nor a person because the body is not an abstract philosophical or legal concept. It is essentially an anatomical and biological reality" (Dierkens, 1966, p. 27). If our existence as human persons has anatomical and biological aspects due specifically to the body, it follows that the existence of statutes with respect to the body are the consequences of existing law.

2. *Legal Nature of the Human Body.* In contrast to the anatomical and biological body, the legal nature of the human body is of essential importance for legal analysis. "The division of our sensible world by the law into two main categories, persons and chattels, is a major evidence" upon which lawyers rely (Galloux, 1988, tome 1, p. 3). One of the results of this dichotomy is that within the bluntness of the legal realm the diversity of the world can only be approached through classification into one of these two legal categories. The human body fits concisely into neither category. As Dean J. Carbonnier has noted, "the human body never appears as such in the civil code" (Carbonnier, 1950, p. 331).

However, taking into account these two categories, lawyers have always considered the human body as falling under the category of persons. Once again J. Carbonnier provides insight when he points out that "the human body is the substratum of the person" (Carbonnier, 1990, p. 17). Similarly, Dean G. Cornu argues that "the body is not a chattel; it is the person itself" (Cornu, 1988, p. 165). Furthermore, as a necessary "substratum of the person, the body is what exists before the mind becomes stronger and what remains when the mind is diminished" (Cornu, 1988, p. 615).

Human beings are persons under the civil law of France, wherein "person" is primarily understood as pure mind. The philosophy of those who inspired and drafted the civil code was spiritualist in the sense that they considered personality to be more a matter of will than of body. But this primacy of the will – of the mind over the body – does not exclude the presence of the body in the law. The legal and moral aspects of both marriage and procreation are not possible without the physical incarnation of bodies. And that which is the case with applicable family law is also true concerning property and contract law. Human activities, which are appropriated under the law, exist only because there exists, in fact, a body to be the instrument of such activity. The body and the mind constitute a whole that testifies in all activities to our common humanity.

Biomedical technologies today, however, have begun to question this circumstance since "procreation and death are manipulated, while new infringements to the body integrity are accepted and even claimed by physicians" (Harichaux, 1989, p. 131). Has the human body moved from the legal category of person into the legal category of chattel? Has the human body lost its sacred character before the law?

B. The Human Body as Protected by its Legal Qualifications

Because the human body historically was regarded as a person, it benefited from the protection granted to the human person; i.e., it was protected against violations of integrity, and against various immoral arrangements.

1. *The Rule of "Noli Me Tangere"*. The primacy of the rule "noli me tangere" is fundamental both in private relationships and in the public relationships which exist between each human person and the society and state. In some countries, such as France, the rule is so absolute that it is usually impossible for a civil court to force someone to submit to medical treatment, or even to coerce blood sampling in a paternity case.

Recent legal treatments concerning the possibilities provided by new scientific techniques utilizing biotechnological techniques suggest considerable change in the legal approach to such questions. For example, can the right to know his/her parents (a right protected under the United Nations Convention on the Rights of Children) be regarded as a justification for compulsory recourse to coercive use of such techniques to determine paternity? In other areas, such as organ procurement and transplantation, legislatures in many countries have already accepted and regulated practices concerning the procurement and transplantaion of organs from deceased and living donors. Such medical interventions are quite obviously a violation of the integrity of the donor for the benefit of the health or life of the recipient.

With regard to the protection of the individual from state interference, the category of personhood has, since the Enlightenment revolution, been extended to those who previously were considered merely slaves or to be civilly dead. Such egregious categories have disappeared and those once concerned now benefit from the protection of the "noli me tangere" rule. While it is true that infringements to this rule have always been possible when they could be justified by legitimate reasons, such as concern for the public health or the prevention of criminal offenses, in democratic states these exceptions are limited and controlled by the judiciary.

Strong scientific incentives in medicine and the biotechnological industries have motivated the biomedical research that has opened new possibilities, primarily through genetic engineering, even to reshape the human body. Moreover, there are immense possibilities for the control of both procreation and heredity that could affect not only each and every human being as an individual, but also the very nature of the human species as a whole. The current legislative trend in states where such techniques are practiced is to allow greater and greater intervention into the human being and the human genome. The effective counterweight for the protection of the human body is no longer "noli me tangere", but rather the principle of autonomy as recognized through the free and informed consent of individuals to receive such treatment.

Yet, is it acceptable for the principle of autonomy to be the exclusive protection? Is it permissible for the state to reach into such circumstances to protect the individual, even against his will?

2. *The Limits of the Right to Autonomy.* Existence of a legal right for a person to dispose of his own body implies the recognition of the possibility of separating our will (i.e., the person) from the object on which the will is applied (i.e., the body). However, in the case of the human body, the person and the body are a whole. For this reason, in many countries the law claims that public order demands that the individual, including his body, should be protected even against the person's own will. This is, for example, the meaning of the "res extra commercium" rule which is incorporated into the civil code: the human body should not be submitted to contract law.

Of course, exceptions and clarifications have usually been admitted in order not "to push too far the fear of 'sacrilege'", as Professor Carbonnier has argued (Carbonnier, 1990, p. 19). On the one hand, this clarification concerns the illicitness of contracts concerning the human body. Contracts, for which the body plays a passive role but is moved by a will, such as contracts for the carriage of passengers, are not illegal. On the other hand, a few exceptions have been admitted when there is a special social interest at stake, such as paying a woman to give her milk to a child, or when the violation is very limited, such as selling hair (Meulders-Klein, 1982, p. 240, n. 31).

The problem is that the biomedical revolution has so largely extended the cases for which it is possible to use human body parts, tissues and other elements that we begin to wonder what is left for such rules to regulate (Harichaux, 1989, p. 130). Elements of the body, such as blood, placenta, gametes, and genes, have become matters of scientific and medical research; they even tend to be parts of industrial applications (Hermitte, 1988, p. 338). Human organs are used regularly, including the harvesting of a single kidney from a healthy living donor, and the collecting of a piece of liver from a living person for graft onto a recipient. In some cases, parents even decide to procreate for the specific purpose of creating a compatible donor for a child waiting for a transplant. In surrogacy arrangements, it is the body as a whole which is solicited for the specific function of reproduction. Research in drug experimentation requires the use of the entire body as well.

Hidden by the law under the concept of a person, but empirically revealed by our biomedical techniques, will the human body come to have the status of a thing to be utilized, to be dismembered for the benefit of the individual or the community, each claiming rights in it? Will the law, and especially the legislation recently adopted in the biomedical field, dig the grave of the living human body? Will it become a thing? I am fearful of what the law will eventually allow, but as J.P. Kauffmann has argued "my body, this is the mind of my life" (Kauffmann, 1993). Therefore what kind of legislation should exist to help define and protect the human body?

II. QUESTIONS ABOUT THE NECESSITY AND FEASIBILITY OF ESTABLISHING LEGISLATION CONCERNING THE BODY

Bioethics debates often afford an excellent opportunity to remind us to search for and validate the foundational values of our lives and of our society. In particular, can we find in our legal tradition elements which will help to solve the following questions? Are individual rights opposed to the respect of the human body and to fundamental human dignity? Is the public interest in using the human body a cause for less liberty and a washing away of our human condition? Or, are we simply becoming cannibals as J. Attali has suggested?

Given that I am a lawyer, I would like to focus on two fundamental risks: the risk inherent in rejecting a legal system founded on the respect of individual rights, and the risk of considering our law as being unable to renew its concepts, notions, methodology, or systematization.

A. Protecting and Extending the System of Individual Rights

Critical arguments have been engaged against the concept of individual rights. However, are such arguments pertinent enough to justify the destruction of the whole system of rights?

1. *The Strengths of the Critical Arguments.* Is it true that the new biomedical techniques because they are challenging fundamental notions of individual autonomy, human personhood, and the human species, are destabilizing our legal system (Bourgeois, 1988)? The fact is that with the biological revolution "individual rights are no more those of the citizen but those of a man as a being made of flesh and blood. Thus it is impossible not to raise the question: Is there any legal idea of what a human being is at the end of the 20th century, I mean a man in his globality?" (Conseil d'Etat, 1988). The reason for this question is that in this century science has appeared to give primacy to technology. Part of this primacy, though, is the inherent consequent ideology devoted entirely to the concepts of utility and fructification. Global society is then conceived of as no more than a rational enterprise whose objective is its own satisfaction (Edelman, 1989, p. 166).

In this circumstance individual rights are perverted through the context of narcissism, wherein it is the right of the woman to have a child, the right of a person to utilize his body as he wishes, the right of the dying to choose his own death, etc. At its very best, the law in such a context, is the law of an artificial world which exists only to satisfy our desires. At its worst, the law then becomes the instrument of the exploitation of the great majority by the "right" of the minority (i.e., the minority currently in legal power). Such appears to be the case in the commercial trafficking of human organs. Of course, there is some reality to this argument. However, it depends primarily on the concept of individual rights amplified and given content within the largely individualistic ideology dominant in our industrial democracies. In contrast, in an overcrowded society rights to procreate might appear to be more limited, and in a poor society a prohibition on the sale of human body products may appear to be tyrannous nonsense, insofar as it restricts families from a vital means of staying alive (Cornavin, 1985, p. 103).

This means that if we think of individual rights as an isolated and universal legal concept we will coercively force other societies to conform to our own and in doing so may very well either contribute to the social problems associated with over population, such as starvation, or remove from its poorest members their only practical option for survival. While this circumstance is certainly the case, it only serves to underscore the pressing need to understand the conception and content of basic human rights as well as how best to ground their protection.

2. *Legal Universalism.* It is true that for Western developed nations the idea that human rights would incite a kind of legal selfishness is absurd (Ferry, 1991, p. 48). Conversely, the philosophy of human rights implies a sort of universalism which grasps at rights that are not merely the expression of an isolated individual but rather that draw the universal limits between the public and the private. Economic and social rights, in contrast, are recognized for the benefit of each individual and do not question this approach because such rights have their own inherent limits.

The concept of universal human rights, though, promotes the emergence of a common interest while still separating the public realm from the private. This implies another advantage enjoyed by the system of democracy and human rights when developing an understanding of the legal and moral status of the human body: all human bodies are different, yet, at the same time, all human persons enjoy equal rights.

Furthermore, legal universalism is not necessarily opposed to the knowledge brought to us through the progress of biology. Scientific knowledge is itself a form of universalism (Langaney, 1988). Consequently, it incites us to be aware that "any regulations adopted domestically to establish limits to new techniques" have very little chance to be efficient or effective if they do not force us to look for international principles to guarantee basic human rights.

One final reason why I am very reluctant to imagine that we could rid ourselves of the system of human rights as the best way to protect the human body is that I do not see the immediate possibility of setting up any alternative system. Alternative systems would surely beg the question in favor of content not available through human reason alone. Reason alone cannot provide the philosophical crossing point that would be necessary to serve to determine the direction and content of that system. Such an approach would not be compatible with the institution and structures of a modern society which decides for itself the rules that must be followed and thus the rules that can be modified (Manetakis, 1992, p. 297).

B. Reconstructing the Concepts

1. *Individuality.* Biomedical technology acutely raises the issue of the legal nature of the human body because it offers the perspective of reconstructing the human body. The question, then, is whether the irreversible effects that biotechnology risks with the intelligence and behavior of the individual, as well as on the genetic characteristics of the human species, lead to a definition of genetic identity. Also, legal protection may focus on the protection of the individual as such, or on the individual as a member of the human species, or merely on protection of the individual's particular assay of genetic components. Legally, the concept of a person has been historically relatively precise: the person is a legal subject, which means that in order to know if the human body before life or after death is a human person we should wonder if such a body is the substratum of a legal subject. In my view, it is difficult to draw from the present state of the law an absolute and positive answer to this question. We should therefore acknowledge that lawyers together with the whole community must face new responsibilities for finding an adequate system of protection for the human person which respect human dignity (Byk, 1992).

2. *Family.* We all know that during the past decades the sociology of the family has changed considerably. Divorce has become a normal end of marriage and non-married couples, as well as single parents, have become "normal" families. The widespread use of reproduction techniques is of particular importance for understanding this evolution because it focuses on the importance of the individual to will to establish families. Biological and genetic parentage is no longer an absolute condition for establishing such a link. The difficulty of the issue in determining family importance for moral or legal concerns is that there is no longer a single family model. But the liberty acquired by the individual to develop his own family projects should have some limits. The interests of children

and the necessary respect for others, such as gamete donors, should be respected.

3. *State and Society.* Should we be afraid of the possible alliance between existing biological mastery and the coming neurobiological mastery? In fact, the prospect of a very different society – either more binding or more democratic – looms on the horizon as a result of bioethical institutions and the internationalization of bioethical issues. It is quite sensible to be concerned about the decision making processes in bioethics: societal control seems inefficient, regulation is often futile and confused, and ethical bodies are often too institutionalized.

Too often bioethics appears to confuse ethics and law. Thus, the risk exists that particular individual ethics will prevail on matters that should instead focus on social rules for a whole society. Another danger exists in the promotion of a dogmatic moral order coercively imposed on all because of some ethereal notion that the ethics developed out of "bioethics committees" is somehow superior to the legal rules developed through the society as a whole.

The modern state ought to find a balanced position which encourages what I call a pedagogy promoting law and values which gives to each component of society, individuals, professional bodies, ethics authorities, and public authorities, its role and responsibility in the collaboration of political choices in agreement with democracy and in respect for human rights. Such a state could not ignore the necessity for international cooperation in these fields as the present European approach of adopting conventions on biomedicine and human rights demonstrates.

Would a good methodology, an efficient policy reinforcing the rights of the individual, protecting the family, and offering opportunities for social changes, be sufficient to give legal and institutional answers to the fears raised in society by biomedical issues and their potential impact on our human nature? Let me simply assert that such a task can only find its meaning if it reinforces our responsibility and solidarity towards others and particularly the most vulnerable among us. This cannot be done without the memory and consciousness of the part played by those many men who throughout our history have sacrificed everything for the sake of human dignity.

University of Poitiers
Paris, France

BIBLIOGRAPHY

Bourgeois, (ed.): 1988, *Collectif, l'Homme, la nature et le Droit,* Paris.

Byk, C.: 1992, 'Conclusion,' in *Les Journées Juridiques Franco-Helléniques,* Journées de la Société de Législation Comparée.

Carbonnier, J.: 1950, 'Terre et ciel dans le droit français du marriage,' *Et. Ripert LGDJ* 1, Paris, 331.

Carbonnier, J.: 1990, 'Droit civil, 1, Les personnes,' *Themis PUF,* n.4, Paris.

Cornavin, T.: 1985, 'Theorie des droit de l'homme et progrès de la biologie,' *Droits, PUF,* Paris 2, 103.

Cornu G.: 1988, *Droit Civil, Introduction,* Domat-Monchestien, Paris.

Counseil d'Etat: 1988, *De l'ethique au Droit,* La documentation française, Paris.

Dierkens, R.: 1966, *Les Droits sur le Corps et le Cadavre de l'Homme,* Masson, Paris.

Edelman, B.: 1989, *Sujet de Droit et Techno-Science,* Archives de Philosophie du Droit, Sirey, Paris.

Ferry, I.: 1991, 'L'Humanisme juridique en question: Réponse a B. Edelman,' *Droits, PUF,* Paris 13, 48.

Galloux, J.C.: 1988, *Essai de Definition d'un Statut Juridique Pour le Matériel Génétique,* thèse, Université de Bordeaux 1, tome 1.

Harichaux, M.: 1989, 'Le corps objet in 'Bioethique et droit,' PUF, Paris, 131.

Hermitte, M.A.: 1988, *Le Corps hors du Commerce, hors du Marché,* Archives de Philosophie du droit, Sirey, Paris.

Kauffmann, J.P.: 1993, 'Mon corps,' l'Evenement du Jeudi, n. 455, 22 au 28/7.

Langaney, G.A.: 1988, *Les Hommes: Passé, Present, Conditionnel,* A. Colin, Paris.

Lombois, C.: 1993, *La Personne Corps et Âme, Journées,* R. Savatier, Poitiers.

Manetakis: 1992, in Les Journées Franco Helléniques, *Journées de la Société de Législation Comparée,* p. 297.

Meulders-Klein, M.T.: 1982, 'Le droit de disposer de soi-même,' *Xe journée d'études juridiques,* Jean Dabin, Bruylant, Bruxelles, p. 240.

SECTION FOUR

THE BODY FOR PROFIT:
ORGAN SALES AND MORAL THEORY

H. TRISTRAM ENGELHARDT, JR.

THE BODY FOR FUN, BENEFICENCE, AND PROFIT: A VARIATION ON A POST-MODERN THEME

I. INTRODUCTION

Do persons have a secular moral obligation to live more healthful lives? And if they do, to whom do they owe the obligation? And under what circumstances can the obligation be defeated? If the obligation exists and people appear bent on violating that obligation, what sanctions or constraints may be imposed? Further, what secular moral constraints exist upon the decisions of individuals to rent, lease, or sell their body parts to others? In particular, if all the parties involved consent freely, and the exchanges are to the financial and health benefit of those concerned, what grounds could there be to object? It is this last question that is most troubling, because it focuses obliquely on the secular specification of the morally proper use of our bodies. More broadly, it involves an understanding of the good life and of canonical human well-being.

Whether one may sell, rent, or use one's body as one will, is closely tied to traditional discussions of perversion, that is, the misuse of one's body alone or with others. The abandonment by the American Psychiatry Association of the concept of homosexuality as a mental disorder, while maintaining as disorders other sexual life-styles, such as shoe fetishism, discloses the post-modern tension. On the one hand, enough of the Christian moral narrative regarding proper sexuality has been lost so that, in a secular context, much of our language regarding sexual perversion no longer has a general meaning. On the other hand, there is the nagging sense that there are impoverished, distorted, misguided, and indeed perverted life and sexual orientations. Many would like at least to be able to show why shoe fetishism as a sexual orientation is deficient when compared with the normative heterosexuality of the West (Spitzer, 1987).[1] The question is, can one from a purely secular moral understanding show that any one use of the body or of its parts is intrinsically better than alternative uses? Or to focus on the specific issues at stake, many would like to show in general secular terms why selling

M.J. Cherry (ed.), Persons and their Bodies: Rights, Responsibilities, Relationships, 277–301.
© 1999 *Kluwer Academic Publishers. Printed in Great Britain.*

one's kidney is morally opprobrious. However intense, indeed religiously well-founded, such moral sentiments in this regard may be, the arguments in secular terms appear inconclusive. Philosophy does not establish a single unambiguous content-full moral account by which to resolve such controversies. There is no generally established understanding of the moral significance of our bodies. Non-agreement-based, secular, moral constraints regarding how we may use them do not appear to be available.

This essay will begin by distinguishing the more manageable moral doubts about the moral use of the body from those that remain as insoluble controversies from a secular perspective. The manageable concerns involve the use of human bodies with the consent of their owners. Also, insofar as common criteria for comparing consequences are accepted (e.g., the monetarization of consequences), one will be able in principle to determine which practices involving human bodies (e.g., the failure to gain timely medical care, indulgence in pleasurable but health-destroying past-times, and the renting, leasing, or selling of body parts) are to be recommended and why. The insoluble controversies will be those that involve claims that the dignity of humans (or of their bodies in particular) morally forbids the renting, leasing, sale, or abuse of body parts. The insolubility of such disputes stems from posing what are religious questions (or quasi-religious questions in the sense of presupposing particular commitments to a set of moral premises) and then attempting to answer that question in general secular terms. After examining the intractability of secular controversies regarding the dignity of the body, the 1987 Office of Technology report regarding the ethical foundations of law and public policy in this area will be explored. The goal will be critically to assess a failed attempt to justify restrictive law bearing on the rental, lease, or sale of human body parts in general secular moral terms. This essay concludes that, in secular pluralist societies, there will not be generally justifiable moral bases for categorically forbidding (i.e., morally warranting coercive state force to prevent) indulgence in self-destructive behavior or in the renting, leasing, or selling of body parts, unless prior agreements are violated or innocent third parties are harmed.

A secular statement of moral responsibility regarding one's own body and those of consenting others will be sparse at best. This sparseness should serve as a warning that secular moral language is a vehicle for communication among moral strangers and that concrete understandings of human well-being and responsibility must be found in concrete moral communities. The content-less character of secular morality should also underscore the considerable distance between what can be morally established in general secular terms and what can be known within an authentic Christian moral perspective. It is from this latter source, however distant, that the Office of Technology Assessment (OTA) draws the moral sentiments in its 1987 report regarding the ownership of human tissues and cells (Office of Technology Assessment, 1987), so as to give an account of the secular moral appropriateness of the sale, lease, or rental of human organs and body parts. As will become clear, this attempt to derive a secular morality from theological roots fails. Those who would wish to know more about the proper use of their bodies are ill-advised to seek such guidance from general secular morality or from partially secularized theological accounts, they will need to seek it in authentic religious terms (Engelhardt, 1995). Through an examination of the arguments advanced by the Office of Technology Assessment, this essay will in particular show the difficulties of deriving a content-full secular moral account of the significance of the sale, leasing, or renting of human organs.

II. THE PROBLEM IN MANAGEABLE AND IN TROUBLESOME TERMS

An element of the puzzles about the moral obligations we have regarding the use of the human body can be accounted for in terms of duties to third parties or in terms of interests in realizing certain goods or values. Insofar as the moral significance of living human bodies depends on explicit actual agreements, the matter will be fairly straightforward. One will need to determine what the agreements are, their intended scope, and the circumstances under which they may be nullified. Implicit actual agreements will be somewhat more difficult of exegesis because the elements of the agreements are not laid out. But the underlying issues will be manageable because one will be able to trace to the source of the moral constraints, namely, to the promises or the contracts of the participating parties. In such circumstances, the moral force of constraints on the use of living human bodies (or for that matter, of the obligation to place living human bodies at the disposal or others) can be derived from the agreement of those participating (e.g., X may not sell his kidneys to Y because ownership in them has already been transferred by prior agreement to Z). One will not need to appeal to any particular ranking of values, canonical moral intuitions, sense of the sacredness of human bodies, or endorsements of ideal hypothetical decision makers, to determine what is morally at stake. Rather, the practice of resolving moral controversies, in the absence of canonical rankings or orderings of values, by an appeal to mutual agreement will be sufficient to justify the web of moral obligations. This justification can be accomplished without metaphysical appeals or difficult philosophical arguments that hinge on establishing a canonical vision of human well-being.

If one knows how (or agrees how) to rank or aggregate the consequences of particular human practices and/or actions, the matter will also be straightforward. For example, if one focuses only on monetarizable costs and benefits, individuals who begin to smoke and drink in later life but eschew driving while intoxicated will likely save resources on sum for society because they will typically not draw excessively on Medicare and social security. At the very least, if controversies can be put in such generally accepted terms, one can resolve most of the issues by developing practices through which individuals indemnify third parties or agree not to draw down on communal resources, should their use of their bodies engender a cost for others.

But much of the controversy regarding obligations of individuals not to misuse their bodies seem deeper than questions regarding the proper calculation of costs and benefits and the exact nature of agreements with third parties. This is particularly apparent when one considers the question of the morality of renting, leasing, or selling body parts, a question which arises, for example, in surrogacy for hire or the selling of a kidney. Beyond the matter of agreements and characterizable costs and benefits, such proposals appear for many to involve an intrinsic moral wrongness.

The special moral hesitations many feel regarding the renting, leasing, selling and abusing of bodies can only adequately be understood by the continued force of the Christian worldview on modern moral sentiments. The countries and societies that emerged to dominate the modern era colonially, economically, and militarily sprang in great measure from Christian cultures that carried Christian sentiments into the modern world, although those sentiments have often become disarticulated from their origins. The result is that one finds both individuals in the history of modern philosophy as well as contemporary individuals seeking philosophical arguments for moral premises that were once secured within a Christian moral context. The Western Christian self-understanding embedded matters of faith within rational arguments in an attempt rationally to justify through natural law arguments many conclusions that are credible only within a particular worldview with premises such as those in the Judeo-Christian tradition.[2]

The modern era emerged from a millennium of established monotheism with the intellectual task of maintaining the general lineaments of Christian morality on the basis of reason, but without reliance on a particular set of religious moral commitments. Recoiling from the Reformation's fracturing of a half millennium of Roman Catholic hegemony, secular reason was invoked to maintain the catholicity of moral and political foundations. This reaction was heightened by the Thirty Years' War and the Civil War. It was expressed in the Enlightenment aspirations of Locke, Kant, and Bentham to establish a content-full, unified moral account with little reliance on, or in some cases in the absence of, the Christian God. The hope was for a rationally purified, but still content-full, single morality without the full-blown commitments of Christian monotheism.

At times, the attempts are tantamount to a self-parody, as when Immanuel Kant endeavors on the basis of reason alone to demonstrate the fundamental immorality of the sale of a tooth or of masturbation.[3]

> However, it is not so easy to produce a rational demonstration of the inadmissability of that unnatural use, and even of the mere unpurposive use, of one's sexual attributes as being a violation of one's duty to himself (and indeed in the highest degree where the unnatural use is concerned). The ground of proof surely lies in the fact that a man gives up his personality (throws it away) when he uses himself merely as a means for the gratification of an animal drive. But this does not make evident the high degree of violation of the humanity in one's own person by the unnaturalness of such a vice, which seems in its very form (disposition) to transcend even the vice of self-murder (Kant, 1964, §425).

Many who find Kant's arguments regarding masturbation implausible still hope for analogous arguments that can secure the moral impropriety of surrogacy for hire or the selling of organs on the grounds that such activities involve a violation of the dignity of humans or an exploitation of humans despite the presence of consent.[4]

The attempt to secure many of the sentiments against the renting or selling of organs by an appeal to deriving them from the notion of man's creation in the image and likeness of God (i.e., the concept of the imago dei) represents an intermediate step between traditional Christian understanding and an attempt to provide a secular philosophical ground for such sentiments. Such theological understanding was married to philosophical arguments by the Roman Catholic church. In particular, within the Roman Catholic church, one finds the clearest natural law condemnation of the view that people may use the organs of their bodies as they wish.

It is to be observed also that even the individual human being – as Christian doctrine teaches and the light of reason clearly shows – has no power over the members of his own body except so far as he uses them for their natural purpose; he cannot destroy or mutilate them, or in any other way render himself incapable of his natural functions, except where there is no other way of providing for the welfare of the body as a whole (Pius XI, 1960 [1930], p. 34).

Outside of a particular religious framework, this conclusion, which would be sufficient morally to include forbidding the misuse of one's organs or the practice of surrogate motherhood for hire, does not appear easily defensible. It depends on a notion of natural biological-moral ends, which is highly problematic in secular terms that must construe human nature as the result of evolution, not design. One is placed in the position of Kant with his desire to supply rational arguments to sustain his condemnation of self-abuse. One firmly holds moral convictions, which therefore appear so self-evident that it seems implausible that rational arguments would not sustain them. Yet the requisite rational arguments are not straightforwardly available.

We have what Alasdair MacIntyre and other cartographers of the post-modern era document (MacIntyre, 1981). One encounters moral sentiments and conclusions, which once functioned as integral parts of a coherent whole, but which now are isolated fragments in search of a justification that cannot adequately be restored outside of an act of faith. Here it might be useful to remember how Christianity first approached the integration of faith and reason. Most initially did not assume, as did Thomas in the 13th century, that reason would be sufficient to secure a content-full understanding of proper human conduct.[5] Thus, even Blessed Augustine of Hippo suggests that discoursive reason completes faith, "It is not because they have known that they believed, but in order to know, they believed. For we believe in order to know; we do not know in order to believe" (*In Ionannis evangelium tractatus*, 40.8.9).[6] The difficulty is that questions regarding how one ought properly to use one's body were once relatively easy to answer within a religious framework. Also the religious framework that dominated the West for over a millennium developed the peculiarity of claiming that its key moral claims regarding the proper use of the body depend not on faith alone, but that they can be secured independently by rational argument. These Western Christian claims regarding the capacities of reason have continued to be maintained even after philosophy was largely divorced from its previous religious moorings. At the same time, it has become increasingly apparent that one cannot secure a particular content-full understanding of moral reasoning (including the proper ways to use one's body) without first accepting a particular set of initial moral premises (i.e., some thin theory of the good or religious faith).[7] Another way to encapsulate the difficulty is to recognize that a religious (or quasi-religious) question has been confused with one that could be stated and answered within secular moral terms. The question seeks content that secular morality cannot provide, thus engendering a controversy that cannot reach closure by a rational argument framed in secular terms.

III. A FOUNDATION FOR SECULAR LAW AND PUBLIC POLICY

This confusion of the religious and the secular in an attempt to frame a content-full secular morality is well illustrated by the OTA report. The report shows: (1) the impossibility of establishing a general secular moral basis for the state's coercively forbidding a market in organ sales, leases or rentals, (2) the confusion or shards of once-intact religious moral concerns with secular moral interests, (3) the significantly post-Christian character of the secular morality bearing on the licitness of organ sales, leases, or rentals, and (4) the need to look elsewhere than to secular morality for an account of how properly to regard the use of human organs (Engelhardt, 1995). In addressing the moral propriety of selling human cells, tissues, and organs, the report by the *Office of Technology Assessment* for the United States Congress structures its analysis around the three basic ethical principles (i.e., respect for persons, beneficence, and justice) endorsed by the National Commission in 1978. The Report reveals a general difficulty in establishing a moral basis for coercively interdicting a market in human parts.

The Report begins its moral analysis with a focus on the principle of respect for persons.

> First is respect for persons: the idea that trade in human materials ought to be limited to the extent that the body is part of the basic dignity of human beings. If the body is indivisible from that which makes up personhood, the same respect is due the body that is due persons. Conversely, if the body is considered incidental to the essence of moral personhood, trade in the body is not protected by the ethical principle of respect for persons (Office of Technology Assessment, 1987, p. 130).

This gloss on the principle of respect for persons only indirectly trades on a concept of the inalienability of one's rights over one's own body. Instead, the Report portrays the status of rights to sell one's organs as dependent on the character of embodiment. As a consequence, for the Report the first question is whether the body is so integrally part of the person as morally to preclude the alienation of organs. The Report asserts:

> From a Cartesian point of view, human tissues and cells are valuable only to the extent that they provide a temporary substrate or basis for the existence of the human persons. The relationship between the human person and a particular tissue or cell is not essential, particularly if these materials are replenishable. This is not to say that Cartesians would be reluctant to attach a monetary value to such materials. In fact, they may be quite inclined to make tissues and cells the object of commerce because there is no great significance attached to such materials in terms of the human mind, personality, or identity (Office of Technology Assessment, 1987, p. 137).

From the text, it is fairly clear that the term Cartesian is not used in an historically strict sense.

The term Cartesian is not employed in the Report to identify a metaphysical Cartesianism with claims concerning the independent substantial character of the mind. Instead, the term identifies what might be termed a value-theoretical Cartesianism, in which the person as the source of values is contrasted with the body as that which receives values through the goals, designs, and choices of persons. Thus, Cartesian is employed to identify those accounts of persons that contrast the person as a self-conscious moral agent with the person as a physical body. But even so, further distinctions are in order. Given the broad usage of Cartesian in the Report, the term would compass Kant, who contrasts the autonomous duties of persons with their mere inclinations as humans. Yet, Kant provides a paradigmatic example of a moral theory, which precludes the alienation of one's parts because such alienation cannot be understood as following from the notion of a rational moral agent.

A further qualification is therefore in order. The set of accounts of the body which the Report characterizes as Cartesian is not just those that regard the body as receiving its true value from the person. Nor is it the set of accounts of the body which regard the body in instrumental terms, for Kant sees the importance of tending to one's body as a means of fulfilling one's duty to act as a responsible moral agent. Rather, it is that set of accounts of the body which regard the body instrumentally in terms of goals that Kant would characterize as heteronomous: the use of the human body in order to satisfy the desires of persons.

Once the values persons may properly assign to the use of their own bodies are disjointed from a canonical, content-full morality, then persons may regard their bodies as disposable in terms of their own goals, with no constraints in principle set on those goals, as long as those goals do not involve unconsented-to injury to innocent third parties. It is the liberal worldview[8] grounded in the moral authority of actual decision makers, not decision makers constrained by an ahistorical content-full understanding of moral reasoning or correct moral decisionmaking, which renders the body and its parts disposable to the goals of competent individuals and their consenting collaborators. It is this liberal understanding of the relationship of persons to their bodies that the Report characterizes as Cartesian.

But the Report does not adequately develop its account of Cartesianism. As a consequence, the Report in general is blind to the possibility that integral to the moral controversies regarding the probity of selling one's cells, tissues, or organs is the ambiguity of the principle of respect for persons: whether respect for persons depends on respecting (i.e., not forcibly interfering with) the peaceable free choices of persons or whether respect for persons requires ensuring (even through coercive means, if necessary) that persons realize their true nature, dignity, and obligations, including their religious obligations to God. The approach of the Report likely stems from the hesitation of its authors to explore the possibility that persons may alienate their bodies in whole or in part by marriage, joining the military, or entering indentured servitude.[9] However, if the cardinal secular moral significance of persons is their being the source of moral authority in a polity of moral strangers, then respect of persons includes respect for their decision to alienate their selves and their body parts.[10] The Report does not take seriously the possibility that the unique secular moral dignity of persons may derive simply from the capacity of persons to make claims, to give authorization for the use of themselves and their property, and to be imputable for their actions. But if the secular moral dignity of persons is grounded in their unique capacity to authorize the use of themselves and their property, then respecting agreements to rent, lease, and sell bodily parts is integral to respecting persons within the content of secular morality. Instead, the Report assumes, behind its account of embodiment, an inalienability thesis with some analogue to the Lockean account of natural rights, where ownership of human beings by God sets limits to the contracts into which persons may enter. But the Report substitutes a particular overriding secularized concept of human dignity for the role played by God in Locke's account. One's free choices are limited by a concept of human dignity, which in the past would have been buttressed by religious considerations.

The Office of Technology Assessment, in providing a background account of the principle of respect for persons (and its implications for buying and selling bodies and their parts), gives special attention to the work of four individuals, beginning with the work of the Protestant theologian, Paul Ramsey.[11] The Report sees Ramsey as holding that "human beings exist in their bodies and that respect for the body is indivisible from respect for the person" (Office of Technology Assessment, 1987, p. 131) because of his belief in the dignity conveyed by God's creation of man. However, the Report does not recognize the cardinal significance of this theological premise for claims regarding the inalienability of one's rights over oneself and one's body, because of the Report's attempt to fashion a secular moral account of embodiment.[12] Again, the Report assumes that a secular notion of personal dignity can do the work previously done by a concept of God's ownership.

The Report is correct regarding Ramsey's concerns about dualism. But his concern is directed not against a metaphysical dualism. He addresses instead a value-theoretical dualism that contrasts persons with their bodies as objects for technological manipulation. Moreover, he sees the difficulty raised not just by the sale of organs, but by organ transplantation itself.

> Our culture is already prepared for technocratizing the bodily life into collections of parts in which consciousness somehow has residence for a time, and for calling by the name of "the direction of mankind toward God" what can better and exhaustively be described in secular terms as the onward thrust of technological progress. Therefore, the use of hearts from living donors can come to be regarded as a proper administration or stewardship of a man over his organic life. The contagious dualism of modern culture has already placed him, as a spiritual overlord, too far above his physical life. To most of us a part of the body or the bodily life as a whole is already only a thing-in-the-world, not to be identified with the person (Ramsey, 1970b, p. 193).

It is the objectification of the human body that Ramsey regards as a moral affront to human dignity and a perverse self-understanding. He objects to the mentality that would regard the body as an object under the control of persons through technology.

> If we wish to avoid being inundated by either horticultural or engineering metaphors, that will depend in great measure on the recovery of a proper sense of the integrity of man's bodily life, as against the Cartesian dualism and mentalism of the modern period which rejoices without discrimination over every achievement or intervention or design which shows that the body is only a thing-in-the-world to be subjected to limitless control (1970b, p. 209).

For Ramsey, regarding the human body as an object to be technologically manipulated in order to satisfy the desires of persons is a violation of the dignity of persons, of which the sale of organs is a special grave instance.

This concern with illicitly objectifying the human body through making it an object of technological manipulation is also the focus, it might be noted, of the condemnation by the *Congregation for the Doctrine of the Faith* of homologous in vitro fertilization and embryo transfer even without surrogacy or zygote wastage.

> Such fertilization entrusts the life and identity of the embryo into the power of doctors and biologists and establishes the domination of technology over the origin and destiny of the human person. Such a relationship of domination is in itself contrary to the dignity and

equality that must be common to parents and children (Congregation for the Doctrine of the Faith, 1987, p. 30).

But, the affirmation of human dignity by the Roman Catholic church over and against a reduction of the significance of the body to the ways in which it can be used to fulfill the desires of persons, does not in its terms preclude the sale of tissues and organs. Once the transfer of tissues and organs has been independently justified (e.g., in terms of concerns for preserving the totality of another person's organic life), there is no objection in principle to payment. Thus, Pope Pius XII argues:

> Moreover, must one, as is often done, refuse on principle all compensation? This question remains unanswered. It cannot be doubted that grave abuses could occur if payment is demanded. But it would be going too far to declare immoral every acceptance or every demand of payment. The case is similar to blood transfusions. It is commendable for the donor to refuse recompense: it is not necessarily a fault to accept it (Pius XII, 1956, pp. 381-382).[13]

Unlike Immanuel Kant, who would forbid all gifts or sales of human cells, tissues, or organs, and unlike Paul Ramsey, who would forbid all sales, Pope Pius XII embraces the position that some payments could indeed be justified.[14]

Pope Pius XII takes this position despite his view that the body has special dignity. Once the obligations that arise because of that dignity have been satisfied through an adequate justification for the transfer of cells, tissues, and organs, the acceptance of payments is not, in and of itself, forbidden. To employ Kant's terminology outside of its theoretical ambit and Kant's own views, the sale of organs does not necessarily involve heteronomy. This is principally the case for Pope Pius XII because the required moral side-constraints (i.e., the principles of totality and charity) and the concept of just compensation prevent human cells, tissues, and organs from being viewed merely as commodities.

The Report also gives an inadequate account of the position of Leon Kass, because of the accent it places on Cartesianism. What is at issue is not whether Kass is a Cartesian in the sense of seeing a stark separation of mind as a *res cogitans* from body as a *res extensa*. It is rather that Kass holds that physicians "respect and minister to bodily wholeness because they recognize, at least tacitly, what a wonderful and awe-inspiring – not to say sacred – thing the healthy living human body is" (Office of Technology Assessment, 1987, p. 133). It is not so much the necessity of embodiment, but the sense of sacredness that appears essential to Kass's arguments. The difficulty is giving an account of the sacred in secular terms. For Kass, the sacredness to which he speaks identifies a commitment to a particular sense of the wholeness of the human person, mind and body. Moreover, this view is combined with an understanding of medicine as directed to meeting the somatic needs, not the mere desires, of patients. But this secular account of human dignity is but one among many alternative proposals for properly understanding the importance of the human body.

The difficulty lies in selecting any one proposal as justifiable for all in general secular terms. In particular, there is the difficulty of excluding on general secular principles the practice of buying, selling, and renting organs as one would other commodities (given adequate appropriate safeguards against fraud and other forms of overreaching). Here, one can join with the issue of persons using their bodies as they may wish in order to meet their own special desires or special life projects (e.g., hang-gliding, driving a motorcycle without a helmet, driving without the use of seat- and shoulder-belts, smoking four packs of Camels a day, or attempting to vault over a canyon on a motorcycle). Given a free choice on the basis of one's own balancing of benefits and harms, and given that one protects innocent persons from unconsented-to costs and damages, it will be difficult to establish a particular canonical moral understanding of the proper use by persons of their own bodies (or the bodies given, rented, leased or sold to them).

In merely secular terms, the human body, as we find it, is the result of evolution, chemical constraints, blind accidents, and other natural events. The human body could in many ways have been otherwise, and it is only "natural" for humans as self-reflecting agents to judge the body's virtues and vices critically. As Sir Peter Medawar noted, " ... nature does not know best" but has left us with a " ... tale of woe [including] anaphylactic shock, allergy, and hypersensitivity" (1960, p. 100f). Indeed, one can outline various ways in which the body can be improved and might in time be improved with germline genetic engineering.

The body has already generally become a taken-for-granted object of technological manipulation. The primordial experience of the body being manipulable by drugs, potions, and alcohol has been developed with the advent of drugs of well-known and targeted action. The use of hormonal contraceptives are, as the Roman Catholic church rightly notes, a step toward constraining the human body not to follow the rhythms given it by nature, but instead the desires imposed by persons. Moreover, the availability of cells, tissues, and organs, as well as prosthetic valves and joints for transplantation, has altered our everyday sense of the body. The body is now experienced as repairable by the insertion of new organs when the old ones fail. The body has taken on the sense of an object available for technological manipulation in the way the Paul Ramsey decried.

This view of the body as constituted out of replaceable organs makes difficult a secular notion that the original organs have any particular significance or any special bond to the secular moral dignity of persons. In this respect, the Cartesian thesis of the Report is true. Though the body maintains its special uniqueness as the condition for the individual's continued survival in the world, and as the unique presence of the person in the world, it loses much of its mystery, such that the replacement and use of its parts becomes a matter of manageable, prudential, secular calculation. Therefore, the sale of a kidney becomes one way among others of accumulating capital. It must be compared with others such as long stressful days and nights of hard labor, perhaps associated with special occupational risks. The decision then becomes a choice among possible injuries to the body (and other costs), one sort arising from anesthesia, surgery, and the reliance on only one kidney, and the other arising from particular, accepted, occupations. Unless a concept of human dignity can be established to preclude in principle one of the options, the decision for the individual is most plausibly tipped in favor of the choice with the most benefits and the fewest costs. If one concludes that the sale of a kidney will expose one to fewer costs than the kinds of labor one would have to undertake to accumulate equivalent capital, and one has no special religious or other commitments that foreclose the sale of a kidney, then its sale becomes the most reasonable choice.

Even if one enters into the project of selling kidneys, one does not become a metaphysical Cartesian. The body can still be understood not just as a necessary condition for being alive in the world, but as one's very presence in the world. It is simply that this presence is manipulable and that not all tissue in one's body is integral to one's own personal embodied presence in this world. But this sense of the body as manipulable and as having replaceable or salable parts is compatible with regarding the body as "sacred" in the secular sense of being the intimate presence in the world of one's acts of love and hate, mercy and vengeance, etc. It is just that one will have concluded that renting or selling a particular organ will not undermine life or one's sense of the sacred (should one have such a sense.)

The matter of the secular moral licitness of using one's body as one wishes turns then not on the extent to which one is Cartesian, but on the extent to which authorization by actual individuals is sufficient to transfer cells, tissues, and bodies to other persons, despite countervailing interests. This is not the place to attempt a theory of ownership, but insofar as ownership is grounded in respect of persons (so that products become property insofar as labor makes them an extension of the producer), then one's self, cells, tissues, and organs will be paradigmatic examples of personal property (Engelhardt, 1996). Moreover, insofar as controversies regarding different visions of the proper use of human bodies are irresolvable, as is the case in secular pluralist societies (which do not possess a way of canonically discovering which preferences are legitimate or illegitimate, or of justifying a unique canonical ordering of social desiderata), the more it will become plausible to resolve such controversies in ways that depend upon mutual agreement.[15] This will place the market in central stage. In addition, it will be difficult in such a context to secure general secular grounds for holding that there are sufficient moral reasons to defeat the free agreement of persons, which free agreement provides the one way for moral strangers peaceably to resolve controversies with secular moral authority in the absence of a common understanding of moral reasons or a common set of beliefs.[16] Again, this will have implications not only for the sale and rental of organs, but for the rights of individuals to use their own bodies to achieve their desires, even if this decreases life expectancy and increases the likelihood of injury, disability, or disease.

The OTA Report, after its examination of the principle of respect for persons, turns to the principle of beneficence.

> The second moral principle is beneficence. Would commercialization of human materials (perhaps of specific kinds) be more beneficial than a ban on such commercialization? Proposals for markets in human tissues, for example, could be justified on the grounds that they would lead to a preponderance of good results over bad. On the other hand, objections to the same markets could likewise be couched in consequentialist (outcome-oriented), beneficence-based terms (Office of Technology Assessment, 1987, p. 130).

The Report then proceeds primarily to examine the likely risks or costs associated with a market in human cells, tissues, and organs.

The Report summarizes the criticisms of a market in human parts by first raising grounds for being skeptical that individuals would maximize their own well-being in such a market, and then second by exploring wider alleged negative effects of commercialization of human body parts. Under the first rubric, the Report lists four major grounds for hesitation.

1. The question is raised whether the market would function well, since the presumption of a well-functioning market is that individuals are rational. However, since a wide range of individuals display irrational behavior, this is taken to be a justification for considering markets inadequate to the task of distributing human parts. However, absent angels or other species of superior rational beings administering governments or other bureaucracies, the same concerns about rationality and venality that one would need to raise with respect to the market would be equally appropriate with respect to non-market solutions. In part, the final answer would be an empirical one based on studies of the efficiency of market versus government distribution of resources. In part, the answer would also turn on the right of individuals to exempt themselves in whole or in part from the control of governments over the use to which they would want to put their bodies or their body parts.

2. The second source of concern turns on whether to allow children and the mentally infirm to contribute organs to the market. This area of concern does not give rise to objections in principle for which there may not already be a cure at common law, where minors and the incompetent are already barred from commercial contracts. And again, questions would need to be raised regarding the ground for holding that governments would be less likely to be exploitative than the market.

3. The third source of concern is that fraud, misrepresentation, coercion, and other overreaching in the market will bring more harm than good. This concern is reasonable only if one holds that the risk of such fraud, despite reasonable means of containing it, will be greater than the harm done to individuals by: (a) the fraud, misrepresentations, and other overreaching that may be a part of non-market organ distribution systems; and (b) the use of coercion to deprive individuals of the opportunity freely to buy and sell organs. Moreover, the history of fraud, misrepresentation, and coercion by governments has been dramatically worse than what has been seen in the market.

4. Finally, the Report suggests that there may be a discrepancy between what people desire and what they need, and that the free market will not provide what people need, only what they desire. This critique is plausible only if one believes: (a) that it is possible to discover true needs and distinguish them from desires, and (b) that governments are more likely to provide effectively for the true needs of individuals, rather than for their desires, than would a market.

The Report then turns to possible macro-level criticisms of a market in human organs.

1. The Report raises the prospect that "commercialization of the body will lead to disrespect and devaluation of the human body in general" (Office of Technology Assessment, 1987, p. 135). This concern is reasonable in secular terms only: (a) insofar as one can make out a coherent, canonical, general secular moral sense of the proper respect or value due to the human body, apart from how the body or its component parts meet the desires of competent individuals in a context of free consent, and (b) insofar as the disrespect to persons consequent on coercively not respecting their free decisions to sell or rent their organs by forbidding such markets, is less than the harm from undermining the proper respect and value due to the human body as a consequence of such markets.

2. The Report raises the specter that commercialization will threaten important ideals of equality. The premise is that purchasing organs will allow the rich to have access to treatment not available to the poor. But surely this problem can be met by governments and charities purchasing organs for the poor (e.g., perhaps providing organ stamps on the model of food stamps). Moreover, forbidding a market in organs is a way to maintain financial inequality, because it deprives poor individuals in America as well as in the Third World from easily acquiring capital from the rich with little risk to the poor and thus without coercion decreasing the disparity between the rich and the poor.

3. The third criticism begs the question: "Moving from the concept of gift to a market in human tissues and cells carries with it such important losses to the common good that they will, on the whole, outweigh the immediate benefits" (Office of Technology Assessment, 1987, p. 135). It is surely open to be argued as to which is more noble, altruism or the free market. The Western Middle Ages accented altruism and constrained the market, while the modern era has unleashed the market and produced riches. It is the existence of international free markets that has created the

contemporary breadth of wealth which allows a significantly wide proportion of individuals to be altruistic in a substantial fashion. The moral genius of the market is that it allows moral strangers to exchange goods and become rich, though they may not share common basic metaphysical or moral assumptions.

4. The final criticism is a special application of the previous and is as fraught with unsecured moral and empirical assumptions.

> In the special case of human biological materials donated for research to nonprofit institutions (e.g., university-based biomedical research), the shift from a gift to a market basis could have damaging consequences in the cost and availability of such materials, public perception of and generosity toward biomedical researchers, and increased suspicion of health providers (Office of Technology Assessment, 1987, p. 135).

But of course, if the market created greater access to more of the desired human cells, tissues, and organs, public perception would be positive. More fundamentally, any definitive answers would require understanding how to balance all the kinds of consequences that will result from any particular policy, including consequences for liberty.

The difficulty is to discover how to measure consequences, how to compare consequences, how to know which outcome is best, how to know which program is most beneficent. To answer such questions, one would have to solve the root challenge to any proposal of a canonical, secular, content-full account of morality: there appears to be no way to identify which hierarchy or way of coordinating the pursuit of values is correct or how to rule out of consideration preferences that many hold to be unacceptable. All content-full accounts of proper moral action, which appeal to a canonical sense of moral reasoning, moral sense, moral intuitions, hypothetical decision-makers, or hypothetical contractors must in the end beg the question by endorsing a particular moral sense or thin theory of the good in order to determine which preferences should count for how much and why. One needs already to have a particular value sense to know when to give what priority to considerations of liberty, equality, prosperity, or security. This intellectual crisis is acknowledged *en passant* by the Report, when it notes the insoluble difficulty of identifying any particular account of distributive justice as canonical. "Since our society appears to subscribe to several, sometimes incompatible ideals of justice, there will be no easy way to list the ethical implications of commercializing human biologicals from a 'correct' theory of justice" (Office of Technology Assessment, 1987, p. 135). And the problems recognized regarding accounts of justice plague those of beneficence as deeply. In summary, if one cannot provide grounds for embracing one account of beneficence over another, one account of justice over another, how will it be possible on secular moral grounds to establish whether it would be good or bad to sell, rent, or "abuse" one's organs or those of consenting others? The ethical analysis provided by the OTA Report leaves us, in the end, bereft of guidance, but better acquainted with the character of morality at the end of the Christian age.

IV. SOLVING PROBLEMS THAT ARE INSOLUBLE

We find ourselves in a world without arguments sufficient to establish in general secular terms: (1) whether it is good or bad to have a market in which one can rent, lease, or buy human body parts, or (2) whether it is morally improper for individuals to use their bodies in ways that are likely to foreshorten their lives or to lead to disability, deformity, or disease. The difficulty is, as Marjorie Grene puts it, "there is no nous." At least, we find ourselves with our nous darkened so that we do not have the capacity veridically to discover through secular reason alone the goods and proper directions underlying nature and the human body.[17] In epistemology and science, this decrepitude of reason paradoxically does not cause as much difficulty as it does with morality and public policy, in that science does not usually involve the coercive imposition of a particular orthodoxy.[18] As a consequence, individuals can establish by implicit or explicit agreements sufficient *ceteris paribus* conditions, so that they can engage together in a particular community of investigation. Those who decide not so to engage can go their own way, pursuing their own understandings of knowledge (though from the perspective of others, they will be understood as grossly deviant and misguided). But, since morality can promise foundations for coercive political structures, even peaceable moral deviance can become the object of coercive state intervention. The problem, then, is to provide a moral justification for a secular state enforcing a particular content-full account of proper organ use, when such a particular content-full account cannot be discovered as canonical.

This moral decrepitude and the consequent contrast between the views of particular moral communities and what can be established generally through secular reason is troublesome here, as has been noted, because general secular questions about the proper consensual use of organs cannot be answered abstractly, that is apart from any particular tradition of understanding the significance of the body and the importance of caring for one's own physical well-being. Traditions, histories, particular moral communities convey to their participants or members concrete understandings of the importance of life and the significance of particular actions with and through their bodies. The failure of the Report to answer satisfactorily the question about the proper use of organs derives, as the third section of this paper shows, from such answers being available only within particular moral traditions. But the answer is sought in universalist terms, because the public policy debate is occurring in a large-scale secular nation compassing many moral communities. The *Office of Technology Assessment* poses a question that can only be answered within a particular moral tradition and then attempts to answer it outside of any particular tradition, that is, within the context of a secular pluralist society that compasses many traditions but cannot have a particular tradition, absent a coercive imposition of that tradition. The Office of Technology Assessment seeks to derive content-full moral guidance as from nowhere, so as to claim moral authority everywhere and for all moral communities (i.e., across the bounds of a pluralist society), when such guidance always comes from a particular somewhere.

A. Why Hegel Was Right and MacIntyre Nearly Right

In his *Philosophy of Right*, Hegel criticizes Kant for his failed attempt to derive from rationality itself a concrete understanding of duties and rights. As Hegel shows, moral reasoning, apart from any tradition, can justify general practices, such as promise-keeping, but it cannot establish what individuals should promise. Reason can disclose the *nisus* of morality toward the achievement of the good, but it cannot specify the good in any concrete fashion, outside of a particular moral community and its history. Considered abstractly, outside of the context of any particular moral community, duty is left without content.[19] It is within a particular community that one understands in a concrete sense the goals and obligations of life.[20] But "there are no tradition-independent standards of argument by appeal to which they can be shown to be in error" (MacIntyre, 1988, p. 403).

The alternative to life within a particular tradition with its robust disclosures of the purposes of bodies and the concrete directions for the good life is what MacIntyre describes as the life and language of cosmopolitans.

> [T]he condition of those who aspiring to be at home anywhere – except that is, of course, in what they regard as the backward, outmoded, undeveloped cultures of traditions – are therefore in an important way citizens of nowhere is also ideal-typical. It is the fate toward which modernity moves precisely insofar as it successfully modernizes itself and others by emancipating itself from social, cultural, and linguistic particularity and so from tradition (MacIntyre, 1988, p. 388).

Those who would span the particular moral communities in secular, pluralist societies in order to provide the peaceable fabric of a common polity must assume the stance of cosmopolitans, or in fact be cosmopolitans. They must, to follow MacIntyre, act "in detachment from all substantive criteria and standards of truth and rationality" (1988, p. 384). But the Hegelian solution, unlike the solution proposed by MacIntyre, does not require a choice between the cosmopolitan perspective and the perspective of a particular moral community. As always, the Hegelian perspective does not require an either/or, but a both/and. It presupposes the plurality of concrete moral communities and offers the perspective of a state that brings them together in a web of formal rights and duties (see, for example, Engelhardt and Pinkard, 1994).

B. Liberty and Responsibility in the Post-Modern World

If individuals do not share a common tradition or set of beliefs, and if by secular reason one cannot establish a particular canonical moral vision then, if one wishes to resolve controversies in public policy by appealing to authority that can be commonly understood by moral strangers in a secular pluralist society, other than by appeal to force, authority for common action must be derived from the actual consent of the actual individuals involved. In such circumstances, rights to forbearance as the conditions for mutual, peaceable negotiation in a secular, pluralist society will provide the common fabric of a polity. From commonly owned resources, refuseable welfare rights will be legitimately fashionable by whatever processes have been established for the deployment of such properties. Because it will be impossible to presume how individuals will wish to use their own bodies, and the bodies of others put at their disposal, rights to privacy will loom large in the area of the sale of bodily parts, and the "abuse" of bodies (and in such circumstances, insurers should be free to set rates to meet the risks that free individuals assume by their actions, e.g., special policies for helmet-less motorcyclists).

Because limited democracies with rights to privacy are morally unavoidable, given the limits of moral reason, the actual lines of freedom and responsibility in a secular, pluralist society will be determined by the actual agreements of actual persons. Those to whom the grace of God has given a concrete understanding of the proper use of human bodies will be equally at liberty to condemn as perverted, misguided, improverished and sinful those who pursue the lives of liberal cosmopolitans. The moral geography of the post-modern world is one in which, with the recognition of the false expectations of a concrete universal secular moral narrative, one is at liberty from a general secular point of view to do what one wishes with consenting others, even when it is wrong, because concrete understandings of responsibility and individual realization will be discloseable only within a concrete moral community.

Center for Medical Ethics and Health Policy
Baylor College of Medicine
Houston, Texas

NOTES

[1]   See, for example, the taxon "Fetishism," 302.81, in American Psychiatric Association, *Diagnostic and Statistical Manual of Mental Disorders* (DSM-III-R and DSM-IV). DSM-IV places "fetishism" under the category of "Sexual and Gender Identity Disorders" (see pages 526f).

[2]   Traditional Christianity recognizes that one can only learn the proper use of one's body through turning to God. This requires turning to Him in worship, almsgiving, fasting, and other practices of asceticism, through which one will experience how one ought to employ one's body. In the absence of these conditions, all is in danger of being perverted, as St. Paul warns (Rom 1:22-23) and as out contemporary secular society shows. This essay defends not what ought to be, but rather shows what is by default the condition of general secular morality.

3   "To deprive oneself of an integral part or organ (to mutilate oneself), e.g., to give away or sell a tooth so that it can be planted in the jawbone of another person, or to submit oneself to castration in order to gain an easier livelihood as a singer, and so on, belongs to partial self murder" (Kant, 1964, *Akademietextausgabe* IV, 425).

4   The bioethicist Thomas H. Murray is quoted as claiming that "We may be more than mere protoplasm, but we're nothing without our bodies (at least in this world). Putting a price on the priceless, even a high price, actually cheapens it. So we don't approve of selling our body parts: and the body isn't quite property" (Office of Technology Assessment, 1987, p. 127).

5   The strong reliance on reason eventually led to Thomism becoming the official philosophy of the Roman Catholic church (Pope Leo XIII, 1879).

6   "Non quia cognoverunt crediderunt, sed ut cognoscerent crediderunt. Credimus enim ut cognoscamus, non cognoscimus ut credamus."

7   An important acknowledgment of the limits of the Kantian endeavor to discover in reason the content for proper moral conduct is found in John Rawls, 'Justice as fairness: Political not metaphysical,' (1985). There he recognizes the limits of hypothetical decision theory by noting its dependence on particular intuitions: " ... since justice as fairness is intended as a political conception of justice for a democratic society, it tries to draw solely upon basic intuitive ideas that are embedded in the political institution of a constitutional democratic regime and the public traditions of their interpretation" (p. 225).

8   I use liberal worldview here in a very narrow sense to indicate a notion of the relationship of persons to each other and to the world, such that the only constraints are those set by mutual agreement. In particular, the liberal worldview with regard to nature is taken to deny moral constraints set by nature on the actions of persons.

9   The Report characterizes Engelhardt as permitting "indentured servitude, as it exists, for example, when one receives support for education in exchange for a commitment to military service" (Office of Technology Assessment, 1987, p. 134). The writers of the Report do not envisage that indentured servitude may involve rights over the very person, not merely the body of the servant.

10   If persons are the source of secular moral authority in a polity of moral strangers, it will follow that persons may agree to have their lives ended or their attitudes changed under certain circumstances. That is, they may alienate themselves by giving up rights to their own life and dispositions. One might imagine an individual contracting to work for an intelligence agency of a government engaged in a just struggle with an evil aggressor. The agency offers certain select employees the option of an additional $500,000 per year to be paid to a secret Luxembourg banking account. For this payment, these employees agree to commit suicide when requested or to have their attitudes refashioned so as to meet the needs of the agency. Thus, one might imagine the following two scenarios. Scenario 1: "Excuse me, Joe, but did you receive a copy of the ultra-secret memo XQ-47, dated December 7?" "Yes, I did. It couldn't have been meant for me." "Joe you are right. And I am so sorry. We really had not wanted anyone to know about this memo except for the Director and me. Would you mind terribly stepping into the suicide room and terminating yourself. As you know, we cannot take the risk of a security leak, should you be caught and tortured." Scenario 2: "Excuse me, Joe, but recent performance reports show that you do not approach the garrotting in back alleys of enemy agents [who have been found guilty of war crimes in absentia and sentenced to execution] with the zest and zeal that we expect from our top employees. Please report next weekend for an attitude-retraining session. We have developed new reconditioning techniques. When we get through with you, you will derive much more satisfaction from

your work (i.e., you will have a brand-new positive attitude regarding the garrotting of individuals in back alleys.)" In general secular terms, no more will be able to be said, though much more should be said.

11 The OTA explores the principle of respect for persons in terms of the implications of the accounts given by Paul Ramsey, Joseph Fletcher, Leon Kass, and H. Tristram Engelhardt, Jr., with respect to the proper use of human bodies and their parts.

12 Ramsey, for example, emphasizes that "We need rather the biblical comprehension that man is as much the body of his soul as he is the soul of his body" (1970a, p. 133).

13 One might note that the provision of blood for transfusion or skin grafts for transplantation are considered unproblematic because they involve a renewable part of the body. Therefore, their removal does not threaten the integrity of the body as a whole. "E contra mutilatio non habetur, si v.g. homo parte cutis vel portione sanguinis privatur (ad transplantationem vel transfusionem in corpus alterius aegroti), quia corpus manet integrum et hae partes mox restaurantur" (Noldin, 1938, p. 312).

14 Very likely, the Pope brings to the notion of payments in this regard not the idea of a market-value, but of a just compensation, a concept rooted in special considerations of Roman Catholic moral theology.

15 For example, one may grant that liberty, equality, security, and prosperity are the major social desiderata. However, society will have a radically different character, depending on how these desiderata are ranked. One will not be able to appeal to consequences in order to determine which structure is preferable, because one will already need to know how to rank or aggregate liberty, equality, security, and prosperity consequences. Nor will appeals to moral reasoning or hypothetical decision theory help unless one already imports into the notion of moral reasoning or one's notion of a canonical, hypothetical decision-maker a particular moral sense, ranking of values, or thin theory of the good.

16 As the OTA Report phrases it: "Others, such as ... Engelhardt, argue that commercialization must be tolerated as part of recognizing the limits of governmental authority to interfere in private choices, even on behalf of important goals or special beliefs certain groups may have about the sacred character of body parts that individuals may freely wish to sell" (Office of Technology Assessment, 1987, p. 143).

17 "Since the world we live in is infinite and open, not finite and closed like Aristotle's, there is no way for us to get an ultimate grasp on first principles literally stateable for a given fixed region of the real" (Grene, 1979, p. 196).

18 For some studies of the roles played by negotiation even in empirical science, see Engelhardt and Caplan (1987).

19 "Duty itself in the moral self-consciousness is the essence or the universality of that consciousness, the way in which it is inwardly related to itself alone; all that is left to it, therefore is abstract universality, and for its determinate character it has identity without content, or the abstractly positive, the indeterminate" (Hegel, 1965, §135, p. 89).

20 "In an ethical community, it is easy to say what man must do, what are the duties he has to fulfill in order to be virtuous: he has simply to follow the well-known and explicit rules of his own situation" (Hegel, 1965, §150, p. 107).

BIBLIOGRAPHY

American Psychiatric Association: 1987, *Diagnostic and Statistical Manual of Mental Disorders*, 3rd edition, revised, American Psychiatric Association, Washington, D.C.

American Psychiatric Association: 1994, *Diagnostic and Statistical Manual of Mental Disorders*, 4th edition, revised, American Psychiatric Association, Washington, D.C.

Augustine of Hippo, St.: *In Ioannis evangelium tractatus,* 40.8.9.

Congregation for the Doctrine of the Faith.: 1987, *Instruction on Respect for Human Life in its Origin and on the Dignity of Procreation [Donum Vitae]*, Vatican City.

Engelhardt, H.T., Jr. and A. Caplan (eds.): 1987, *Scientific Controversies*, Cambridge University Press, New York.

Engelhardt, H.T., Jr. and T. Pinkard (eds.): 1994, *Hegel Reconsidered: Beyond Metaphysics and the Authoritarian State,* Kluwer Academic Publishers, Dordrecht.

Engelhardt, H.T., Jr.: 1995, 'Moral content, tradition, and grace: Rethinking the possibility of a Christian bioethics,' *Christian Bioethics* 1, 29-47.

Engelhardt, H.T., Jr.: 1996, *The Foundations of Bioethics*, second edition, Oxford University Press, New York.

Grene, M.: 1979, 'Comments on Pellegrino's anatomy of clinical judgments,' in H.T. Engelhardt, Jr., S.F. Spicker, and B. Towers (eds.), *Clinical Judgment*, D. Reidel Publishers, Dordrecht, pp. 195-198.

Hegel, G.W.F.: 1965, *Hegel's Philosophy of Right*, T.M. Knox (trans.), Clarendon Press, Oxford.

Kant, I.: 1964, *The Metaphysical Principles of Virtue,* J. Ellington (trans.), Bobbs-Merrill, Indianapolis.

Leo XIII, Pope.: 1879, *Aeterni Patris, 4 August 1879.*

MacIntyre, A.: 1981, *After Virtue*, University of Notre Dame Press, Notre Dame.

MacIntyre, A.: 1988, *Whose Justice? Which Rationality?* University of Notre Dame Press, Notre Dame.

Medawar, P.: 1960, *The Future of Man*, Methuen, London.

National Commission for the Protection of Human Subjects of Biomedical and Behavioral Research: 1978, *The Belmont Report*, U.S. Government Printing Office, Washington, D.C.

Noldin, H.: 1938, *Summa Theologiae Moralis iuxta codicem iuris canonici, vol. 2, De Praeceptis dei et ecclesiae*, Felix Rauch, Leipzig.

Office of Technology Assessment: 1987, *New Developments on Biotechnology*, vol. 1, *Ownership of Human Tissues and Cells*, Government Printing Office, Washington, D.C.

Pius XI, Pope: 1960, *Casti Connubii, 31 December 1930*, translation from *Papal Teachings: The Human Body*, St. Paul Editions, Boston, pp. 25-36.

Pius XII, Pope: 1956, 'Address to a group of eye specialists,' May 14, *Papal Teachings: The Human Body*, St. Paul Editions, Boston. pp. 381-382.

Ramsey, P.: 1970a, *Fabricated Man*, Yale University Press, New Haven, Conn.

Ramsey, P.: 1970b, *The Patient as Person*, Yale University Press, New Haven, Conn.

Rawls, J.: 1985, 'Justice as fairness: Political not metaphysical,' *Philosophy and Public Affairs*, 14, 223-257.

Spitzer, R.L.: 1987, 'The diagnostic status of homosexuality in DSM-III: A reformation of the issues,' in *Scientific Controversies,* H.T. Engelhardt and A. L. Caplan (eds.), Cambridge University Press, New York, pp. 401-416.

DONNA C. KLINE

DESPAIR, DESIRE, AND DECISION:
A FUGAL RESPONSE TO ENGELHARDT

In 'The body for fun, beneficence, and profit: A variation on a post-modern theme,' (herein after referred to as 'The body for hire') H. Tristram Engelhardt (1999) offers witty and scholarly arguments for conclusions with which many of his readers will sympathize; i.e., that certain important decisions about the uses to which one's body is put ought to be left to individual design, free of state coercion, so long as innocent third parties are not harmed. We, the sympathetic readers, therefore, accord his arguments considerable latitude. I will argue that unfortunately, those appealing conclusions do not follow from any non-tautologous reading of the essay's premises. More importantly, however, if Engelhardt is correct about the impossibility of demonstrating moral propositions and if he is correct in the criteria that he sets out for adequate justification, then it follows that no moral proposition, appealing or repugnant, whether concerning state actions or the actions of individuals, can be justifiably asserted, at least against those who disagree with it. And, I argue, it further follows that Engelhardt's own conclusions rest unsupported upon moral propositions as unjustifiable as the canons that he attacks. Thus, this essay is almost entirely directed at Engelhardt's justification for his own affirmative position and less at his elegant critique of competing theories, including his review of the Office of Technology Assessment's report.

Indeed, this essay will argue that the post-modern tension, as described by Engelhardt, not only results in our being pulled between cultural diversity (our recognition of the apparently irreconcilable variations in the values held by various groups in our society) and enlightenment rationality (our desire to categorize actions on an objective basis, so that we can both make decisions about what action for the society to take against its members in regard to their uses of their own bodies *and* justify those decisions to those who would disagree with us) but makes rational dialogue on a solution to the dilemma impossible. That tension is, indeed, painful to anyone who reflects on the issues confronting society in regard to bioethical issues of organ transplantation and related matters and who also wishes to constrain state coercion to actions in which the coercion

M.J. Cherry (ed.), Persons and their Bodies: Rights, Responsibilities, Relationships, 303–330.
© 1999 *Kluwer Academic Publishers. Printed in Great Britain.*

can be justified. But, I will argue, the post-modern metaphysics and epistemology described by Engelhardt in fact make the very discourse put forward by Engelhardt impossible. Having run the fragile ship of rationalism onto the reef of nihilism, one cannot reasonably expect to salvage a few consoling planks from the wreck, like some modern day philosophical Crusoe constructing an acceptable *modus vivendi* from some salvaged odds and ends of modernism.

I. SYNOPSIS AND FORMAL RESTATEMENT OF ENGELHARDT'S POSITION

Engelhardt addresses two primary issues: (1) the overall justification for state action regulating the uses of persons' bodies and (2) a specific policy which he recommends we adopt to govern the sale of body parts. The two positions are intertwined because the former is basically a meta-ethical question while the latter is a specific question of ethics or political philosophy which needs to be answered within the framework of the more general theory. The meta-ethical issue is, in summary, "How does one justify public policy in the absence of consensus of the citizens?" and the political philosophy question is "What public policy with regard to the sale of body parts can be so justified?"

These are issues with which Engelhardt has wrestled at greater length elsewhere. In *Bioethics and Secular Humanism: The Search for a Common Morality* (hereinafter *Secular Humanism*), he says:

> I begin with a confession. This volume is meant as much to address the fundamental philosophical and cultural challenges of the Post-Modern age as to give an account of bioethics or to explore the significance of secular humanism. ... the challenge is defined by the failure of religion or reason to establish a canonical account of justice or morality. Secular humanism is the attempt to articulate what we as humans hold in common without special appeal to religious or other particular moral or metaphysical assumptions (1991, p. xiii).

Engelhardt concludes that secular humanism, religion, and reason cannot provide substantive moral guidance but that secular humanism can articulate sufficient common grounds to provide a framework for "a content-less perspective for moral negotiation among moral strangers" (pp. xvii, xviii).

This essay responds specifically to 'The body for hire' which argues for a radical conclusion based on sweeping meta-ethical premises. In *The Foundations of Bioethics* and *Secular Humanism,* Engelhardt presents more limited, and I would argue more plausible, claims; although I am far more inclined to agree with those elegantly reasoned views, I believe that the arguments set out in this essay apply to any Engelhardtian meta-ethical claim to the effect that either of those works can avoid the post-modern meta-ethical quandary described by Engelhardt (1986; 1996).

In 'The body for hire,' Engelhardt describes his undertaking as follows:

> This essay will begin by distinguishing the more manageable moral doubts about the moral use of the body from those that remain as insoluble controversies from a secular perspective. The manageable concerns involve the use of human bodies with the consent of their owners. Also, insofar as common criteria for comparing consequences are accepted (e.g., the monetarization of consequences), one will be able in principle to determine which practices involving human bodies (e.g., the failure to gain timely medical care, indulgence in pleasurable but health-destroying pastimes, and the renting, leasing, or selling of body parts) are to be recommended and why (1999, p. 278).

It soon becomes apparent, however, that Engelhardt is neither going to list the principal rights or duties of an individual with regard to that individual's own body nor provide us with a guide as to which specific practices (insofar as individuals are concerned) are to be recommended and why. Indeed, as will be discussed more fully later, 'The body for hire' essentially argues that no justified, meaningful, guidance, at least by a moral stranger, can be given to the prospective donor who asks "To what procedures should I consent?" At best, under Engelhardt's formulation of the post-modern predicament, the most one can do is advise the inquirer to check with that community with which he feels the closest ties or to negotiate for whatever deal he thinks he might like the best. That silence, however, does not indicate a mere lapse of attention on Engelhardt's part; the abdication flows necessarily from his profound skepticism about the possibility of justifying moral choices, a skepticism that I conclude is so profound as to descend to a nihilism on moral choice. Engelhardt cites Hegel for the proposition:

> Considered abstractly ... duty is left without content. It is within a particular community that one understands in a concrete sense the

goals and obligations of life. But "there are no tradition-independent standards of argument by appeal to which they can be shown to be in error," (1999, pp. 296-297).

One should not underrate the power of that statement that "duty is left without content" because, as we shall see more fully later, it leads to two radical positions with regard to the individual who needs to choose what to do with his own body. Those are (1) that one may do as "one wishes with consenting others" and (2) that no justification can in principle be given for any particular individual action, except within the traditions of the community to which the actor belongs. The question of what one may (or ought) agree to is not simply left blank; it is declared to be forever a blank which can only be filled in from the values of the community to which one chances to belong (or from whimsical personal preference) and never more meaningfully than that, never with a confident sense of a moral position rationally arrived at. Engelhardt argues that:

> Philosophy does not establish a single unambiguous content-full moral account by which to resolve such controversies [about what to do with our bodies]. There is no generally established understanding of the moral significance of our bodies. Non-agreement-based ... constraints regarding how we may use them do not appear to be available (1999, p.278).

* * *

> The insolubility of such disputes stems from posing what are religious questions (or quasi-religious questions in the sense of presupposing particular commitments to a set of moral premises) and then attempting to answer that question in general secular terms (1999, p.278).

Engelhardt does not seem to feel the same profound skepticism with regard to the possibility of providing moral guidance, derived from philosophical analysis, in the area of political action; 'The body for hire' itself is an effort to delineate the circumstances in which state coercive action can legitimately be taken against individuals to control the actions which they may take with their own bodies. Engelhardt is quite clear that philosophical reflection demonstrates to us that there are limitations on what we can justifiably do when we act as agents of the state or, to put it another way, what the state can justifiably do when it acts for us.

We are all aware that profound disagreements apparently exist between people of moral seriousness and good intentions. Moreover, we know that decisions about matters upon which disagreement exists need to be addressed because the existence of technology, which can satisfy certain strong desires, such as the desire to have a child, forcibly bring the issue to our attention. That is, doctors know how to transplant organs from one human being to another, at least under certain circumstances, and people who desire the outcome of such a transplant, such as those who suffer from diseases which can be cured or ameliorated by such transplants, would pay considerable sums of money for the organ necessary to that cure. For example, a couple who desire to have a child which is genetically theirs but who cannot have such a child in the biologically ordinary way would be willing to pay a woman to carry such a child for them. The coexistence of desire and technology means that decisions as to the use of the technology will be made by someone and in some fashion.

Now, we also know that such decisions can be made in a variety of ways and at different levels of decision making in society. That is, as a society, we can pass laws which specifically prohibit or require a particular action, or we can leave the matter to individual decision (which amounts to the society permitting the conduct in question). Although this answer is obvious, it is a significant, but somewhat hidden, aspect of Engelhardt's paper that the question he is answering concerns the position or role that society should take rather than the action that an individual should take. That is, a rich person who asks himself whether he may morally buy a kidney from a poor person (or a poor person who reflects on whether it would be a morally permissible – or even laudable – action to sell the kidney) will not get much guidance from Engelhardt's paper. More specifically, he will get essentially non-guidance: since Engelhardt argues that no moral theory can be proven, he would leave the decision to the individual, without further moral guidance. The person who holds a particular moral theory, such as a Roman Catholic, may presumably continue to follow that theory, but the person seeking the best guidance for his actions will presumably be left by Engelhardt to make his way as best he can among the thickets of competing theories, after noting in passing that, in fact, there is no way out of the maze. The tension between the perception that no moral theory has been acceptably demonstrated and the continuing desire to have some moral guidance may be taken as the core of the first part of Engelhardt's paper.

That tension merits closer description because of its prominence in 'The body for hire' and in contemporary policy discussions. Engelhardt uses the American Psychiatric Association's demarcation of some sexual preferences as normal and some as abnormal as an example of the patent impossibility of rationally and objectively distinguishing between the normal and the abnormal, let alone the right and the wrong (1999, p. 277).

Later in the paper, he offers an eloquent description of the post-modern dilemma, as follows:

> The failure of the Report [of the Office of Technology Assessment] to answer satisfactorily the question about the proper use of organs derives, as the third section of this paper shows, from such answers' being available only within a particular moral tradition. But the answer is sought in universalist terms, because the public policy debate is occurring in a large-scale secular nation compassing many moral communities. The Office of Technology Assessment poses a question that can only be answered within a particular moral tradition and then attempts to answer it outside of any particular tradition; that is, within the context of a secular pluralist society that compasses many traditions but cannot have a particular tradition, absent a tyrannical imposition of that tradition (p.296).

That is, the desire for answers remains even when one is convinced of the impossibility of obtaining such answers on the grounds on which one would like to have them.

It is worth returning to the psychiatric analogy for a moment; the point of that example, in my understanding of it, is twofold. First, we do continue to separate the sane from the insane, even if we perceive that no clear dividing line can be drawn, even if we perceive that cultural norms are inextricably built into the enterprise. Second, however, if no clear dividing line can be drawn between the normal and the abnormal, then the whole enterprise of psychiatry is drawn into question, and an enterprise that we want to characterize as a science is in fact reduced to the mere tyrannical implication of societal preconceptions about behavior. That is, the impossibility of objectively distinguishing between the sane and the insane arises from the nature of the world, not from a failure of the science. Therefore, we cannot simply hope that the progress of knowledge will erase the problem by providing us with more accurate observations or more powerful theories; there is no objective information to be had. Any action which relies for its justification on such a distinction, therefore, must remain forever unjustified. Thus, not only does the science of the mind lose its objective or empirical character, but any action taken in the name of that character shows itself as tyrannical. Engelhardt makes the analogy specific, as follows:

> [In the realm of scientific inquiry] individuals can establish by implicit or explicit agreements sufficient *ceteris paribus* conditions, so that they can engage together in a particular community of investigation. Those who decide not so to engage can go their own way, pursuing their own understandings of knowledge (though from the perspective of others, they will be understood as grossly deviant and misguided). But, since morality can give foundations for coercive political structures, even peaceable moral deviance can become the object of coercive state intervention. *The problem, then, is to provide a moral justification for a secular state enforcing a particular content-full account of proper organ use, when such a particular content-full account can not be discovered as canonical* (p.295).

Engelhardt, in the service of that goal, offers to us the moral stance of Hegel which "presupposes the plurality of concrete moral communities and offers the perspective of a state that brings them together in a web of formal rights and duties" (p.297). Achieving that goal and, along the way, sketching out the dimensions of that state's control over the human body are among Engelhardt's enterprises in 'The body for hire.' One dimension of such a state is shown as follows: if individuals do not share a common tradition or set of beliefs, and if by secular reason one cannot establish a particular canonical moral vision, then, if one wishes to resolve controversies in public policy by appealing to authority that can be commonly understood by moral strangers in a secular pluralist society, other than by appealing to force, authority for common action must be derived from the actual consent of the actual individuals involved (p.297). Another dimension is drawn in as follows:

> This essay concludes that in secular pluralist societies, there will not be generally justifiable moral bases for categorically forbidding (i.e., morally warranting coercive state force to prevent) indulgence in self-destructive behavior or in the renting, leasing, or selling of body parts, unless prior agreements are violated or innocent third parties harmed (p.298).

Engelhardt's logical strategy, then, is that if all of the other possible sources of justification for state coercive actions can be shown to fail necessarily to provide such justification, then it will follow that there is only one principle which will be involved as justification for state action (the enforcement of agreements among all the parties concerned directly in the transaction).[1] I have tried to clarify Engelhardt's arguments against state regulation of traffic in human bodies (which I shall refer to here under the general rubric of "prostitution") by compiling them into the forms of premises and conclusions.[2]

(1) *The Justification Assumption*: Coercive state action requires justification by an appeal to an authority.[3]

(2) *The List Assumption*: There are four (and only four) sources of authority for common (i.e., societal or state) action, which are force; universal agreement on moral rules (usually in the form of shared traditions or religious values); consent of the person against whom the coercive action is taken; and philosophical proof that the coercive action is morally correct.[4]

(3) *The Absence of Authority Argument*: Three of the four possible sources of authority are unavailable to justify coercive state action in the area of medical decision making, as shown by the following arguments:

(4) *The Force Assumption*: Mere force is not a sufficient moral justification for state action.[5]

(5) *The Unprovability Argument*:

(A) Efforts to provide a secular, content-full, moral code by means of philosophical argument (i.e., one which is not derived from either religion or tradition) have all failed. Or, it is not possible to demonstrate a content-full moral code by means of philosophical argument because all substantive claims about the uses of the human body are necessarily derived from religion or tradition (pp.283-295).

(B) Therefore, philosophy also fails to provide any justification which will suffice to authorize coercive state action (pp.283-295).

(6) *The Lack of Consensus Argument (Re: Religion and Tradition)*:

(A) Traditions and religions vary in modern society, and no tradition or religion has any more claim to being true than any other one, and many of us think that they are historical artifacts which have no hold on us today.

(B) Therefore, tradition and religion do not suffice in modern society as justification of coercive state action in regard to prostitution (pp.283-295).

(7) *Engelhardt's Conclusions*:

(A) Therefore, only the actual consent of the parties concerned can justify any coercive action by the state in regard to prostitution (p.297).

(B) Therefore, all actions involving the use of the body must be permitted by the state, provided that (a) the parties thereto have agreed (or have not agreed to do otherwise) and (b) no innocent third party is harmed (p.278).

II. REFUTATION OF ENGELHARDT'S REASONING

It is my position that Engelhardt's reasoning fails on four grounds: (1) that it is internally inconsistent; (2) that it begs the question twice (once by smuggling the concept of a social contract or the consent of the governed into the argument and once by assuming that consent confers legitimate authority); (3) that neither of the conclusions flows from the argument, even if the above flaws did not exist; and (4) that the conclusions will not provide the moral guidance for the situations which Engelhardt claims it covers.

A. The Inconsistency Criticism

The first topic of discussion is the position in 'The body for hire' with regard to the meta-ethical issues of the sources of moral theory and the justification required of a moral theory about state action. As set out above in the *Justification Assumption*, Engelhardt postulates that state intrusion into our decisions about what to do with our bodies requires justification. Furthermore, Engelhardt tells us that force is not sufficient justification. In fact, it is implicit in his work that the justification be stronger than that conferred by the usual democratic processes. In other words, there would not be much point in Engelhardt's paper if he were content to hold that no restraint should be imposed on people's use of their own bodies unless the law imposing that restraint were adopted in the constitutional manner required by a particular state.

Propositions about justification, however, are moral propositions. Engelhardt's argument will not work unless he can begin with the major premises that justification is required for (legitimate?) state action and that force is not a sufficient justification. He must begin with a moral proposition in order to complete his argument from post-modern despair about the nature of moral propositions to a state policy of leaving matters to private contract. *To put it another way, the absence of persuasive justification for coercion only matters if one believes that justification is required.*

The problem with these propositions is that Engelhardt's argument also turns on the lack of consensus in modern society (the *Lack of Consensus Argument*) and the inability of philosophy to establish the truth of any particular moral proposition (the *Unprovability Argument*); so, then, where do we obtain these propositions?

If we agree with Engelhardt (on the *List Assumption*) that there are only three legitimate sources of authority, which we may expand to be proven philosophical truth, religion or tradition, or perhaps universal consent, we can see that none of these can be relied on by Engelhardt to establish the *Justification Assumption* or the *Force Assumption*, for the simple reason that to rely on any of the three legitimate sources would be inconsistent with Engelhardt's major thesis of the post-modern dilemma of diversity and lack of faith.

As quoted above, Engelhardt argues eloquently about the failure of philosophy:

> Philosophy does not establish a single unambiguous content-full moral account by which to resolve such controversies [about what to do with out bodies]. There is no generally established understanding of the moral significance of our bodies (p.278).

Thus, Engelhardt cannot say that the requirement that the use of force be justified is established by philosophy.

Similarly, Engelhardt argues that the intellectual status of religious constraints on prostitution is the same as "quasi-religious" views about the sanctity of the body in that both types of views hinge on "presupposing particular commitments to a set of moral premises." Analogous to the American Psychiatric Association's quandary over the definition of "normal," the disputes involving these premises are insoluble because the notion of the sanctity of the body necessarily depends on *a priori* assumptions, which cannot be adequately defended on purely secular grounds. In other words, I read Engelhardt as saying that neither the existence of a particular God nor the truth of particular moral claims about our persons can be demonstrated without appealing to some premise which is itself not demonstrable in secular terms, a premise which must, in other words, be taken as a given or be based on some authority which depends for its authority on the views being discussed. An example of this would be the familiar argument that any moral principle in the Bible must be true because the Bible is the word of God; on Engelhardt's analysis, our being persuaded of the authority of the Bible turns on our antecedently being persuaded that God exists and that he inspired the various writers of the Bible. And, Engelhardt argues, we are not so convinced. On these grounds, then, he cannot turn to religion or tradition to establish that justification is required for the use of coercion or that force is not sufficient justification.

One might be tempted to say that "surely Engelhardt is entitled to these innocuous assumptions because in fact we all do agree with them." But, of course, it is Engelhardt's post-modern point that we do not necessarily agree on any moral premise.[6] Moreover, we know from the such contemporary events as the abortion controversy and the constraints imposed by various governments on sexual conduct that a large number of people do not feel that any justification is required beyond that which they already have access to in their own religious views. One is tempted to say that persons still comfortable with those allegedly religious or quasi-religious sources need read no further because they may be content within their traditions and simply refuse to believe that their cherished proofs are not sufficient.

III. CIRCULARITY AND THE ROLE OF AGREEMENT IN MORAL DISCOURSE

The post-modern moral thinker, on Engelhardt's view, will be satisfied only with state coercion only if it is justified by a moral system which is provable philosophically or rests on a universally accepted religion or tradition.[7] I will argue in the following section that what appears to be a mere description of the post-modern situation actually involves an ethical premise relating to state action, which is that no coercion by the state is permissible unless there is universal agreement with the principle upon which the state is acting. Since Engelhardt's conclusion is that state coercion is warranted only to enforce agreements, I submit that this moral postulate both is inconsistent with Engelhardt's post-modern thesis and also question-begging with regard to his conclusion.

A review of Engelhardt's description of the post-modern dilemma, particularly his concept of a secular argument and his description of the failure of tradition, shows that he places two stringent requirements upon the meta-ethical proof of any moral theory which is invoked to justify state coercion, which are (1) that the proof be sufficient to convince everyone, no matter from what premises they start, and (2) that the theory itself proceed from no moral starting point. In other words, it is not sufficient to justify state coercion that the state be acting on a moral theory which is in fact true: in order for coercion to be justified, the proof of the morality being enforced must be so lucid, so unequivocal, that everyone is convinced. Presumably, such a proof would be so lucid that, for every moral agent in the society, either that moral agent would concede the moral correctness of the coercion or one would be entitled to proceed anyway on the grounds that the person was irrational or perverse.[8] Moreover, for Engelhardt, each and every moral proposition in the system must be itself demonstrable without recourse to any undemonstrated starting point.

As Engelhardt puts it, the question of what uses one may morally make of one's body seeks content that a secular answer cannot provide, thus engendering a controversy that cannot reach closure by rational argument. A secular argument is one which does not involve accepting a particular set of initial moral premises (i.e., some thin theory of the good or religious faith). In other words, 'The body for hire' advances the thesis that the above conditions cannot be satisfied in a pluralistic society

In defense of this point, which I call the *Unprovability Thesis*, Engelhardt offers two basic arguments: (1) that persistent disagreement exists and no way has been shown to prove (or even to convince everyone) that any one of the competing theories is superior and (2) that, necessarily, all moral theories about prostitution must involve a "quasi-religious" proposition, i.e., must include some moral premise, which cannot itself be demonstrated and which is not universally accepted.

But, what does the existence of disagreement actually show us? Mere disagreement does not show us that none of those theories is true, of course, because those who disagree with any given theory could simply be wrong. The fact of disagreement or criticism of other theories does not really prove to us that no theory is demonstrably true; we might be wrong in our criticism or responses to our (apparent) criticisms might exist which have simply not been pointed out or discovered yet. What the existence of disagreement (even disagreement as to every theory) shows is simply that there is disagreement, i.e., *not* that no moral theory is true (or has been proven) but that no *universally convincing* proof of any moral theory has yet been given.

Perhaps Engelhardt's effort at establishing the *Unprovability Argument* might be characterized as an informal inductive argument, along the lines that "if so many approaches have been tried and failed, then it is unlikely that any effort to prove a moral theory will succeed." Even so construed, however, the argument will not get Engelhardt to the conclusion that "no theory can be proven" but only to the proposition that "no moral theory can be proven in such a manner that all of us will be convinced by the proof."

One might ask why it matters that people disagree as to whether a given moral theory is true (or proven). This question will be discussed below, but it is worth noting here that agreement only matters if the truth of a moral theory is not sufficient alone to entitle the state to act coercively in accordance with that theory. Obviously, any number of groups and individuals disagree violently with that view, feeling that it is the truth of their beliefs (such as beliefs about the sanctity of fetuses) which justifies the state acting coercively (to make abortion a crime, for example) regardless of the disagreement of numerous other individuals in the society. To put it another way, if Engelhardt's argument turns on the importance of moral consensus, then it fails on his own ground of profound moral disagreement in the society.

Let us look further at the relationship between justification and agreement which is implicit in 'The body for hire.' It will, I submit, turn out to be murkier and more complex than it appears. As suggested in the discussion above, there is, I think, something of a slide from "demonstrable" to "established" to "commonly understood" or "accepted" in the structure of 'The body for hire.' The significance of the ambiguity is that it leaves room for at least two different construals of the argument. One may read it as including an implicit premise that *volenti non fit injuria*, so that people who agree with the state's position should not object to having it enforced against them, or one may read it as proposing that only propositions which have been demonstrated (to everyone's satisfaction) to be true may legitimately be enforced by the state. One might argue that if Engelhardt meant that the state would be justified in coercing people to comply with a moral theory with which they agreed, then his view itself assumed a moral premise of the form "it is not wrong to coerce those who wish to act contrary to a view which, if rational, they would agree was true." Presumably, the notion is that it is permissible to force someone to do something which they would admit is the right thing, even if, for example, selfish motives momentarily inclined them to do something else.

But that notion, that moral premise of the moral significance of consent or agreement, is of the very kind that Engelhardt has indicated cannot be utilized in a secular argument, and thus Engelhardt's position again appears either circular or inconsistent. If, hypothetically, Engelhardt wished to argue that *this* one premise was itself universally accepted or could be proven to be true, he would be violating his own factual premises about the divided nature of our secular pluralist society and the impossibility of demonstrating the truth of any content-full ethical theory, as well as begging the question.

Here we may wish to consider briefly Engelhardt's arguments to the effect that all positions, such as that advanced by the *Office of Technology Assessment*, that place limitations on the proper uses to which the human body may be put involve religious or quasi-religious premises. If we assume for the moment that he is correct that all propositions about the legitimate use of the body derive essentially from religious notions of the dignity of man, let us consider Engelhardt's position on the significance of such derivation.

In order to avoid commission of the genetic fallacy (i.e., that a given proposition is false because it comes from a particular source, such as religion), Engelhardt's point cannot simply be that the history of ideas shows that all moral constraints on, for example, prostitution derive from religious considerations. The mere fact that a moral premise is derived from religion or some other source, like an otherwise discredited moral theory, is not sufficient to show that the premise in itself is false.

Engelhardt's point, therefore, may be either that (1) all theories about the moral status of prostitution and the sale of organs (other than the one put forward by Engelhardt) are based on religious or quasi-religious views about the proper use of the body and not all of us (or, at least, not all of those to whom the essay is addressed) believe in those underlying views any longer or (2) all such propositions (other than the one put forward by Engelhardt) are based on religious views about the proper use of the body and those theories are demonstrably unprovable in the sense set out above. For the same reasons set out above, however, neither of these construals of his position can save it; the notion of unprovability will not work because it either smuggles in a moral concept of agreement or fails logically.

A person who wished to offer less controversial support to the conclusions in 'The body for hire,' (which would still be consonant with Engelhardt's position) might argue that the role of agreement (on the political level) in the argument is only to secure a peaceful solution to the question of what the state should do about prostitution.[9] In other words, we might say that people will agree to be free from coercion, even if they don't agree on the moral right, for example, to that freedom. That is a quasi-Rawlsian argument, akin to the argument actually advanced in support of the First Amendment's freedom of religion by the Founding Fathers, i.e., given that no one can be sure that his group will have the political power to enforce their own religious views, each should be willing to agree that no one will have the right or power to do so.

There are three objections to Engelhardt's taking this road. First, he seems to want to offer more than simply a way of reaching agreement; he wants to answer the "manageable" questions about the proper use of human bodies. Second, although I personally find the Rawlsian approach helpful on the issues in question, it still presumes a degree of cultural uniformity which does not seem consistent with the Engelhardtian post-modern view of the world; why should we think that there would be rational agreement on the desirability of avoiding forceful imposition of one tradition when reading the daily paper will bring numerous instances of violent intolerance to our attention? Third, that interpretation of 'The body for hire' as advocating a particular pragmatic political consensus (e.g., if I can't be sure that I will prevail politically, I ought to agree to a system in which no one has the right to impose their views on me), requires consensus at least two moral propositions, *i.e.,* (1) that it is morally permissible to agree to live in a society which tolerates acts which one knows to be immoral (at least if one is justifiably afraid that one might not prevail in the political decision-making as to those acts) and (2) that the consent of the parties concerned is morally relevant. It is not at all clear, however, that there is widespread agreement on either view; to take only one example, the current abortion debate suggests that such a version of political tolerance is not accepted at all by those who oppose abortion. To take another example, the historical debates over religious difference show many more centuries of bloody warfare than years of rational peace. Indeed, there were (and obviously are today) moral or religious theories which hold that it is obligatory to be a martyr in the struggle to establish a religious state, if one cannot win the fight to bring the state in line with one's religious views.

Another option would be to argue that Engelhardt aims only to provide that guidance which he believes is available to us without adopting any particular moral theory, and such guidance consists in the formal structure of a state delineated without reference to any disputable moral constraint. As is argued further below, however, a consent-based system necessarily calls upon moral propositions and culturally based perceptions about the world, both of which are called into question by the post-modern world view.

In 'The body for hire,' we can see that agreement plays a moral role on two levels. First, Engelhardt's position rests on the moral postulate that if consensus as to general moral principles about prostitution existed, the state could legitimately enforce those rules. Thus, consent or agreement would justify state action, if it were universal. Second, however, since consensus does not exist on substantive rules, Engelhardt argues that we should rationally leave the matter to agreement on the individual level. In other words, according to Engelhardt, the only position we can agree on (pragmatically, if not as a matter of moral truth), is that questions about prostitution ought to be decided by the parties to the particular transaction contemplated. I will argue in the next section that the above position does not follow from Engelhardt's premises.

IV. THE FAILURE OF ARGUMENT FROM DISAGREEMENT TO
CONTRACT AND THE FAILURE OF CONTRACT AS A MORALLY
NEUTRAL DECISION MAKING STRUCTURE

Assuming, however, that we accept the *Unprovability Argument* and that we, like Engelhardt, desire to find a peaceful way to coexist with our fellow sojourners and that we agree that some justification is necessary for state action, the question then is whether it follows that the only acceptable position on prostitution is to leave matters to the private agreement of the parties immediately involved, to what the Office of

Technology Assessment discusses in terms of the market. What argument gets us from the unprovability theorem to a private contract position?

It is worth repeating here the core of Engelhardt's position which is that if individuals do not share a common tradition or set of beliefs, and if by secular reason one cannot establish a particular canonical moral vision, then, if one wishes to resolve controversies in public policy by appealing to authority that can be commonly understood by moral strangers in a secular pluralist society, other than by appealing to force, authority for common action must be derived from the actual consent of the actual individuals involved (p.297). As noted above, his arguments which purport to eliminate philosophical proof and tradition as sources of authority do not succeed without the addition of moral premises which are inconsistent with the description of the post-modern dilemma. Even if, however, the general conclusion (that "authority for common action must be derived from the actual consent of the actual individuals involved") followed from the premises about unprovability and consent, that conclusion is not identical with (and, without more argument, does not entail) the freedom of contract or market view.

One reason for this conclusion is that the level of consent varies; on the one hand, we are talking about the authority for joint, coercive, action against an individual, while, on the other hand, we are talking about the individual's consent to a particular transaction. The proposition that consent is the only basis for the imposition of state authority does not logically entail the proposition that private contract is the only justifiable social structure; among other views, the social contract theory (under which consent to the structure of government is the only legitimate basis for state authority) is equally consistent with the proposition that, given consent to the social structure, social force may legitimately be employed to enforce moral strictures such as prohibitions on the sale of organs. One might, to put it more simply, consent to a structure of government which imposed paternalistic limitations on the sort of contracts one might enter.

Indeed, it may be argued that some form of implicit social contract theory is at work in 'The body for hire;' unless Engelhardt envisions contracts which themselves contain provisions for their enforcement by the state, consent to the particular transaction does not alone entail consent to be coerced by the state. That observation leads us to the larger point, which is that leaving matters to individual decision in particular transactions still does not justify coercive intrusion by the state, at least without several additional moral claims, including such claims as "The

state may justifiably use force to require someone to keep a promise."
The mere fact that two or more people have reached an understanding on
something does not entail that the state may or should enforce that
understanding.[10]

Two further comments may be made relating to Engelhardt's solution;
first, that it is not a solution and second, that it cannot be a solution
because it violates Engelhardt's conditions for a satisfactory solution.
These two points are closely related: (1) the relegation of decision making
to private contract is not a solution because it leaves too much
unanswered, and (2) answering the unanswered questions requires moral
premises which in turn violate Engelhardt's other constraints.

Looking back at the argument above, a moment's reflection reveals
that Engelhardt's concluding position (to the effect that all agreed upon
transactions about bodies must be permitted so long as there is agreement
and no innocent third person is hurt) itself contains moral terms, such as
"innocent," which are not in the preceding premises. Where did that
limitation on the power to regulate our lives by agreement come from?
How can such propositions coexist logically with the premises involving
the unprovability of moral propositions. Clearly, unless the exclusion of
contracts which harm the innocent or which are not entered into freely,
for example, can be supported by reasoning which, in Engelhardt's
words, does not involve "religious or quasi-religious" premises, then that
proposition cannot consistently be the conclusion of the argument. In
other words, the argument will fail to meet the criteria which Engelhardt
himself has set if it necessarily includes reference to moral notions.

But, what else are propositions that require that a contract be freely
entered into in order that it be binding or that innocent third parties not be
hurt, if not moral propositions about the dignity of persons? The question
is whether in fact agreement is such a secular value that it does not
involve us in the same kind of religious or quasi religious thinking about
the body that the theories rejected by Engelhardt do. First, we may ask,
"Doesn't everyone agree that at least matters of agreement have moral
standing?" And, to that inquiry, of course, the answer is a flat "no," since
many people (such as those who believe that voluntary prostitution or the
voluntary sale of organs should be illegal) do not believe that agreement
has sufficient moral force to determine the issues about the use of the
body. Second, we may ask what moral force the notion of agreement does
have. Where else would the idea come from that the consent of the owner
of the body, so to speak, had anything at all to do with questions of what

might properly be done with the body, except from antecedent notions of rights in the body?

One might say that the concept of contract is a non-religious legal or secular fact, but to do so would be to slip from a conceptual point about the nature of certain claims to a mere cultural artifact. The concept of the meaningfulness or relevance of consent is not secular in the sense of being unrelated to the idea that one person's consent with regard to a particular body (i.e., the inhabitant of that body at the time) is more relevant to the question of what to do with that body than the views or desires of any other person, such as doctors or the person who would like to take a kidney out of the body in question.

One also may ask "Whose consent counts?" Shall we not say that every party to the procedure must consent to be involved, specifically, both the doctor and the laboratory and the hospital (and perhaps the taxpayers whose money supports the hospital) and the nurses?

But that is not all, for Engelhardt says also that innocent third parties must not be harmed. Who are those innocent third parties? Do they include the poor, but honest, scholar who desperately needs the same kidney that is being sold for millions? If a woman agrees to bear a child for a childless couple, who may legitimately complain? Her husband who will be emotionally shattered by the choice? He may be innocent, and he is certainly a "third party" to the transaction. Her parents who cannot bear to think that "their" grandchild will be raised by "strangers"?

Moreover, we may ask, when does the binding choice, if there is one, take place? That is, when does the contract become binding? Suppose, for example, that Poor Person from the third world has a kidney of the rare tissue type desired by Rich Man. Poor Person agrees to sell his kidney to Man for a considerable sum, but Person later changes his mind. Is there a point at which Man can justifiably demand that the state enforce the contract? If so, when? After the money has been paid? When Man has justifiably relied on Poor Person's promise and foregone another opportunity to get a kidney?

It should be held in mind in considering this issue that none of the traditional justifications for enforcing promises can be used in support of Engelhardt's position, for all such arguments (according to Engelhardt) turn on some moral notion, whether utilitarian or deontic or rights based; if any of those justifications worked for the adoption of rules governing state enforcement of contracts, they could be tried out on the issue of the legitimacy of the sale in the first place, the place where Engelhardt maintains they do not and cannot suffice to provide answers. Therefore, neither the social utility of having contracts be enforceable nor the notion that it would be unjust or inequitable to permit people to walk away from their freely assumed "obligations" nor the notion that people simply ought to keep their word can be relied on in Engelhardt's desert context.

Now one might object, on Engelhardt's behalf, that surely people can agree that a commercial world without contracts would be impossible and that, in their own self interest, if nothing else, they would all agree that contracts should be enforced. But, Engelhardt's theory raises serious objections to such a response. First, it is the *sine qua non* of his argument that substantive agreement on any basis cannot be reached; once one postulates that people would agree to one rule or the other, there is no reason to think that they would not agree to a rule about hiring bodies. For example, if one says, *pace* Rawls, that people would all agree that contracts should be enforced, why should one not say that people would agree that no one should not be permitted to sell their organs; put in reverse, why would someone (who believes passionately that organs should not be sold) agree with a shrug that, well, if one is going to permit them to be sold, then of course, a contract to do so must be enforceable with all the coercive power of the state?

Engelhardt's argument, however, rests on two more fundamental assumptions, so deeply laid in our tendency to agree with him as to be invisible. The first of these is a moral boundary line that the Western liberal tradition draws about our bodies. If anyone is to decide what is to be done with this particular pound of flesh, we hold, then it should be the person who inhabits it. Yet, if, following Engelhardt's criticism of the Office of Technology Assessment's view, we truly let go of any concept of the dignity of personhood, upon what foundation would that boundary line be drawn? What is the strong feeling that the fate of a body ought to be decided by the mind to which it is attached, if not itself a moral preconception based, perhaps, upon the dignity of man or some irreducible notion of personhood? Yet, of course, there could be thinkers who would argue that dominion over the body could rest with others, for example, with those naturally entitled to dominion, such as masters or fathers or husbands or priests, or with those who are best able to decide what should happen to each member of the community or family, such as doctors.

Secondly, 'The body for hire' assumes a division between the individual and society, which is itself indigenous to only certain societies.

Now, Engelhardt could quite plausibly respond to this criticism that highly socialized tribes or peoples could simply be left to work things out for themselves, and that his view is only intended to address the moral issue in the reasonably cosmopolitan, developed world, the society of alienated, atomistic individuals. But, the post-modern view itself includes the notion that social reality is highly constructed and that such a construction is shaped by unobserved preconceptions about the relationship of a person to herself and to others. The relationship of individual to the state or community is, after all, a post-modern thinker would say, itself a matter of to what community one belongs – Engelhardt's argument assumes a certain view of the community which is itself malleable. Although I frankly believe that it may be a sufficient response simply to limit one's argument to its intended audience, nonetheless, one should not underestimate the corrosive power of views that nothing is known, that all the world is phenomenologically malleable, and that our perception is fatally skewed by the community in which we live. The reason that the deeper issue arises at all is that the very question which Engelhardt seeks to answer turns on at least some agreement among his hypothetical debaters as to what the issue is and the constraints upon the answer, i.e., that the state must have some justification for interfering in people's autonomous state. A view of the world in which there simply was no perception of the need for justification nor sense of alienation from the needs of the community theoretically would not find it necessary to pose the question and would not embark on Engelhardt's road.

The observations above were largely phrased in theoretical terms. Those who subscribe to the view that all matters should be regulated by the marketplace may also be asked some questions, however, which stem from the elements of contract law, assuming that by the "marketplace" one means some version of a capitalistic society regulated by contract. The first question is, shall these prostitution contracts be enforcable? The answer may seem obvious on Engelhardt's view, since he refers to agreement on the measurement of consequences by monetarization, but it is not. The law distinguishes between contracts which are illegal and those which are merely unenforceable. It is not lawful to enter a contract for paid sexual services; in some situations, it constitutes the crime of prostitution or pandering. Similarly, it is not lawful to enter contracts to pay a judge to make a particular decision nor to pay someone to kill another person, although slang, in its gallows humor, refers to the latter as "putting out a contract on someone." The mere entering of such agreements is itself criminal and the state affirmatively takes action to preclude either the making or the carrying out of such agreements. On the other hand, there are contracts which are not enforceable but which are not themselves illegal. For example, if I promise to become a Christian or to quit smoking if you pay me $1 million, you cannot recover significant damages if I fail to perform, although you can probably get your money back. More specifically, in the so-called personal service contracts, which are contracts for one particular person to perform a particular action for another, the contract is enforceable to the extent that damages will be awarded for non-performance, but the courts will not award specific performance, meaning that they will not compel the worker to perform the job. So, if I contract to paint your portrait to your satisfaction and decide that you are a despicable dictator and ought not be memorialized, you can recover money damages for the loss of the painting, but the court will not exercise any other power to compel me to paint. Moreover, the courts will not enforce excessive penalties for failure to perform.

Therefore, to agree to leave matters to the agreement of the parties does not fully answer, in the present legal system, the question of what contracts will be enforced by the state, if any. Looking back at 'The body for hire,' we do not find any ready answers and for a very good reason – the "love for sale" view is an entirely negative one which focuses on the exclusion of the state from a particular role in our lives. It does not focus on the state's compelled intrusion into our lives. Thus, for example, on Engelhardt's view, if the state may validly enforce freely entered contracts and if only freely undertaken obligations may be so enforced, what is the obligation of the sheriff to carry out the enforcement of such contracts? Must his duty turn, in each case, on his actual agreement with the proposed transaction? If not, why not? What is the source of his duty which does not turn on actual consent to a particular transaction? We could say that he consented to assume a particular role but only if, as noted above, we import the existing machinery of courts and contract law and lawyers and sheriffs into our consensual state. Suppose that an Engelhardtian kidney transplant sale contract provides for forcible enforcement by private thugs. Ought the sheriff to protect a welshing party to the contract from the thugs who show up at his door, bent on dragging him unwilling to the operating room? Well, one may say, "Surely the state can also act to prevent people from being forced by other individuals to do or not do something." But, what moral principle legitimizes the state on the post-modern view in coercing someone not to coerce us? There are perhaps two ways to get there – one way that appears free of most moral presuppositions is to adopt a little Kant and hold that one who treats himself as a free moral agent must logically allow others the same freedom (and then, of course, one needs the premise that "it is morally permissible for the state to prevent someone from failing to act as a consistent moral agent.") But, Engelhardt has told us that Kant's view is laughably connected to prohibitions on masturbation, and therefore not plausible as a positive content-full moral theory. Besides which, such quasi-Kantian reasoning would not apply to someone who did not accept the moral premise that the state may coerce people who do not respect other persons as moral agents. It does not need a detailed reference to the history of ideas to observe that one of the many things that people have failed to agree on over time is who counts as a moral agent, a person entitled to be treated like oneself – blacks, Indians, "natives" of all sorts, women, and persons under the age of 18 or 21, being only a few examples of excluded groups that come immediately to mind.

Now, the question of whether the state can morally coerce people to abide by their contracts under Engelhardt's view arises again. The answer is again in doubt and appears to be "no." First of all, of course, no good justification can be given for any moral content-full moral proposition such as "promises ought to be enforced." And, of course, some such moral premise is necessary to justify the state in acting coercively.

Note that it does not matter that one of the persons contracting agrees, implicitly or explicitly, to state intervention, because the propositions "promises ought to be enforced" or "consent licenses coercion" likewise are content-full moral propositions as to which there is no agreement. Engelhardt tries eloquently to derive some guidance from the absence of guidance, to make the absence of universally accepted justification for coercion into a moral imperative for tolerance up to the margin of contract. The meaning of "everything" is, unfortunately, everything; if everything is permitted, oppression and tolerance stand on an equal footing.

In the context of social policy relating to the sale of body parts, Engelhardt describes our position as being that we must decide both as to our own conduct and as to the policy of the state toward ourselves and others, if only to avoid the coercive power of the state being used against us in a way to which we object, while we cannot justifiably impose any particular theory on others. In other words, because we consider here a social policy, clearly we are deciding at least for everyone in our society. Engelhardt, however, makes a heroic effort to point out to us a third choice, a choice that is not an unjustifiable imposition on other persons, a choice that avoids, therefore, the dilemma of either not choosing or of choosing tyrannically. The only option which will avoid that tragic dilemma, according to Engelhardt, is to choose not to choose, that is, to choose to allow persons to do as they please and to refuse to guide them in that decision. Thus, one can try to avoid some portion of despair by finding a path which does not lead to tyranny, at least.

V. ENGELHARDT'S RESPONSE IN OTHER WORKS

In *Foundations* and *Secular Humanism,* Engelhardt offers a somewhat more modest, but also more complex and powerful, account of bioethics, drawing on philosophical strategies of Kant, Hegel, and, I would argue, Rawls, Nozick, and Hobbes. His hope in *Secular Humanism* is that "Secular humanism is the attempt to justify and elaborate a common moral framework grounded in what we share as persons" (1991, p. 138). In *Foundations*, he asserts that "Mutual peaceable negotiation emerges as the linchpin of public authority in general and of authority in health care in particular" (1986, p. 44). He later claims that

> Even with the failure to establish generally the authority of a particular moral sense, one can still have authority for common actions in pursuit of particular moral goods. One can secure a justification for moral judgments for certain biomedical policies. The authority is that of common assent (1986, p. 45).

Although this essay is directed at the structure of 'The body for hire,' I would assert that my arguments apply equally to *Foundations* and *Secular Humanism* to the extent that those works attempt to claim that the specific conclusions rest on premises which are exempt from the post-modern dilemma postulated by Engelhardt in 'The body for hire.' To the extent that *Foundations* or *Secular Humanism* rest on the assumption of a community of persons who have some underlying values (or at least desires) in common, such perhaps as the desire for the peaceable community and the desire to accord to others the same liberty of decision-making that one accords to oneself, then the arguments in this paper would apply only in modified form. For example, for anyone who has worked in concrete bioethics contexts of actual decisionmaking, it is hard to disagree with Engelhardt's assertion in *Foundations* that

> Patients as persons thus meet others who as persons possess different views of proper conduct and of the good life. Patients, physicians, nurses, and other health care professions must decide how they will cooperate with each other in common understandings and undertakings. The context of health care is an arena where an important community of understanding must be fashioned (1986, p. 243).

However, to the extent that Engelhardt claims that his principle of justice (as including the principles of autonomy and beneficence) can be sustained as "principles for resolving moral disputes among individuals who do not share a common moral vision" (1986, p. 85), I continue to disagree, not with the substantive content so much as with the meta-ethical claim; those principles (and the felt need for them as justification for social action) rest, I would argue, expressly upon common moral understandings.[11]

VI. CONCLUSION

In essence, then, it may be said that the Engelhardtian post-modern view is to philosophy as the mythical all-dissolving acid is to chemistry. The old joke shows a white coated scientist holding a flask of "universal solvent" or "all-dissolving acid" and asks "What is wrong with this picture?" What's wrong, of course, is that the universal solvent or all-dissolving acid would destroy its container. Flask, hand, scientist, white coat, table, and laboratory would all be eroded if the discovery were truly made; similarly, with regard to the post-modern skepticism, the philosopher cannot contain this perverse philosopher's stone in a glass-walled vial of almost invisible preconceptions and hope. It should not be overlooked that Engelhardt expressly disavows not only the desire but the possibility of providing any answers to the prospective parties to a contract about what, if any contract, they ought to enter. If the person desiring to buy a kidney asks, "But is it right?" the only answer is "No answer can be given." Notice that the response is not "Whatever you choose to do freely is right" or "Whatever involves two freely consenting human beings is permissible." The only response that issues from the mouth of the post-modernist can be, at most (and only if Engelhardt is correct), "Whatever involves two freely consenting human beings ought to be exempt from state coercion." After that, we are left to our own devices or lack of them. Either we turn to a tradition, which can have no more rational standing than that which we are accustomed to give it, or we abandon all our futile attempts at arriving at the morally correct decision and base our decision on desire or chance or whatever comes to hand. In other words, our despair of acting, in our own personal lives (as opposed to as members of society), as agents both moral and rational, then is the true post-modern despair, the grandnephew of the Sartre and Camus' existential despair. We desire to act rightly, to think that our actions have moral significance, but if we are Engelhardtian post-modern rational agents, we should know that they do not or, at least, that if we do, we shall never be able to convince those who *a priori* disagree with us. If every proposition in Engelhardt's paper were true and were fully demonstrated, the best we could hope for is that we can sometimes in our public lives act morally; the policeman who refrains from stopping the surgeon who wishes to perform a kidney transplant for hire has acted, one may assume, morally or, put another way, has refrained from acting immorally. But, I argue, the respect for free choice, the impulse to limit state coercion to those circumstances where we can justify it to all rational members of the community, the sense that mere power is not a sufficient justification for the use of social force to constrain individuals in the use of their own selves, even the preference for a peaceable consensual community over a state of social conflict are all moral or ethical choices, in whole or part.

In 'The body for hire,' Engelhardt first draws for us a dismal map of the post-modern labyrinth, complete with dead ends, pitfalls, and darksome tunnels, where every turn seems to lead the well-meaning wanderer deeper toward the fearsome lair of moral nihilism, a monster begotten by cultural diversity upon rationalism. Engelhardt tells us that there is no weapon in the intellectual armory which will slay that monstrous nihilism. Just as we are about to despair of ever resuming to the sunlight of any clear moral decision, Engelhardt holds out to us the Ariadne's thread of consent, promising to lead us surefootedly out of the labyrinth. We want to cling to that thread because it promises to lead us out of our dilemma, to allow us to establish some societal norms without unjustifiably imposing our own merely personal moral code on unwilling others or having them impose theirs on us.

I have argued in this paper that the thread is too frail a marker and there is no such easy path, that the notion of consent is itself a choice of a particular moral and societal point of view. To leave decisions to the marketplace is to pretend that one is not choosing, that the market is a handy social mechanism that exists without our contrivance and to which we may entrust decisions that we decline to make. In fact, that mechanism is a choice of its own and our visible hands are constantly adjusting and guiding it. If that is so, then, I would say that we cannot help but decide and that we must decide as best we can, using our moral intuitions, our religious concepts, our commitments to moral propositions about tolerance, consent, and freedom, and our perception of the social reality, as carefully as we can, not in the placid certainty of being right but in the knowledge that our decision is necessarily an uncertain one.

Houston, Texas

NOTES

[1] It is not an important part of my arguments in this paper, but I believe that the principle of avoidance of harm to innocent third parties may be an implicit corollary of the enforcement of agreements premise: *ex hypothesi*, an innocent third party is one who is not a party to the agreement and, based on Engelhardt's other arguments, no action can be taken against or by those who are not third parties.

[2] Although I believe that the above schema is fair to Engelhardt's reasoning, to save space and focus on substance, I have not elaborated on my reasons for attributing these views to 'The body for hire,' but have simply cited the text on which I relied.

[3] His goal will be critically to assess a failed attempt to justify restrict law bearing on the rental, lease, or sale of human body parts.

[4] Only four options are considered in 'The body for hire' and Engelhardt does not suggest any others (see 1999, pp.277-278).

[5] "If individuals do not share a common tradition or set of beliefs, and if by secular reason one cannot establish a particular canonical moral vision then, if one wishes to resolve controversies in public policy by appealing to authority that can be commonly understood by moral strangers in a secular pluralist society, other than by appealing to force, authority for common action must be derived from the actual consent of the actual individuals involved" (1999, p.297).

[6] In fact, I would say that Engelhardt finds himself precisely in the position occupied in the 1970's by liberal opponents of Critical Legal Theory. When a person adhering to a strong version of critical legal theory advocated, for example, the use of force to dampen dissent (or cynical political maneuvering to gain power in a law school faculty), the liberal (who

simultaneously espoused tolerance and governance by rules but believed in the post-modern position that no moral view, including tolerance and governance by rules, can be demonstrated) was at a loss. Engelhardt's attempt to extricate our hypothetical liberal from his dilemma gives his essay a larger importance than its narrow topic might suggest.

[7] Or, of course, which follows from Engelhardt's reasoning.

[8] It may be remarked that this demand for "provability" is like the skeptic's demand for certainty and subject to the same objections.

[9] Indeed, conversation with Dr. Engelhardt suggests that he is sympathetic to this interpretation of his position, and it is supported by the condition precedent in his argument that force is not an adequate justification.

[10] The missing proposition is, of course, "The state may justifiably use force to coerce obedience to contracts."

[11] I do not consider here whether such views could be grounded on rational self-interest in the Hobbesian or Nozickian manner. I would note though that even those arguments rest upon the individual ranking self-preservation or self-determination ahead of other values such as the value of a religious state.

BIBLIOGRAPHY

H.T. Engelhardt, Jr.: 1986, *The Foundations of Bioethics*, Oxford University Press, New York.

H.T. Engelhardt, Jr.: 1991, *Bioethics and Secular Humanism*, Trinity Press International, Philadelphia.

H.T. Engelhardt, Jr.: 1996, *The Foundations of Bioethics*, second edition, Oxford University Press, New York.

H.T. Engelhardt, Jr.: 1999, 'The body for fun, beneficence and profit,' in M.J. Cherry (ed.), *Persons and Their Bodies*, Kluwer Academic Publishers, Dordrecht, pp. 277-301.

THOMAS J. BOLE, III

THE SALE OF ORGANS
AND OBLIGATIONS TO ONE'S BODY:
INFERENCES FROM THE HISTORY OF ETHICS

The sale of organs for transplantations raises thorny moral issues. Are there substantial obligations to one's body which render this sort of commercial activity morally unjustified? Or are we free to sell our organs? I restrict myself to the simple case of the sale of a functioning organ that is not needed by the seller, such as a kidney. The question then becomes: "Are there rationally conclusive reasons to think that I have obligations to my body that should constrain me morally from selling a kidney?"[1]

The Federal Government says so, and it assumes that it has moral warrant to enforce these obligations by prohibiting the purchase of kidneys, and other specified human organs, for transplantation.[2] Moreover, a 1987 report to the U.S. Congress by the Office of Technology Assessment assumes that, if one must manifest rationality in and through the body, then there are certain obligatory ways of doing so, and these prohibit commercial sales of body parts.[3] Both the law and, to a lesser extent, the OTA's advice, make assumptions of what is morally good that claim to be cogent simply in terms of secular reason. Otherwise, the law could not claim what it does claim, viz., to be normative for citizens across the spectrum of a secular, pluralist society.

Are these assumptions viable? Are there rationally conclusive reasons to constrain ourselves from selling organs that we do not need to function? I shall argue in the negative. I frame my argument in general philosophical terms, but I shall examine the moral philosophies of Aquinas, of Kant, and of Hegel. Thomas and Kant provide the argumentative frameworks for reservations about and for outright opposition to, marketing body parts. Attending to the defects in their arguments points to the need for an alternative account, which Hegel provides, and which allows us to endorse in principle a market in kidneys. Contemporary discussions about the moral licitness of organ sales usually do not seriously engage figures from the history of ethics. My focus will, I hope, show both the theoretical and the cultural relevance of classical moral thinkers upon this contemporary bioethical issue.

M.J. Cherry (ed.), Persons and their Bodies: Rights, Responsibilities, Relationships, 331–350.
© 1999 *Kluwer Academic Publishers. Printed in Great Britain.*

I. AQUINAS

For Aquinas the human being is a rational animal who can only exercise practical reason in and through the body. One is therefore obligated to respect the body's integrity and to develop it so that it, in its appetites and desires, conduces to moral virtue. The basis of these obligations is a notion of objective good: the intelligible structure within being,[4] which is articulated with respect to human beings as the natural law (*Summa Theologica* I-2, q.94, a.2c; cf. *ibid*, q.91, a.3c, and q.58, a.4c.) The natural law is what the contemporary Thomist John Finnis calls the set of "conditions and principles of practical right-mindedness, of good and proper order among men and in individual conduct" (1980, p. 18). It is constituted by the first principles of good for human practical reason.[5]

I contend that no conclusive reasons of a general non-sectarian kind can be given to agree to such a principles of practical right-mindedness, because no such reasons can be given for any determinate content of the good life. If this contention is correct, no notion of objective good can provide a rationally conclusive basis, a basis for the moral authority of law across a secular and pluralist society such as ours, for obligations to one's body. Before examining this contention, however, consider first the relation in Thomism between natural law and obligations to one's body.

The principles of the natural law are supposed to be self-evident, underived and indemonstrable, the primary ends of action, the ultimate premises to which reasons must appeal (cf. *Summa Theologica*, I-2, q.91, a.3c: per se nota). Finnis elaborates them as the irreducible and equally fundamental highest values to be achieved in human action (Finnis, 1980, pp. 90-92; cf. pp. 59-90): life, knowledge, play, aesthetic experience, friendship, practical reasonableness, and religion. If the moral man is supposed to act to achieve good ends, they are these ends. They are what it is good to do. They are the activities or functions which one should perform well, and of which one should develop the virtues.

Aquinas thus gives[6] the determinate content of the good life for man. These are not themselves moral principles, but "pre-moral"[7] first principles of practical reasonableness. They are the ends to which the man of morally right practical reason, Aquinas' prudent man, must be well disposed in order to act morally.[8] If one is properly disposed, he will realize that he is always morally obligated to act in ways which respect these first principles of practical reason.[9] Most basically, then, evil is to be avoided; so any action whatsoever that is directed against these principles in men is absolutely forbidden. But also, good is to be done and pursued; so all actions that promote these principles in men are to be encouraged.[10]

On such an account, how is one obligated to one's body?

Since human life is one of the alleged primary goods or highest values for practical reason, and one lives in and through one's own body, one has obligations to preserve that body's organic functioning so far as possible. The individual organ has value insofar as it is useful for the good of the organic whole. The good of the organically whole body is the regulative norm for what it is morally licit to do with individual organs. One may, Aquinas thinks, amputate a limb for the sake of the health of the whole organism; but one may not cut away a sound bodily organ. It is "altogether illicit" to "mutilate" one's body (*Summa Theologica* II-2, q.65, a.1c).

On its face, Aquinas' reasoning seems to prohibit the sale of one of our two healthy, functioning kidneys. However, medical science now knows that not all of one's organic parts are necessary for the body's organic functioning, and that one of our kidneys is normally dispensable. It is not the case, as seemed to Thomas, that the good of the organic whole requires both kidneys; and the danger of one failing, so that the other is needed, is relatively minor. If one kidney is not needed for bodily functioning, it seems to follow that one could, for sufficient reason, such as to provide it to someone who vitally needs it, have a sound kidney removed. As the contemporary Thomist Joseph Boyle concludes, it is morally licit to "give up a kidney which is not needed" and assume the risk that the one remaining kidney will continue to function normally, "for the sake of helping another," because one is not thereby sacrificing function (Boyle, 1999, p. 134). One can distinguish the good of the organically functioning whole from the good of the intact whole with redundant organs.[11] The Thomist position in principle requires one to respect the former, not the latter.

With respect to a commercial market in kidneys, then, if one can give away a kidney for sufficient reason, one can in principle sell it for sufficient reason. As Boyle points out, the potential kidney donor may have morally compelling reason to sell, e.g., to get the financial resources to purchase life-saving treatment for a family member; so that his sale of the kidney saves not only the purchaser's life, but another's. And the purchaser will usually have sufficient reason, e.g., an additional period of good-quality living with the transplant, to justify the purchase (Boyle, pp. 135-138). The Thomist has no intrinsic objection to marketing kidneys.

Nonetheless, I want to object, why is the organically functioning whole to be respected, i.e., why ought it to be maintained by one with practical reason? Is it an obligatory good, to be sustained by others even if the human being is incapable of consciousness, as in the case of permanently vegetative patient or the anencephalic? Is life an obligatory good if function is radically diminished? Is one forbidden to improve functions by technological manipulation, e.g., via drugs or hormones, or by genetic engineering? Or radically to alter this or that function? No rationally decisive answers seem to be available to these questions, and reasonable people can choose to differ based on different assumptions and goals. What practical reason dictates in point of content as obligatory on anyone with it, is not clear. *A fortiori*, what it dictates as obligatory conduct towards one's body is not clear.

My point is that Aquinas cannot provide a rationally conclusive basis for obligations towards one's body. The first principles of practical reason – life, knowledge, play, aesthetic experience, friendship, practical reasonableness, and religion, as the supreme values to be pursued by practical reason – are supposed to constitute irreducible determinations of the objective good for practical reason. That they – or, indeed, any canonical ranking of goods – are what any rational person ought to pursue, is not self-evident. Why should we not be able to trade off one of these determinations for others, or all for something else? Why should one not, if he is going to die soon, give, or sell, both kidneys, if friends or family will be decisively helped thereby, even though he thereby shortens his life? Again, why not commit suicide rather than endure further life with, e.g., too much pain to function productively, even though surcease of pain is not one of Aquinas' *per se* goods? Aristotle, like Aquinas, thinks that life is a self-evident good. But he also allows that pain can be evil.[12] Imagine a life with so much pain that the person can no longer function in the state or in worshiping the gods. In this case, Aristotle's framework may allow – at least in a secular pluralistic cosmopolis – the person to commit suicide or commission euthanasia. It is not self-evident that the determinate contents of the good for practical reason are what Thomas alleges them to be.

I want now to contend that there are no sound arguments that can be given in behalf of any canonical ranking of objectives goods, i.e., in behalf of any particular ordering of objective grounds of moral obligation, not only in Aquinas' case, but in general.[13] If there is no canonical ranking of objective goods, then there is no rationally conclusive ground for a society to rank and institutionalize the goods of, e.g., liberty, equality, security, and prosperity relative to each other. More radically, if my argument goes through, there would then seem to be no moral authorization for a society to impose one good or set of goods rather than another upon members who peaceably dissent and are innocent to any prior commitment to whatever organized society authorizes.[14]

One cannot appeal to natural law, or indeed, to any structure of reality, even one conceived teleologically by analogy to the practical reason and purposive behavior of man or of a divine creator.[15] To do so, one must know whether each of these purposes is to be viewed as morally instructive, and how instructive (e.g., as to be eschewed, or to be pursued as obligatory, or as supererogatory), and why instructive. And to know these things, one must have a standard by which to judge these purposes morally good or bad. But if one needs a standard by which to judge natural law (whether *in toto* or in any part), natural law cannot provide that standard. Nor can appeals to self-evidence. What is claimed to be self-evident is supposed to be intuited as so, and intuitions may differ. To appeal to intuition, then, is to forsake giving argument that what is intuited is in fact as one's intuition delivers it; and where the deliverances of intuition can differ, sound argument is called for. Thus, it is not clear why one should think that all of the primary determinations of Aquinas's good are equally fundamental and irresolvably co-obligatory, and that certain considerations such as permanent absence of any cognitive function, or too much pain, can never licitly outweigh the *prima facie* good of life. Intuition cannot establish an objective ground for moral obligation, or for obligations to one's body in particular; arguments are needed.

However, the arguments cannot appeal to the consequences of certain sorts of action as the basis judging their morality. For the consequences would themselves have to be ranked in their importance for morality. And such a ranking presupposes the very standard of moral good which the consequences were to establish.

Nor can one appeal to the idea of an impartial judge, whether the knower of the true idea of justice such as Plato's philosopher-king, or the modern version, which stresses lack of bias, such as Rawls' hypothetical contractors. Either there is some particular ranking of the goods or a particular morally obligatory action or set of actions that the judge has in mind as the standard of moral obligation, or there is not. If there is, this ranking is assumed, not established, as the canonical standard; if there is not, the impartial judge has no basis for judging whether actions are obligatory. If the hypothetical contractors take some "thin theory" of the good to be canonical, this canon is assumed, not established; if they do not, they have no basis for making further specifications of the contract.

I infer that no view of the objective good – no determinate content of what is good for practical reason, and therefore for anyone with practical reason – can be established in rationally conclusive terms that do not themselves appeal to some determinate content. No view of the objective good can provide a rationally cogent basis for moral obligation. Consequently, no notion of obligations to one's body can be given in terms that are morally authoritative for a non-sectarian pluralist society, if the basis for these obligations is a particular notion of objective good.[16] (I am referring, of course, to obligations to one's own body. I am abstracting from obligations I may have concerning others' bodies, e.g., not to assault them, as well as from actual, contractual obligations I may incur. These obligations, I shall argue in the next section, can be conclusively established, but only in terms of the subject *qua* moral agent, not in terms of objective good.)

II. KANT

Kant's moral philosophy is worth examining at this point for two reasons. First, he successfully explains moral obligation in terms not of an objective good, but of the subject's reason insofar as it can control the will. Second, he thinks that this explanation implicates obligations to one's body. I shall contend that Kant is successful in the first endeavor but not in the second.

Kant explains moral obligation by making it coincide with the principle of the obligated subject's free will, i.e., with his reason. With this coincidence one can not understand why the subject is in principle free with respect to what he is morally obligated to do, rather than, as in the case of an objective ground of moral obligation, being unable to explain the compatibility of moral obligation and the obligated subject's freedom. Kant explains moral obligation in terms of reason, insofar as it can control the will. We are morally obligated precisely because we have free will.

What is it, then, that we are morally obligated to will (and do, because the will is the principle of responsible action)?

For Kant we are morally obligated to will ends that can as well be willed as "universal law", i.e., as normative for any rational agent (*Grounding for the Metaphysics of Morals*, AK402). The "fundamental law of pure practical reason", Kant tells us, is "so [to] act that the maxim of your will could always hold at the same time as a principle establishing universal law" (*Critique of Practical Reason*, AK30). The maxim[17] of action must be defensible in terms of reason, and hence in terms which are in principle accessible to any rational agent. Otherwise, Kant thinks, one is consigned to the sensibly experienced world, the *mundus sensibilis*, which is completely determined causally without room for freedom, or therefore for moral obligation.

Kant is wrong to think that the nexus of physically caused events cannot be used by rational agents to achieve their own, freely posited ends. I do as a matter of fact coordinate, discipline, order, and sublimate my impulses and inclinations; so I can use them to achieve ends I set for myself, e.g., to appreciate art or to achieve salvation, or to master a theoretical discipline.[18] The ends that I will are indeed, to some extent, irreducibly constituted by these impulses and inclinations, at least as coordinated, ordered, and/or sublimated. They are ends from my particular point of view. But then they cannot in principle be willed as a universal moral law, because no particular point of view can be soundly argued to be normative for any rational agent. Conversely, an universal moral law would have to abstract from all particular points of view, and so would not have any content.

If, however, no contentful moral law can be shown to be obligatory upon all rational agents, how can Kant have successfully explained moral obligation? Not by articulating a positive blueprint for generating moral laws, because he cannot positively determine content that would be normative for the will of any moral agent. Rather, it is by making reason respect that will, i.e., respect the freedom of any moral agent who has not consented, and thereby obligated himself, to some arrangement, and is innocent of having set a precedent for the use of force against him, e.g., by himself using force against an unconsenting innocent moral agent. That is, unless I can give sound argument for constraining another moral agent, I must respect that person's freedom.

But what objective content does this constrain me to will? One cannot say, with Kant, that you should "so act that the maxim of your will could always hold at the same time as a principle establishing universal law," for one cannot establish universal law apart from particular contents willed; and one cannot show that any particular content is normative for all moral agents. Can one then say with Kant that one should act so as to treat any and every actual rational being, as Kant defines the person (*Grounding for the Metaphysics of Morals*, AK428), whether oneself or another, "always at the same time as an end and never simply as a means"?[19] Yes. Concerning other persons, I am obligated to respect their freedom. I cannot, e.g., injure, steal from, or coerce unconsenting innocent persons. If I coerce unconsenting innocents, I would be using them for my end in a way that violates their status as free and morally responsible, because I cannot give cogent reasons why they should have their freedom forcefully subjected to my will. My end would thereby also violate the principle – reason as controlling will – in virtue of which I can set my own ends and be responsible for doing so. It would violate the logical condition of free and morally responsible willing (and acting). I am not thereby positing freedom as a positive good; there is no constraint upon one freely indenturing himself, e.g., to the French Foreign Legion. Rather, I am observing the precondition for one being morally responsible for what he or she does. The only content-filled ends that morally constrain innocent moral agents are those to which they consent (perhaps as a result of mutual negotiation).

   This conclusion dovetails with the argument of the previous section that no moral end can be established as unambiguously canonical, morally authoritative for innocent moral agents who have not freely consented to it. One cannot, then, say with Kant that as moral one belongs to a kingdom of ends, because no moral authority can be given to the sovereign of such a kingdom (cf. *Grounding for the Metaphysics of Morals*, AK433-435). Kant's kingdom of ends must be a cooperative community if it is to have moral authority. For no content can be established by reason as morally authoritative, if no end can be established by reason as canonical. Nor, indeed, can any procedural morality be thus established; for such procedures are designed to reach certain end-states, and no end-state can be established. Both content and procedure are particular actualizations of free will. What I am obligated to respect, however, is that reason's control of will makes freedom possible. How I ought to use it in my case cannot be said *a priori*: it may even be that in some circumstances I can, or even ought, to extinguish it, e.g., by suicidal sale of my body parts. But what can be said *a priori* is that I must respect the freedom of other moral agents, if they are innocent and cannot be presumed to consent to what I would do with them.

In general, then, I can know what I ought to will and do is to respect freedom in myself and in others. But this amounts to a second-order determination of what I ought to do; it means only that I must respect other moral agents insofar as they are innocent and unconsenting to my ends, and I must respect myself as morally responsible because free. In general I should keep promises, fulfill my obligations, and strive after the good; but what particular promises to make, what particular obligations I have, and what particular goods I should seek, cannot be articulated apart from the particular social contexts I happen to find myself in.[20]

Kant, however, thinks that particular obligations do follow from the categorical imperative, including obligations to one's body: "Man is not a thing and hence is not something to be used merely as a means; he must in all his actions always be regarded as an end in himself. Therefore, I cannot dispose of man in my own person by mutilating, damaging, or killing him" (*Grounding for the Metaphysics of Morals*, AK429). Kant's reasoning here is similar to that concerning duties towards others. I have the obligation to value as a positive good, and therefore as an end to be willed, the intact existence of persons, whether in the case of myself or in the case of others.[21]

The similarity, however, trades upon the ambiguity between freedom as the logical condition for recognizing moral obligation, and freedom as an obligatory good, a positive content of the good life.[22] The former is the logical presupposition for one to be held morally praise- and blame-worthy; it is the condition for anyone to be held morally responsible for his agency. Freedom as a positive good, by contrast, is claimed to be obligatory upon any who exercise moral agency. Kant establishes freedom as the condition for understanding moral responsibility. But his derivation of obligations toward the body depends upon freedom as a positive good.

Specifically, Kant thinks that the condition of being moral obligates one to will to live freely rather than commit suicide to avoid pain, or sell a kidney to obtain the resources for pleasure. For in these cases, Kant thinks, I am letting myself be used as a mere means to avoid pain or achieve pleasure. "To deprive oneself of an integral part or organ (to mutilate oneself), e.g., to give away or sell a tooth so that it can be planted in the jawbone of another person, or to submit oneself to castration in order to gain an easier livelihood as a singer, and so on, belongs to partial self-murder," he says (*Metaphysics of Morals. Doctrine of Virtue*, AK423), for such an act is supposed to involve the same contradiction as self-murder or suicide, viz., "to destroy the subject of morality in [one's] own person as a mere means to some end of one's own liking" (*ibid.*). Moreover, Kant implies, if one sells any body part, even one that is not necessary to one's organic functioning, one is using his person as a mere means for financial gain. "[S]elling one's hair for gain is not entirely free from blame" (*ibid.*), Kant remarks. Even though only one functioning kidney is necessary to a human's organic functioning, we are still supposed to be obligated not to sell the unnecessary kidney.[23]

Kant's contention is that there is a certain end, or content of the good, that I cannot contravene without violating the very condition of moral obligation. In this case, the content of the good consists of continued life as a free and organically functioning rational animal. But Kant here conflates freedom as a positive content of the good life, with freedom as the presupposition for being held morally responsible for one's actions. I am constrained by freedom as a presupposition from acting so as to interfere with the freedom of other innocent persons. But I am not constrained to recognize freedom as a positive contentful good, because no such good can be rationally established as obligatory.

Kant is even wrong to think that, if I consent to allow myself to be used as a means, e.g., to provide someone else with a kidney for a suitable price, I am a means merely. For if I consent to allow myself to be used as a means to an end, I am to that extent assenting to the end and setting it as my own. Consequently, if I will an end that does not embrace freedom as a positive good, I am not violating the condition in virtue of which I and every person is free and morally responsible. We would all still be a part of Kant's 'kingdom of ends', i.e., the peaceable community of persons who respect each other as moral agents even though we do not recognize the same goods.

Moreover, if you act on the basis of the conflation, and thereby think that you are morally authorized to force me to will freedom as an end, the use of force violates the condition of moral responsibility, because it forces another free, innocent moral agent to do as you will. Since you have no rational compelling argument for the particular good, you have no basis to require me as rational to will it. You have no moral justification to force me.[24] Morally, it is only licit to try to persuade me, peaceably.

Kant thinks that he can derive particular obligations toward one's body from the general principle of moral obligation, because he conflates freedom as a side-constraint with freedom as a positive value. Respect of, and constraint by, the freedom of any innocent person is the condition for recognizing a community of persons, a peaceable moral community. But this constraint is quite different from being constrained to will freedom as a positive good. One should not encumber Kant's discovery, the discovery of the harmony between moral obligation and the subject's freedom, with his unsuccessful attempt to derive therefrom particular obligations to the body.

III. HEGEL

Hegel is of interest, because he accurately appreciates both the strengths and the weaknesses of Kant's position, and because he thinks he can nonetheless develop content-filled notions of objective good from the notion of the person as free and yet responsible for the morally good. I shall argue that he succeeds in doing so, but in a way that requires not rationally conclusive arguments, but only the free contractual arrangements of the market. He thus provides the rationale by which market arrangements for kidney sales are shown to be not intrinsically immoral.

Hegel's *Philosophy of Right* gives a theoretical reconstruction of crucial explanatory concepts – categories – for 'Objective Spirit', i.e., for the world constituted by humans' exercise of free will. Hegel may be read here as showing logical inter-relationships among concepts that explain the will as both really free, i.e., as more than mere wishing, and as rational, i.e., concerned with normative practical goods.

Kant's principle of morality is a concept of will that is rational but only subjectively free; it is rational insofar as it can query the moral rightness of any willed good; but the query remains within the reflection of the subject. Such a principle presupposes a stage at which the will can be exercised and realized in the world. For, unless one's will can be understood as enacting freely chosen ends, and thereby claiming a right to the means to achieve these ends, one cannot understand how there can be conflicts between one's claimed rights and what is morally right. At this stage individual practical agents are considered as moral strangers who realize their freewill by claiming and using things – and most immediately, their bodies (*Philosophy of Right* ##47-48) – as means to achieve their ends. They would be acting legitimately, so long as the things cannot set ends for themselves and have not been claimed by another. The practical agent's appropriation and use of them for his ends seems to establish his right to them. Hegel here treats the concepts of property, individuals' claims of rights, conflicting claims of rights, violations of another's right by, e.g., taking his property, the general question of right vs. wrong, and the question of what is right in itself, independently of my will, i.e., of what ought I to will This question permits one to understand the Kantian notion of the will as free not insofar as it is realized in objective particulars but insofar as it is expressed in the subject's rational reflection upon what ought to be done morally.

Having treated the concepts presupposed by Kant's account of morality in terms of free will as rational but only subjective, Hegel then points out why the Kantian account must itself be complemented. Rational will cannot be understood as sheerly subjective; it must also be understood as objectively realizable; otherwise it would be no different from reason's wishes. Moral obligation as such is abstract. At the level of Kantian morality one knows that he is obligated to recognize the legitimate rights of others, but he cannot know what these are in any particular case. One knows that he ought to do what's good, but what that concretely involves, he cannot say. "Because every action explicitly calls for a particular content and a specific end, while duty as an abstraction entails nothing of the kind," Hegel observes of Kantian morality, "the question arises: what is my duty? As an answer nothing is so far available except: (a) to do the right, and (b) to strive after welfare, one's own welfare, and welfare in universal terms, the welfare of others" (*ibid.*, #135). But this determines no normative good at all; and unless normative goods can be determined, Kant's moral philosophy would be vacuous, as would its centerpiece, Kant's explanation of the harmony of moral obligation and freedom in terms of their common root in the will as rational.

Hegel therefore points out that various social unities, such as the family, civil society and the state, are constituted to realize certain practical ends, or goods. Within these unities particular rights and duties arise and are given content, contractual agreements and obligations are enforced, and welfare and beneficence are specified. And, certainly in cases where the unities are comprised of persons who freely subscribe to the ends which the unities are designed to realize, these ends would be normative for those persons. Hegel thus complements Kant's account with categories of societal groupings that are unified to realize certain practical ends. At this stage Hegel thinks that he can articulate structures of free will that are both objective and rational. He first categorizes the family and particular communities which share a particular, content-filled notion of the good, because it is within these groups that particular obligations and rightful claims occur. He then gives categories of civil society, construed as the market economy and the institutions it generates for its effective operation, e.g., civil courts and enforcement procedures. The market allows cooperation for mutual benefit among groups that need not have a shared notion of the good. Finally, Hegel recognizes the state as that societal structure within which freedom is said to be concretely realized (*ibid.*, #260).

The state may be viewed as providing a neutral framework within which various individuals and communities can, even if they are moral strangers to each other, live peaceably with one another. It would derive its moral authority not from a disputable claim to embody a common good which claims to obligate innocent persons who have not assented to it, but from the procedural framework it provides for adjudicating disputes and settling grievances between various communities and individuals, and distributing resources owned in common, and for providing Constitutional safeguards to privacy.

Hegel, in the Aristotelian tradition, wants more. The state is supposed to institutionalize a political union for securing the common good (*ibid.*, ##257, 258, 260).[25] I have suggested that no particular content can be established with a rational conclusiveness that would authorize enforcement of such a good across a secular pluralist society constituted largely of moral strangers, because no rationally conclusive ordering can be given for a canonical ranking of the priority relative to one another of, e.g., liberty, equality, security, prosperity, and any other likely candidate for the objective good. The state's enforcement of a particular notion of the objective good, therefore, even if it is authorized by a clear majority, would seem to violate the moral autonomy of unconsenting, innocent moral strangers. These considerations suggest that the state has moral authority only to be amoral, i.e., not to enforce any particular moral standpoint. Civil servants should be the "universal class", providing services that are neutral to the particular moral vision of those it serves.

Nonetheless, Hegel indisputably does show how the market economy and its arrangements provide a morally authoritative unity to a secular pluralist society of communities which are in ever so many other ways moral strangers to each other.[26] The market not only provides the efficient satisfaction of needs and production of wealth. It also secures the means for social interaction between parties who do not share the same moral framework. Each party, in pursuing its own separate aims in agreement with some others, binds the society together in a morally authoritative whole. For even though the economically cooperating parties share no common moral vision, they freely agree to the structures of the market arrangements, and thus morally authorize their enforcement. We no longer have good reason even to be suspicious of a free market in kidneys; quite the contrary.

The market can facilitate the realization of objective goods by the peaceable cooperation of moral strangers, i.e., by all the constituents of a modern secular pluralist society, not because Hegel provides the missing arguments for the objective good, but because the various parties cooperating in the market can freely agree to obligate themselves by their economic cooperation. Hegel can thus be seen to provide the conceptual framework within which markets in the exchange of kidneys (and other organs not desired by their owners) can be morally endorsed.

IV. CONCLUSION

My thesis has been that it is morally licit – or more precisely, it is not morally illicit – to sell functioning organs such as one of one's two kidneys; and that, in doing so, parties fashion morally binding structures that unify secular pluralist societies such as those of the United States.

The argument advanced in behalf of this thesis, however, is much more sweeping: that in general terms one cannot specify moral obligation but must respect the moral autonomy, or freedom, of unconsenting innocent moral agents; hence one is morally constrained from interfering with parties contracting to exchange that which is properly theirs. The sweep of the argument comes from its negative thrust: showing the inability of reason alone to provide conclusive arguments in general terms for what ought to be done or foregone, and the general inference that one must at least respect the basis of ethics, the freedom of moral agents. More particular inferences should be indicated for many of the subjects touched upon in the body of this paper.

With respect to obligations to one's own body, none can be rationally justified in general terms. There is only the meta-level injunction to use it rationally. This follows from the argument that there is no general, morally canonical theory of the good, or of the just, that can be established by sound argument to authorize a particular obligation towards one's body, valid in all circumstances and for all incarnate moral agents. In terms of pure practical reason – the view from nowhere – we can say that there must be a view from somewhere, but we cannot say what that view itself must be. The first premises from which the content for obligations flows will depend upon what particular moral and religious intuitions, thin theory of the good or notion of rational choice, or ideological opinions, we ultimately rely upon. Regarding those who do not share our ultimate stances, we must respect their autonomy as unconsenting innocents and can licitly only use peaceful persuasion. In Kantian terms, even where we do not ultimately share premises, we can shape a moral order by cooperating on the basis of mutual consent in market arrangements in which we each pursue particular purposes. We each consent to a moral duty to fulfill our part of the market agreement, e.g., to buy, sell, and market kidneys.

From this angle we can see how radically misplaced the Congressional prohibition on the sales of human organs for transplantation is. For the market is in principle a voluntary arrangement. Moreover, the ban on kidney sales deprives the needy with the one readily available source of wealth they all have. One wonders whether, if the writers of the 1987 OTA report had been more sensitive to the history of ethics, they would have seen how defensible, in terms of theoretical reasons as well as of practical consequences, a market in body organs is.

St. Anne Institute
Tulsa, Oklahoma

NOTES

[1]   Distinguish this from the issue of whether, if such duties can be rationally established, the state has the moral authority to enforce such duties and constrain me from selling my organ; as well as from the further issue of whether, even if the state has no moral authority to constrain me, it may have non-moral authority to regulate commerce in this area.

2   It prohibits livers, hearts, lungs, pancreas, bone marrow, corneas, eyes, bones, and skin (1984
     National Organ Transplant Act [NOTA; Public Law 98-507]). The 1987 Office of
     Technology Assessment report (*New Developments in Biotechnology*, 1987, vol.1) says that
     the law was "was based primarily on congressional concern that permitting the sale of human
     organs might undermine the Nation's system of voluntary organ donation" (p. 76). Procuring
     organs by donation is viewed as obviously preferable to procuring them by sale, and so much
     so as to give moral warrant to banning commercialization lest it inhibit those who would
     otherwise donate organs. That the total amount of organs made available by both the market
     and donation would be no less than are available at present, when donation alone is licit,
     seems indisputable. It is nigh incredible to think that legalizing a market in organ sales would
     dissuade more people who would otherwise have donated organs than it would persuade
     those to sell who otherwise would have been unwilling to proffer organs. It is highly
     implausible, therefore, to think that a market would reduce the supply of organs. I infer that
     official policy assumes the sale of non-regenerable body parts to be intrinsically wrong, even
     if buyer and seller would both benefit by the transaction. The reason it is so, is that
     commercial transactions in such body parts would violate the dignity of the parties concerned
     so severely that they should be legally constrained from engaging in such. Secondly,
     Congress was "driven by concern that the poor would sell their organs to the rich, to the
     detriment both of poor people who might feel economically coerced to become organ
     suppliers and those in need but cannot afford transplantable organs" (*ibid.*). The reason for
     this second concern is that organ sales will make suppliers view themselves in ways that
     violate their status as persons, and will alienate those who cannot afford to buy needed organs
     by encouraging them to think of organs as items that can licitly be bought, if only one has the
     resources to do so. This reason, like the first, is Kantian, reflecting the belief that
     commercialization of body parts violates the dignity of the person whose body it is.

3   The OTA report reaches no conclusions its ethical consideration about marketing human
     body parts (*op. cit.*, at pp. 143-144). But it sums up the dispute about the ethical licitness of
     this market as follows: "The dispute between those who believe that commercialization of the
     human body is justified and those who think it is not seems mostly to be an argument
     between those who accept a dualistic view of the separation between body (material,
     physiological being) and mind (immaterial, rational being), and those who do not" (p. 143).
     Accordingly, if one is not a dualist, then there are certain obligatory ways of manifesting his
     rationality in and through the body, and these ways are violated by commercialization. The
     most coherent view that there are such contentful obligations, quite apart from the question of
     commercialization, is to be found in the natural law tradition of Thomas Aquinas, because it
     gives the most articulate defense of the ways in which actions in accordance with practical
     reason ought to be manifested. (Aristotle's views are less clear, and Kant's are skewed by his
     distinction between the causally determined world of phenomena and the *mundus
     intelligibilis* of rational agents.)

4   This is being's *ratio appetibilis* (*Summa Theologica* I, q.5, a.1c: bonum et ens sunt idem
     secundum rem; sed bonum dicit rationem appetibilem, quam non dicit ens). In ontological
     terms this constitutes the perfection or plenitude of being (*Summa Theologica* 18 I-2, q. 18,
     a.1; cf. *ibid.*, a.5c: unumquodque dicitur bonum, inquantum est perfectum: sic enim est
     appetibilis ... .Perfectum autem dicitur, cui nihil deest secundum modum suae perfectionis).

5   In contrast to Stoic natural law, which rests on a notion of physical nature, natural law for
     Aquinas, after Aristotle (cf. *Physics* 199a9-19: nature's telos conceived by analogy with
     purposive human behavior [pointed out by Finnis, (1980 p. 52)], is a teleological conception

of nature that is grasped by analogy with man's practical reason and purposive action. (I am not sure that the analogy holds for Aquinas apart from nature as God's creation. Ethics and politics in Aristotle are conceived of in terms of man's practical reason. But natural law according to *Summa Theologica* I-2, q. 10, a.1c, and q. 94, a.2c, may not be. The thesis that ens is bonum, is Neoplatonic, not Aristotelian.]

6   I owe this exegesis to Finnis, *ib.* (n. 6).

7   The terminology is Finnis's (*ib.* p. 103).

8   Cf. *Summa Theologica* I-2, q. 57, a.4c; q. 58, a.5c; II-2, q. 47, a.6. This necessity for proper dispositions – i.e., states or habits – reflects the fact that the good for Aquinas is objective, consists of concretely determinate ways of being practically right-minded, and so must engage more than one's reason; it is rational to act to acquire knowledge, only if one has, in addition to reason, also interest, a properly disposed curiosity.

9   This is a blunt rendering of Finnis, pp. 101-133. Finnis (p. 129) credits G. Grisez (above all in 'The first principles of practical reason: A commentary on ST I-2, a. 94, article 2,' [1969, pp. 168-69]) with working out the requirements.

10 Cf. *Summa Theologica* I-2, q. 94, a.2c: [P]rimum principium in ratione practica est quod fundatur supra rationem bonis, quia est, *Bonum est quod omnia appetunt.* Hoc est ergo primum praeceptum legis <naturaliter>, quod bonum est faciendum et prosequendum, et malum vitandum. Et super hoc fundatur omnia alia praecepta legis naturaliter, ut scilicet omnia illa facienda vel vitanda pertineat ad praecepta legis naturalis, quae ratio practica naturaliter apprehendat esse bona humana (my emphasis).

11 Even so, the assumption is that the organically functioning whole is normative, i.e., ought to be maintained by one with practical reason. Why so? Is one permitted to improve function by technological manipulation, e.g., via drugs or genetic engineering? change functions? Is life a self-evident good if function is radically diminished? Is the organic functioning whole good if incapable of conscious direction? The inability to give rationally conclusive answers to these questions reflects the problems advanced in the next paragraph of the text.

12 *Nicomachean Ethics* IX 9, 1170a26-b2: life is *per se* good. *Ib.* X 2, 1173a11-13: pain can be *kakon.*

13 My contention here tracks the argument of H. T. Engelhardt, Jr., who argues that no conclusive set of reasons can be given for any objective moral preferences (1996, pp. 40-65).

14 Since the theme here is the moral ground for organized society's institutional sanctions, one cannot simply retort that society's members must be assumed to assent to its institutions and sanctions. I admit, however, that this theme is too complex for me even to have adequately indicated its main lineaments, much less to have adequately dealt with all possible grounds for dissent from my position.

15 One could conceive of God even within a Christian context as having given man nature to work out his salvation, and to use his reason, *in al.*, to redirect tendencies in nature and reshape functions in himself, in the course of achieving that salvation.

16 On the face of it, either nothing would then be obligatory, and nihilism should follow. Or, second, we would have to assume as a brute datum, without reason or justification the first premiss(es) that give content to morality. Or, third, we can say that, although there is no way to provide rationally conclusive content to morality, one is at least required to respect the form of morality, free will guided by reason; and in the absence of cogent general reasons for any particular content, one is constrained – if force requires justification – to rely upon respect for the freedom of the unconsenting innocent moral agent and peaceful negotiation about what ought to be done. I pursue the third choice in part II.

The problem with the assumption in the second choice is evident in Alan Donagan's formulations of such a premiss. One formulation is in terms of St. Thomas, as the first principle of the natural law, *viz.*, that one should "act so that the fundamental human goods" – what we have, after Finnis called the fundamental values the actualization of which are first principles of practical reason – "whether in your person or in that of another, are promoted as may be possible, and under no circumstances violated" (*The Theory of Morality*, 1977, p. 61). This formulation is objectionable, as we have seen, precisely in its articulation of these goods as primary. Donagan prefers Kant's second formulation of the categorical imperative, that one should "act so that you always treat humanity, whether in your own person or in that of another, always as an end, and never as a means only" (p. 63). In this case the difficulty is in the particular content of Donagan's understanding of what is involved in treating persons as ends. "Since a man is a rational creature who is a rational animal," he infers, "respect for man as the rational creature he is implies respect for the integrity and health of his body. Hence *it is impermissible ... for anybody to mutilate himself at will ...* . [D]espite Kant, it is generally and reasonably allowed to be legitimate to give a bodily organ such as an eye or a kidney for transplantation, in order to save a faculty, or the life, of another. Yet his must not be at the cost of that faculty in the giver, or of his life. One may not blind oneself to save another from blindness" (p. 79). It is not at once clear why one who is aging or near death might not be willing, for reasons that may appear good and laudable, to give up a functioning organ he needs to one who can use it more fully, or to sell it in order to secure decisive material advantages or resources for a loved one. Donagan's specification of obligatory content is not even always plausible, much less conclusive. By contrast, the obligations mentioned in the parenthesis immediately following in the text are so, because they follow from legitimate uses of one's will, e.g., in claiming one's body as one's own, or in entering into contractual agreement.

[17] *Grounding for the Metaphysics of Morals* AK421: "A maxim is the subjective principle of acting and must be distinguished from the objective principle, *viz.*, the practical law. A maxim contains the practical rule which reason determines in accordance with the conditions of the subject (often his ignorance or his inclinations) and is thus the principle according to which the subject does act. But the law is the objective principle valid for every rational being, and it is the principle according to which he ought to act, i.e., an imperative."

[18] Hence, too, I can freely engage in evil ends without being, as Kant's theory implies, swept up in an irresistible causal nexus.

[19] This is the second articulation of the categorial imperative (*Grounding for the Metaphysics of Morals*, AK 421). Note that Kant equates "humanity" here with the totality of persons concerning which one has moral obligations. I avoid the term "humanity", so as not to be sidetracked on issues of whether humans are persons, and whether all persons are human.

[20] It may be worth noting that Kant does show that even in cases in which what I ought to will, i.e. its material content, is very largely controlled by the circumstances in which I find myself, yet *how* I will to do it, its form, is mine to control entirely, because practical reason is mine. In this respect, I am free and obligated to be moral.

For example, it may be that, as the head of a household in a very poor society in which the father must provide the means for the education of his only son, and in which that provision is the father's only means for providing for himself in his old age, he has no option other than to sell a kidney in order to get the requisite resources. But the intent (or in this case, the likely combination of intentions) that is willed, is entirely his to control. He may act out of his own self-interest, or out of instinctive paternal love, or out of paternal duty; and he can either simply accept the moral goodness of any or several of these intents, or question the moral goodness of each and see the ways in which it does and does not comport with what he ought to do. Because he can be wholly responsible for his reasoning, and for determining his will accordingly, he is in principle obligated to be moral even in the face of limited options about what to

[21] For Kant it is in virtue of reason alone that one is free from being caught up in a strictly determined sequence of physical causality (*Critique of Pure Reason*, A553-554/B581-582). Epistemically, this is necessary for his justification of the categories to make sense. Were Kant strictly determined causally to make the transcendental deduction, it would be viewed as a completely determined effect, not as a claim to establish the *questio juris*. Practically, one is free insofar as he wills in a way that escapes his causally determined desires, i.e., uses pure practical reason to set up ends for the will and for oneself as willing. The thread common to both the practical and the theoretical strands of Kant's philosophy is, as H.T. Engelhardt, Jr., has pointed out in conversation, freedom as a project of reason: theoretically, in explaining the cogency of reason's *a priori* demands on theoretical knowledge; practically, in determining one's purposes. In both cases Kant assumes that (i)reason constitutes the universal form for particular objective contents that are given *ab extra*; and (ii)insofar as reason is reflective, it has no objective content. (In contrast to Hegel, reason for Kant cannot self-reflectively rework content so as to rationalize it.) Theoretically, then, Kant has to speak of material content prior to its rational formation by the categories – the matter of intuition – although this makes no sense if rational form must structure all contents in order for them to be intelligible; and then he has to assert synthetic *a priori* propositions in order to maintain that the categories are not devoid of content and rationally form the matter of intuition. Practically, Kant has analogous problems: reason must abstract from particular interests in order not to be causally determined, and yet it must attempt to universalize them in order to get content equivalent to the universal form of pure, law-giving practical reason. This attempt cannot succeed, as I argue in the immediately succeeding paragraphs in the text, unless reason can adopt to its own designs the interests in virtue of which one can will something determinate; and this it does insofar as mutual consent to effect some intersubjective purpose is willed as a moral reason. Then several moral subjects will the moral law, with content as well as form, by shaping it for themselves.

[22] I owe this distinction to H. T. Engelhardt, Jr., *The Foundations of Bioethics*, 1986, pp. 68-71; 1996).

[23] One could, presumably, donate it.

[24] It obviously follows that the government ban on organ sales is morally illicit. One may retort that we have tacitly consented to the moral authority of whatever laws are passed by due Constitutional procedure. But one would have to cash this appeal to tacit consent. This issue is too vast and complicated to be addressed here. I must content my self with the indications in the paper's next section about how I would address it.

[25] Cf. *Politics* I 1, 1256a6. Note too that Aristotle's polis is bound to a common good by friendship (*Nicomachean Ethics* VIII 2, 1155a23-27), which a secular pluralist nation cannot assume.

[26] This point is made by H.T. Engelhardt, Jr., in 'Virtue for hire,' (1991).

BIBLIOGRAPHY

Aquinas, T.: 1981, *Summa Theologica*, Fathers of the English Dominican Province (trans.), Christian Classics, Westminster, M.D.

Boyle, J.: 1999, 'Personal responsibility and freedom in health matters: A contemporary natural law perspective,' in M.J. Cherry (ed.), *Persons and Their Bodies*, Kluwer Academic Publishers, Dordrecht, pp. 111-141.

Donagan, A.: 1977, *The Theory of Morality*, University of Chicago Press, Chicago.

Engelhardt, H.T., Jr.: 1986, *The Foundations of Bioethics*, Oxford University Press, New York.

Engelhardt, H.T., Jr.: 1991, 'Virtue for hire,' in T.J. Bole and W. Bondeson (eds.), *Rights to Health Care in an Era of Cost-Containment*, Kluwer Academic Publishers, Dordrecht, pp. 327-354.

Engelhardt, H.T., Jr.: 1996, *The Foundations of Bioethics*, second edition, Oxford University Press, New York.

Finnis, J.: 1980, *Natural Law and Natural Rights*, Clarendon Press, Oxford.

Grisez, G.: 1965, 'The first principles of practical reason: A commentary on ST I-2, a. 94, article 2,' *Natural Law Forum* (now *American Journal of Jurisprudence*) 10, 1169-1201; reprinted in A. Kenny (ed.): 1969, *Aquinas*, Doubleday, New York, pp. 168-96.

Hegel, G.W.F.: 1991, *Philosophy of Right*, A. Wood (ed.), A. Nisbet (trans.), Cambridge University Press, Cambridge.

Kant, I.: 1993, *Grounding for the Metaphysics of Morals*, J.W. Ellington (trans.), Hackett, Indianapolis.

Kant, I.: 1992, *Critique of Practical Reason*, L.W. Beck (trans.), McMillan, New York.

Kant, I.: 1991, *Metaphysics of Morals*, M. Gregor (trans.), Cambridge University Press, Cambridge.

National Organ Transplant Act: 1984, Public Law 98-507.

Office of Technology Assessment Report: 1987, *New Developments in Biotechnology*, Government Printing Office, Washington, D.C.

SECTION FIVE

PERSONS AND THEIR BODIES:
KEY ARGUMENTS AND CONTEMPORARY CRITIQUES

GERALD P. MCKENNY

THE INTEGRITY OF THE BODY:
CRITICAL REMARKS ON A PERSISTENT THEME IN
BIOETHICS

Perhaps the most fundamental fact about medicine is that it acts upon our
bodies, and the most distinguishing characteristic of modern scientific
medicine is the scope of its capacities to intervene into and reorder our
bodies. Yet surprisingly, the dominant forms of bioethics say very little
about the body. Their lexicon is instead populated with terms referring to
obligations and duties regarding persons, and bodies are usually treated
only indirectly, as they are implicated in choices of or harms to persons.
This lexicon has proven to be remarkably thin in comparison to the robust
views of the body that the West inherited from its Aristotelian, Christian,
Jewish and other sources, which can be seen, in different forms, in texts
as diverse as Moses Maimonides' "Eight Chapters" and Immanuel Kant's
Tugendslehre.
 The thinness of most bioethical accounts, and the failure of proponents
of thicker accounts to persuade others of their validity, has meant that on
the whole the bioethical focus on persons has supported the Baconian and
Cartesian commitments to expand choice and eliminate suffering, but has
been unable to tell us what choices to make or what kinds of suffering to
eliminate. Combined with the ever expanding capacities of medicine, the
results for the body are predictable, and have been articulated in
characteristically wistful fashion by Leon Kass:

> In the twenty-five years since I began thinking about these issues, our
> society has overcome longstanding taboos and repugnances to accept
> test-tube   fertilization,   commercial   sperm-banking,   surrogate
> motherhood, abortion on demand, exploitation of fetal tissue, patenting
> of living human tissue, gender-change surgery, liposuction and body
> shops, the widespread shuttling of human parts, assisted-suicide
> practiced by doctors, and the deliberate generation of human beings to
> serve as transplant donors (Kass, 1992, p. 85).

Kass goes on to say that it is not the list itself but a much deeper issue that
underlies it that captures an uneasiness felt by many even when they do
not find all of the above practices objectionable:

M.J. Cherry (ed.), Persons and their Bodies: Rights, Responsibilities, Relationships, 353–361.
© 1999 *Kluwer Academic Publishers. Printed in Great Britain.*

Perhaps more worrisome than the changes themselves is the coarsening of sensibilities and attitudes, and the irreversible effects on our imaginations and the way we come to conceive ourselves. For there is a sad irony in our biomedical project ... We expend enormous energy and vast sums of money to preserve and prolong bodily life, but in the process our embodied life is stripped of its gravity and much of its dignity. This is, in a word, progress as tragedy (Kass, 1992, p. 85).

It is something like this sense that modern medicine has irreversibly degraded our embodied life that, I believe, leads many reflective people to conclude that standard forms of bioethics have failed to address many of the most important issues raised by the conquest of the body by medicine. This sense in turn leads some bioethicists to propose or retrieve more robust views of the body which they hope will rule out those interventions into the body that they find objectionable. One family of such views appeals to one or another conception of the *integrity* of the body. Usually, the appeal to bodily integrity is directed against the view that the body is the property of the person whose body it is and/or the view that the body is a machine with replaceable parts.

It is notable that four essays in this volume, representing traditions as diverse as natural law and phenomenology, are largely based upon an appeal to bodily integrity over against the views of the body as property and machine, and draw from the integrity of the body conclusions regarding various kinds of intervention into the body. Since the notion of bodily integrity apparently has such wide appeal in Western culture, it is worth considering whether it is capable of enabling us to distinguish permissible from impermissible ways of intervening into and reordering the body and its processes. In what follows, I will examine the various views of bodily integrity and its importance found in the essays by Joseph Boyle, Drew Leder, Thomas Powers and S. Kay Toombs. In so doing, I will try to sort out and evaluate the specific moral judgments regarding organ donation and transplantation made by the four authors in the name of these notions of integrity. Since my verdict on bodily integrity as a usable concept in bioethics will be largely negative, I will conclude by arguing for an alternative way for bioethicists to address the question of the body.

In what does bodily integrity consist? Four different answers are found in the authors of this volume. In the natural law tradition, represented by Boyle, but also by Powers's Kant, integrity is a biological concept. The question is whether it refers simply to the body as intact, so that any form of mutilation (including the removal of organs for transplant) is impermissible, or to organic function, so that some forms of mutilation (including organ removal) may be permissible so long as they do not compromise functioning. Kant seems to have held the first position. His view that in contrast to plants animal parts cannot be removed from their organic systems (Powers, p. 212) and that each part of the body has its own purpose and is necessary for the success of the whole organism (Powers, p. 216) appears to lead to the conclusion, which Powers draws, that mutilation is permissible only in cases when amputation is necessary to preserve the whole body.[1] On the basis of extensive discussion of the well-known principle of totality in Catholic moral theology, Boyle is able to distinguish clearly between the two conceptions, point out the commitment of contemporary natural law theory to the second one, and draw the conclusion that there is nothing impermissible about removing organs for transplant.

Working within the path laid out by Edmund Husserl, Maurice Merleau-Ponty and Jean-Paul Sartre, Toombs argues that the parts of the body form an intentional unity in and through which the subject engages the surrounding world and experiences her body as a whole (pp. 79-81). This kind of integrity, with its irreducible reference to the experiencing and acting subject, clearly differs from both kinds of integrity claimed by the natural law tradition, which are described biologically rather than phenomenologically. Nevertheless there is a remarkable similarity between this view and the view of integrity as the intact body, a similarity that is especially evident in Toombs's concern to distinguish the experiential integrity of the body from the Cartesian view of the body as a machine with separable organs and parts (pp. 84-85). From this perspective she raises the issue of organ transplantation in a new way: the question is not about removing organs from one's body but about how and whether one can integrate or "in-corporate" transplanted organs– which, *contra* the view of organs as replaceable parts, never fully lose the otherness that derives from their having belonged to another human being–into one's own lived body (pp. 85-86). Toombs does not rule out organ transplantation on this ground; she is less interested in issuing permissions and prohibitions than in alerting health care professionals to be aware of the difficulties organ recipients will face in maintaining the intentional unity of their lived bodies.

Leder also opposes the Cartesian view of the body as a machine that can be understood and controlled by being analyzed into its component parts, a view that complements and supports the capitalist view of the body as product or commodity (pp. 238-239, 243-244). But his solution, which is less phenomenological in the original Husserlian sense than that of Toombs, is to understand the lived body as formed and sustained in a broad web of interconnections that constitute its integrity, i.e., the integration of self and body, self and others, self and world that the lived body mediates (pp. 251-255). Like Toombs, Leder does not use this notion of the integrity of the body to rule out organ transplantation. He is more concerned with the broader context of modern medicine that assigns such a high priority to transplantation at the expense of other things. Hence he argues that his view of integrity as interconnection makes it more likely that organ transplantation will yield priority to more primary forms of care while at the same time making it possible to understand transplantation, with its transgression of rigid boundaries between one body and another, as a profound affirmation of the interconnection between embodied selves (pp. 256-259).

A second set of questions concerns the reasons why violation of the integrity of the body is held to be morally wrong. There appear to be three major reasons, all of which Powers finds in Kant's writings (though Powers does not seem to realize he is dealing with different reasons in each case). The first and most common claim is that the integrity of the body is a constituent of the dignity of the person, so that to violate the integrity of the body is to violate the dignity of the person. Boyle and Powers are united in affirming that for their respective moral theories respect for the integrity of the body is a part of the respect owed to ourselves and others as rational creatures (Boyle, p. 119; Powers, p. 219). While Boyle does not derive any precise prohibitions from this claim, according to Powers, Kant opposes the selling of body parts on these grounds, namely that such selling reduces the body and with it the person to a mere means. Interestingly, Boyle, who shares the concern about reducing the body to a mere means, does not rule out the selling of organs. The disagreement between Powers and Boyle is over whether or not commercialization necessarily reduces the body to a mere means. Powers believes that it does (p. 227); in this he is joined by Leder who, while not employing the Kantian schema of means and ends, also views the commodification of the body inherent in organ markets as a degradation of humanity (Leder, pp. 233-264). In contrast, Boyle argues that the body is reduced to a mere means only when something is done to it against one's will (pp. 130, 137). This concurs with Boyle's broader argument that most questions about the treatment of bodies must be left to the discretion of individuals.[2]

There are two problems with this first claim. The first is that it is difficult to account for exactly how respect for the integrity of bodies, as this is understood (though differently) by Kant and the natural law tradition, follows from respect for persons on purely philosophical grounds. Powers simply affirms the connection. Boyle grounds the connection in Christian beliefs about the relation of body and soul, the resurrection of the body and the Eucharist, all of which quite plausibly show that for Christians the body is part of the person (pp. 117-119). However, Boyle feels the need to deny that this claim is strictly religious, yet when the reader looks for a secular argument he finds only an appeal to the authority of Alan Donagan, who himself simply asserts without argument that the nature of human beings as rational animals joins the integrity of the body with the dignity of the person (Boyle, p. 119; Donagan, 1977, p. 79). The second problem is that, as Powers recognizes (p. 220), most questions regarding permissible and impermissible uses of the body do not turn on whether the body is being used as a mere means but whether one is using one's body for an acceptable end. Impermissible ends violate one's humanity while permissible ends do not. But if this is so, it seems that we need a theory to distinguish permissible from impermissible ends in order to determine what counts as an impermissible violation of the integrity of the body.

Phenomenologists have a much easier time making their own version of the connection between bodily integrity and the dignity (or in their terms, the integrity) of the person since their very characterizations of "person" and "body" necessarily refer to each other. It follows, for example, that interventions that produce significant physical changes in the body also threaten to disrupt personal integrity (Toombs, p. 90).

The second reason for regarding violations of the integrity of the body as morally wrong is that integrity of the body is a necessary condition of the fulfillment of human moral and other purposes, which may depend upon the integration of some body part into the whole (Powers, pp. 221-222). Powers argues, plausibly, that Kant would reject living organ donation on these grounds, since one would never know whether or not one might later need, say, a kidney to fulfill the purposes of life. As Powers notes, this makes the integrity of the body instrumental to these purposes, but Powers seems unaware that this instrumental status of the body is at least in tension with the body as a constituent of the dignity of the person. Boyle also recognizes the importance in the natural law tradition of maintaining the body for the fulfillment of self- and other-regarding purposes. But unlike Powers's reconstruction of Kant's likely position, Boyle does not reject living organ donation. Once again the disagreement is not over bodily integrity *per se* but over the acceptability of incurring the risk that one will later need the organ one has donated. Powers argues, quite plausibly, that Kant would look askance at incurring such a risk (p. 221), while Boyle does not find anything wrong with doing so for the benefit of another (p. 134).

The third reason for the wrongness of violations of bodily integrity is the claim that respect for the integrity of the body is a condition for proper moral sensibility. This reason refers to Kant's recognition that while we have duties *towards* persons only, we have duties *regarding* other beings such as animals since, for example, to treat animals cruelly is adversely to affect our moral sensibility. In the case of selling organs, Powers asserts–as a matter that "depends entirely on empirical psychology"–that organ markets would threaten the proper respect for bodies that the dignity of the person requires (pp. 226-227). However, aside from the fact that Powers fails to cite any empirical study that would support his claim, the analogy between selling organs and cruelty to animals is weak (selling, unlike cruelty, is not morally wrong in itself). In any case, Leder and Toombs also present versions of this reason, arguing that the reduction of the lived body to a machine-like entity with replaceable parts that seems to underlie the entire practice of organ transplantation leads to a moral climate in which patients are reduced to biological entities (Toombs, p. 85), the self is diminished, human nature itself is degraded, and whole groups of marginalized persons are exploited (Leder, pp. 240-242).

The conclusion to be drawn from this brief survey of bodily integrity in the four authors is that despite the role they assign to it, the concept is not able on its own to do much work in resolving disputed questions about uses of and interventions into the body. One reason is that there are two many conceptions of integrity to expect much agreement on its nature and normative force. Another reason is that even where there is or could be agreement, crucial moral judgments do not turn on claims about bodily integrity but on other matters, such as whether selling parts of one's body necessarily reduces the body to a mere means or threatens crucial moral sensibilities, whether it is acceptable to risk one's future capacity to fulfill one's duties in order to engage in present beneficent activity, whether a certain end of bodily activity is permissible or not.

Of the four authors Boyle seems to recognize the limitations of the concept of integrity; he accordingly (and appropriately) derives the fewest publicly enforceable moral judgments from it. However, in leaving room for the discretion of individuals he tells us little about what kind of moral formation would produce individuals capable of judging well in the area of discretion left to them. It seems obvious that one formed according to the Christian convictions and practices that led to the positive valuation of the body that found its way into the natural law tradition would decide differently than one formed by a society that impels its members to desire and produce bodies that approximate societal standards of perfection, but Boyle's fear of bringing too much particularity into his "common" morality prevents him from exploring this point. Here Leder is more helpful: he recognizes that only if we form ourselves in a robust view of the body (in his case, that of neo-Confucianism) will we be able to cash out a conception of integrity into a substantive moral stance. But if Boyle is afraid of particularity, Leder's solution would seem to require a mass conversion to neo-Confucianism as a universal philosophy of life.

In short, the concept of bodily integrity seems ill-suited to resolve ethical debates regarding the capacities of medicine to act upon the body. Perhaps the concept has always been an unsuccessful attempt to recover or preserve a premodern approach to the body from modern attitudes and practices regarding the body. The natural law approach, and especially the older strand that identified integrity with intactness, tries to express in more modern terms an ancient sense of the body as sacred–a sense that Kant and natural law thinkers seem still to have but can no longer express in its native idiom. But it has become clear that the new idiom (of human beings as ends, of conditions for human purposes and moral sensibility) is ill-suited to preserve the old sense of the sacred. The phenomenological approach follows several centuries of desacralization in which the body came to be understood through Cartesian medicine in terms of machines, and through liberal political theory in terms of property. It tries to recoup something of a sacred dimension it never knew by preserving the experiential unity of the subject against the alienation it finds in machines and property. But as Leder makes clear over against Toombs, it is not self-evident that alienation is always a bad thing; the alterity involved in giving and receiving organs may be a morally positive thing, and (going beyond Leder now) the alterations of identity that come from interventions into the body may decenter the subject in a way that makes a moral or spiritual transformation possible. The lesson is that for the concept of bodily integrity to work it must be imbedded in a more substantive set of attitudes and practices regarding the body. Within such a set of attitudes and practices one can make determinations, and be formed in a way that enables one to make them wisely, concerning the matters on which, as we have seen, particular judgments regarding integrity depend. In short, the concept of the integrity of the body can not stand on its own but can only receive its content from a much broader view of the place of the body and of our various ways of intervening into it and reordering it in a morally worthy life.[3]

Rice University
Department of Religious Studies
Houston, Texas

NOTES

[1] Technically, of course, even amputation under these circumstances violates the intactness of the body, and if Powers is correct about Kant's acceptance of such amputations, Kant here gestures toward the second view, in which integrity refers to the functioning of the whole body. In any case, Powers does not distinguish the two views, and it seems likely that Kant did not either.

[2] There is an interesting agreement between Boyle and H. Tristram Engelhardt's essay in this volume, since both argue that persons are treated as means rather than ends only when something is done to them without their consent. However, this space Boyle assigns to the discretion of individuals follows not from a libertarian theory but from the irreducible particularity involved in applying a substantive moral vision to an area of life where individuals and circumstances necessarily differ and where personal responsibility must be exercised.

[3] I develop this theme at length in my book (McKenny, 1997).

BIBLIOGRAPHY

Boyle, J.: 1999, 'Personal responsibility and freedom in health care: A contemporary natural law perspective, in M.J. Cherry (ed.), *Persons and Their Bodies: Rights, Responsibilities, Relationships*, Kluwer Academic Publishers, Dordrecht, pp. 111-142.
Engelhardt, H.T., Jr.: 1999, 'The body for fun, beneficence, and profit: A variation on a postmodern theme,' in M.J. Cherry (ed.), *Persons and Their Bodies: Rights, Responsibilities, Relationships*, Kluwer Academic Publishers, Dordrecht, pp. 227-302.

Donagan, A.: 1977, *The Theory of Morality*, The University of Chicago Press, Chicago

Leder, D.: 1999, 'Whose body? What body? The metaphysics of organ transplantation,' in M.J. Cherry (ed.), *Persons and Their Bodies: Rights, Responsibilities, Relationships*, Kluwer Academic Publishers, Dordrecht, pp. 233-264.

McKenny, G.P.: 1997, *To Relieve the Human Condition: Bioethics, Technology and the Body*, SUNY Press: Albany, N.Y.

Powers, T.: 1999, 'The integrity of body: Kantian moral constraints on the physical self,' in M.J. Cherry (ed.), *Persons and Their Bodies: Rights, Responsibilities, Relationships*, Kluwer Academic Publishers, Dordrecht, pp. 209-232.

Toombs, K.: 1999, 'What does it mean to be "somebody?" Phenomenological reflections and ethical quandaries,' in M.J. Cherry (ed.), *Persons and Their Bodies: Rights, Responsibilities, Relationships*, Kluwer Academic Publishers, Dordrecht, pp. 73-94.

STEPHEN WEAR, JACK FREER,
AND BOGDA KOCZWARA

THE COMMERCIALIZATION OF HUMAN BODY PARTS:
PUBLIC POLICY CONSIDERATIONS

I. INTRODUCTION

Given the continuing scarcity of human body parts (hereafter HBPs),
increasing support seems to have emerged for allowing the
commercialization of such "commodities" toward the presumed result
that more will be available for transplantation. Such a sea change in
public policy, at least in the developed countries, of course, runs counter
to various taboos, e.g., putting a price on something deemed sacred (or at
least the repository of the sacred), but as its proponents tend to
emphasize, to outlaw such commercialization infringes personal liberty,
deprives the disadvantaged of a way to lessen their plight, and would
respond to the dire need of all those who wait for transplantation,
watching their bodies failing, hoping against hope, and so forth. But,
conversely, commercialization raises the specter of the rich taking
advantage of the poor in yet another way, in effect making them, as usual,
means to the ends of those more advantaged. Hardly a comfortable
scenario; surely a quite poignant and difficult dilemma.

One of the values of this volume lies in the fact that its contributors are
generally unwilling to allow such a complex societal dilemma to be
decided on mere sentiments or unargued opinions, whether such arise
from taboos, vague discomforts, or unreasoned appeals to traditions that
only claim the allegiance of some of us. In contribution after contribution
this volume's authors seek to report and analyze the moral wisdom of
various specific moral, legal, cultural and religious perspectives toward
coming to terms with such an issue. More specifically, they fruitfully
address the issue of what rights and obligations we all have with regard to
our bodies, whether this be for the sale of HBPs, self-mutilation for
pleasure, the option of voluntarily terminating one's life, and so forth.

Our own contribution to this volume will proceed down a somewhat
different path, one that we see as tellingly absent in an important sense
from this volume. Specifically, we will attempt to provide a moral

M.J. Cherry (ed.), Persons and their Bodies: Rights, Responsibilities, Relationships, 353–383.
© 1999 *Kluwer Academic Publishers. Printed in Great Britain.*

account of the appropriateness of a given nation, or other political entity, presuming to limit, or wholly forbid a practice such as the commercialization of HBPs. This may surprise the reader. Does this volume not address the public policy realm here? It seems that it does repeatedly. We should first be clear about what we take to be the "lacunae" here.

II. THE ENGELHARDTIAN GESTALT OF THIS VOLUME

Only brief attention to the other contributions in this volume would seem to give the lie to our claim here. Obvious public policy documents are repeatedly referenced and critiqued, e.g., the report of the Office of Technology Assessment regarding the commercialization of HBPs (1987). Moreover, all of our fellow contributors to this edition seem variously engaged in identifying and assessing the various considerations that should be ingredient in any public policy debate as to whether to allow such a practice or not (and if so, how so).

Deeper attention to the various contributions supports our view, however, that the specifically pro and con, weighing of benefits and harms, sort of reflection that characterizes public policy debates is absent here. Moreover, the final "message" of the volume seems to be that any such debate can only legitimately conclude that commercialization must be allowed as it would be immoral for any society to infringe upon the liberty of its citizens by outlawing, or even significantly limiting, their access to the market in HBPs. More specifically, we see the contributions in this volume as falling into the following two basic categories: (1) philosophical renditions of what various traditions or perspectives would tend to say about the commercialization of HBPs (among other issues); the natural law tradition (Boyle, 1999), the perspective of ancient Greek philosophy (Hankinson, 1999) as well as that of Locke (Mack, 1999), Kant (Powers, 1999), Mill (Donner, 1999) and phenomenology (Leder, 1999; Toombs, 1999) are arrayed herein, as well as the perspective of specific cultural views, e.g., contemporary French jurisprudence (Byk, 1999). Then, (2) we have two quite forceful renditions of the Hegelian (or post-modern) view, a la Thomas Bole (1999) and H. Tristram Engelhardt, Jr. (1999), that although such perspectives can be forceful for those who voluntarily subscribe to them, they are not morally compelling for those of other traditions or perspectives and, generally, can not be utilized to justify the sort of broad secular public policy position that advocates anything other than respect for the choices of persons in the marketplace and public life. One author, *viz.*, Kline (1999), does attempt to take Engelhardt up regarding his libertarian/minimalist state approach, and another author (Leder, 1999) proceeds to spell out a detailed view of a society that would, in fact, limit such activities. But the former does not appear to us to be successful, at least to the issue that we want to engage, and the latter seems to proceed without responding to the minimalist arguments of Bole and Engelhardt. Such a response is precisely what we shall initially offer.

Bole and Engelhardt forcefully present a viewpoint that should be quite familiar to those who have followed this *Philosophy and Medicine* series over the years, as well as those who have attended specifically to Engelhardt's work, especially as fully articulated in his *Foundations of Bioethics* (1996). As Bole and Engelhardt both point out, this view has its antecedents in Hegel's critique of Kant, and enjoys contemporary support from various well-known and respected thinkers, e.g., Alasdair MacIntyre (1981). The view commended to us in Engelhardt's *Foundations*, and spelled out repeatedly in his prolific writings, is basically as follows: the failure of the Enlightenment project to generate a content filled moral account of the good life, *on the basis of reason alone and binding on all rational persons,* has failed. The situation of post-modern man is one of living within nations which are composed of diverse moral communities and traditions which do not share the same views of what the good life is and how it is to be pursued and sustained. In sum, we live in a world of moral strangers and the only fitting result, morally, is for us to respect each other's personal choices to the extent they differ. We may well have gained or fashioned some content-filled moral vision for ourselves, generated either out of our own experience and thought, or out of whatever moral, cultural or religious community we happen or choose to belong to. We can, likewise, choose to set up moral communities among like-minded individuals that are content rich. But we violate basic moral principles if we choose to forbid or limit the activities and choices of those who disagree with us. In the end, as far as secular reason goes, what is morally binding upon us all is to respect and support the secular peaceable society wherein the free choices and activities of all persons are allowed so long as they do not directly harm other persons. And if, for example, an agreement is reached between competent persons for the sale of HBPs from one to the other, then no one is harmed and it must be allowed. For the state to step in and forbid or seriously limit such transactions would be an unconscionable, morally illicit interference with personal liberty, whatever intuitions one may have about the public policy pros and cons.

That this volume is thus quite Engelhardtian in process and result should be apparent. Most of the contributors, aside from Bole and Engelhardt, basically aim at informing us as to what a given tradition, culture or viewpoint commends regarding our rights and obligations to our bodies. But, almost as if they sense Engelhardt *et al.* in the background, there is little sense that they are doing any more than informing us of what those perspectives come to. Some seem also to commend their views to us, but no more forcefully than in the sense of arraying their arguments to which we might or might not be inclined. The Engelhardtian minimalist state may never have otherwise existed, but it certainly seems to have a robust life in this volume. Correspondingly, the typical public policy sort of debate, aimed at considering the appropriateness of legislation where many pros and cons obtain, and absolute consensus neither exists nor is likely to be forthcoming, is impeached in effect as well as by explicit argument.

We will seek to correct this deficiency, i.e., to provide the grounds by which a secular state may legitimately at least consider forbidding or otherwise limiting the behaviors of its citizens with regard to their bodies without presuming or needing their individual consent. We, at least, take ourselves to be in good company with all those who think that public policy is just such a pro and con, weighing of benefits and harms, sort of enterprise, and are at least clear that they are attempting to decide on matters that will never enjoy absolute consensus in our pluralistic post-modern societies.

III. THE ARGUMENT FOR THE MINIMALIST STATE AND ITS FLAWS

We can save much space and time here by stipulating much of what Engelhardt *et al.* have argued extensively for, *viz.*, the failure of the Enlightenment project. We, in fact, can not find, nor will we attempt to fashion, any content-filled moral view that presumes, *on the basis of secular reason alone*, to be binding on all rational persons of good will. It is the further conclusion that this result leads irreversibly to the minimalist state that Engelhardt *et al.* insist upon that we will take issue with. Let us first review the arguments that purport to support such a state from the stipulated fact of the "post-modern predicament", i.e., the failure of the "Enlightenment project".

The relevant modes of argument of the Engelhardtian juggernaut come in three basic forms: (1) arguments for the post-modern predicament itself, whose conclusion we have already agreed to; (2) arguments that a given public policy conclusion does not enjoy any full consensus and, in fact, that its arguments are insufficient to oblige all rational persons to assent to it; (3) empirical or ad hoc sorts of arguments that attack the public policy conclusion as such, i.e., on its intrinsic merits, beyond any stronger claim that it is rationally obligatory. Now the reader must understand a few things about such argumentation overall: (1) given assent to the "post-modern predicament," the conclusion that any specific content-filled moral view or conclusion is not binding on well-intentioned, rational persons is already made, at least for Engelhardt *et al.*; this is what the post modern predicament means; (2) given this, the further project of showing that any specific moral view or conclusion is not morally obligatory is more like an exemplification of the first point, not a further proof. To some extent one might think that each and every moral view or conclusion must be shown to be rationally non-obligatory, but one senses that Engelhardt *et al.* really believe that the general post-modern critique of the Enlightenment project is decisive by itself; and (3) when Engelhardt *et al.* move to empirical or ad hoc arguments that the specific view or conclusion is actually flawed, inaccurate, intrinsically incoherent, or what have you, one may be sure that the validity of the minimalist state that (1) calls for, and (2) supports, in no way stands or falls on the basis of such further criticism. The jury is already in on the basis of (1), with some unnecessary help, one might assume for those who are unable to appreciate the implications of (1), from the sort of arguments that (2) provides.

Of these three forms of argument, it is thus the first two that must concern us at this juncture, the third is clearly just frosting on this particular cake. This is so as once one has received both barrels from the Engelhardtian artillery, i.e., the global post-modern predicament critique, and the critique that any particular public policy view is not rationally obligatory on persons in the peaceable society, then it is really bootless to fret over how the particular pros and cons of any such view balance out. No such balancing argument can be rationally obligatory and thus public policy must simply accept and support the voluntary choices of individual citizens.

Now it is noteworthy that the terms of proof here are variously expressed, not always with the full sense of the austere standard actually at hand. At times Bole and Engelhardt talk of a given view not being "cogent in terms of secular reason," or not "normative for citizens across the spectrum of a secular, pluralist society" (Bole, 1999, p. 331), or not offering "generally justifiable moral bases" (Engelhardt, 1999, p. 278). But elsewhere both authors make clear the high standard of proof they demand for moral argumentation: it must be "rationally conclusive" (Bole, 1999, p. 331) or "rationally decisive" (Bole, 1999, p. 331) according to Bole; or as Engelhardt insists, such argumentation must "preclude in principle" behaviors it presumes to forbid (Engelhardt, 1999, p.282).

One might first note here that the standard of proof these authors insist on is, at least, not that which those who engage in public policy debates take themselves to be accountable to. There can surely be nothing "rationally decisive" or binding "in principle" from the typical pro and con balancing reflection that is the stuff of the usual public policy debate. One might well ask why one is bound to offer such a level of proof. And, more to the point here, on what grounds do Bole and Engelhardt make this level of proof normative upon public policy?

We might well respond to Engelhardt and Bole's austere standard as follows: you have argued that the Enlightenment project has failed to provide a rationally compelling, content filled vision of the good life, as was its supposed purpose. Many at such a point might well go back to Aristotle, to the extent they still wish to engage in moral or public policy debate, and conclude simply that the Enlightenment project simply broke the ancient rule that one should not require more of a field of inquiry than it can bear. Instead we are left with all sorts of moral intuitions and sensibilities, issuing from various traditions, cultures, and personal experience that will somehow get sorted out in a much less elegant fashion, and derive whatever authority they may merit by some sort of consensus agreement, e.g., in the form of the legislation of a representative democracy. But Bole and Engelhardt do not take this route. *The Enlightenment project may have failed, but the standard of rational decisiveness that went with it is to be retained.*

One might ask why they insist on retaining such an unprofitable, self-stultifying standard. But, of course, the claimed profit will be in the libertarian, minimalist state and ethics which they commend. The result of the Enlightenment failure as to content results in the successful establishment of the moral necessity of such a state and the essentially unbridled liberty that its citizens must be allowed.

But one might well wonder why one would insist on sticking with the "person as rational agent and legislator" that lies at the heart of the Enlightenment project. For this "entity," as we agreed, has come up quite empty handed at the end of the quest it began as it shirked off the dogmas of the Dark Ages and attempted to emerge into the light of reason. Is this not to engage in the same sort of enterprise as those who no longer profess the Christian religion, but want to retain its precepts, i.e., "God is dead, and Mary is his Mother"? Add to all this the post-modern critique of the rational self a la Freud *et al.* and one wonders how the rational self, the person as legislator, or what have you, has managed to retain his crown when the kingdom has admittedly become such a wasteland.

The preceding, of course, is little more than rhetorical, and hardly constitutes a decisive rejection of the standard of proof and libertarianism that Bole and Engelhardt offer. Nor do we have the space to offer any such compelling proof. But two basic points can be made, one specific, and one general, toward providing the basic lineaments of a more forceful and adequate criticism. Specifically, freedom appears to have been made into an absolute trump here. Only those in a given society who consent to a given restriction are bound to it; to limit those who do not concur (and, for example, want to enter the market in HBPs) is morally unconscionable. But the result is that the recommended society can not accommodate the intuitions, experience and traditions of those who do have certain views as to what practices should be restrained, limited or forbidden. To many, who believe that a nation, or other political entity, derives its viability and value from its attention to the conduct and character of its citizens, and stands or falls, at least in the long term, by its ability to support, mold and guide "the better angels of our nature," the recommendation is a recipe for disaster. Similarly, it appears to be a recommendation for something like the "*libertum veto*" of recent Polish history, where one legislator was sufficient to bar any governmental measure. But history seems clearly to suggest that this is a sure way to get one's country repeatedly partitioned by less "circumspect" neighbors.

To turn the tables on Bole and Engelhardt, keeping in mind the much more common preference and support for a truly robust state that does "intrude" to some degree on the liberty of its citizens, we might well ask them to help us understand why they think our freedom is so grievously assaulted or impaired if the state decides, by decision of "the representatives of the people assembled," to outlaw certain practices identified in this volume, e.g., shoe fetishism, mutilating one's body for pleasure, or selling one's body parts? That freedom is intruded upon here is clear, but how much? Grievously so? Grievous enough to justify the absolute prohibition of the formative and regulatory functions of the state that many of us take to be essential to the peaceable (and viable) society that we all hope for for ourselves and our loved ones? It does not seem so to us.

This leads to the suggestion of a more general criticism of the Engelhardtian view. On what grounds is your austere libertarian view "rationally decisive" for the rest of us who feel the basic need for a much more activist state? Because of the failure of the Enlightenment project? One should think that the failure of an enterprise counsels abandonment of the assumptions of that enterprise. Because freedom is shown to be so important and basic to the moral life? This is not apparent to the rest of us who see many other, competing values ingredient in the quest for the good life and the institutions that might support and protect it. Because such a state commends itself to our moral intuitions, particularly in the post-modern period? Never having existed, the rest of us may well wish someone else provide us with the proof in the form of a trial run ... and do so elsewhere.

All societies that we are aware of have sought to restrict, limit and at times forbid various forms of the exercise of liberty of their citizens. Thinking of the American principles of "life, liberty and the pursuit of happiness," countries have routinely drafted unwilling citizens into their armies to the resultant death of many, have restricted their personal and commercial liberties in myriad ways (e.g., anti-trust laws), and restricted their pursuit of happiness (e.g., by criminalizing the usage of recreational drugs). During the War of Southern Insurrection (the American Civil War for those who prefer an inaccurate but neutral "description"), the North quickly opted for a draft while the South hesitated, and Lincoln violated *habeas corpus* for the sake of the national emergency, while the south debated its form of union. Which was correct? We will side with those who think that the long term viability of a society is part of the criterion by which one judges its merits as a "peaceable society." Failure, and defeat, *prima facie* signal flaws and inadequacies, not virtue. The common good, as loosely identified and agreed to by the "representatives of the people assembled" seems a much surer course, for all that we may fear its possible excesses. The alternative makes freedom a fetish and sacrifices the public order, and the good of its citizens, on the altar of an enterprise that, as we have agreed, has produced "a haul which will not bear examination" (Eliot, 1943, p. 32).

IV. THE MORAL LIFE AND LIMITS OF A REPRESENTATIVE DEMOCRACY

If Bole and Engelhardt's minimalist state is not "rationally compelled" for the rest of us as we have tried to suggest, and is, in fact, contrary to the moral intuitions of most, the issue then relates to how robust a state, and in what form, we should subscribe to. Here, more by stipulation than extended argument, we will simply presume to suggest that current American institutions appear to be quite adequate, at least in broad outline. Still sharing the conclusion that the Enlightenment project has failed, and recognizing the broad, robust diversity of views concerning the good life and proper conduct, such diverse views and intuitions can be balanced out within the deliberations of the "representatives of the people assembled." *This is not just another content filled moral view, impeached by the Enlightenment result. Rather it is another form of process by which the fact of diversity, and the moral indecisiveness of secular reason, can be accommodated in a "peaceable society."*

Engelhardt and Bole believe that the individual's consent to any governmental restriction is morally necessary. We commend subscribing to an alternative process of adjudicating differences that does countenance such restrictions, in line with the much more dominant belief (and historical practice) that civil polities have positive and robust responsibilities to guide and, at times, restrict the behaviors of its citizens. This leaves us with the rough and tumble, nothing certain or "rationally decisive" world of public policy which, earlier thinkers believed (or at least hoped), was the "free market place of opinion" where "the best ideas would win out." Here it will be to non-theory driven considerations, the pros and cons that most of us believe we face, that public policy will attend. And coercion of some citizens against their wills will be justified by the findings and legislative activities of the "representatives of the people assembled."

A. The Place of Freedom in Such a Polity

Engelhardt *et al.* have proceeded by enshrining freedom as a side-constraint on the activity of government; the minimalist state is the result as consent is morally necessary and state coercion unacceptable. Rejecting this, we are arguably obliged to countenance whatever arrangements the "representatives of the people assembled" opt for, including, one must admit, some sort of totalitarian state where the individual will is subsumed under the national will. This could be temporary, as in some sort of national emergency, or ad hoc, in the sense that submission of the individual will may have to occur in certain areas, as when one is drafted into the military. Could it be permanent and global? Must we countenance such a possibility?

Toward this issue, we will simply refer the reader to the reflections of another contributor to this volume, *viz.*, Joseph Boyle, who rests much of his contribution on the idea that freedom of individual conscience and activity is, in modern times, increasingly a widely held and primary value (Boyle, 1999). Coming as this does from the natural law tradition, which in its religious antecedents certainly had few if any qualms in limiting freedom (actually seeing this in various areas as one of its virtues), we would tend to hope that we are not subscribing to a polity that will tend to restrict freedom overly. In fact, we suspect that the tendency will be to make freedom of individual conscience and liberty a quite primary value, one that, in effect, places the burden of proof on any who desire to limit individual freedom. But, it must be emphasized, this is to divert fundamentally from the Engelhardtian view where freedom is a side-constraint and consent a necessity. In an important sense, in fact, Engelhardt does not even tout freedom as a value; he would allow all sorts of freedom limiting contracts between consenting individuals. For our parts, and supported by a number of the reflections in this volume, we commend it as a primary value in any public policy debate, and expect, in the post-modern world, that it will tend to be taken as such.

B. The Public Policy Deliberations of
the "Representatives of the People Assembled"

If such a polity, and the public policy debates that will occur within it, are deemed morally legitimate, then a few final points about its actions and ingredients should be made clear: (1) any perspective whatsoever is fair game, and one should expect that some representatives will simply vote their conscience in terms of views that many others do not share. If legislator X, for example, believes that the kidney is the repository of the individual soul (because the God Marduk has instructed him that this is the case), then he or she will be likely to vote against any commercialization of kidneys, whatever the other pros and cons of the issue may be. In our suggested polity, this is to be expected and as reasonable as any other view. If Engelhardt *et al.* are right about the Enlightenment failure, as we believe they are, then ultimately one person's view is as good as any other's. And the virtue of such a polity is that it allows any such view to come into play.

Such beliefs about kidneys, or any other organs however, are rare. More often, we suspect, vague discomforts or latent taboos may tend to be operative. Fine. The trick for those who wish *rationally* to consider public policy options will thus be to array considerations and values that may counter any knee-jerk tendencies on the part of the legislators. This can be done both by general appeal to certain widely held considerations, or to considerations within a particular sub-culture or other national group that happens to be politically robust enough to secure representation in the national assembly. This, of course, yields a much less robust and forceful sense to the reflections of the philosophers. But this is, to our minds, another sense of what the Enlightenment failure *means*. What considerations seem generally significant, beyond the particular views of national groups or sub-cultures? It may well be that a broad consensus will be available in certain areas. To the possibility of such a consensus regarding the commercialization of HBPs, we now turn.

V. GENERIC PUBLIC POLICY CONSIDERATIONS

We will conclude this inquiry into the politics and commercialization of HBPs with a number of basic suggestions, first regarding the renewed significance of the other contributions to this volume within the non-Engelhardtian forum that we have proposed, and then regarding various clinical and societal issues that the policy makers will need to attend to as they consider allowing commercialization.

A. Resuscitating the Contributions to this Volume

It is, first, the case that the contributions to this volume should be seen as "resuscitated" from the nether world that the Engelhardtian critique leaves them in by noting that they will have much to say, at least by implication, to many of the "representatives of the people assembled" who will be the final court of appeal in our alternative polity. Boyle, as noted, goes to great lengths to make clear that the natural law tradition is obliged to set great store in the notion of individual liberty, and thus restrictions upon it must satisfy a robust burden of proof (1999). For those legislators who tend to speak from a particular cultural or religious perspective that has a natural law tradition at its base (e.g., Roman Catholicism), such arguments must be attended to. Similarly, for those who come from a more "pragmatic" or consequentialist orientation, Donner's reflections on Mill's view about our liberties, with particular reference to our bodies, should be instructive (1999), just as those with more deontological tendencies would profit from Powers' reflections on what Kant's view might be in this area (1999). Finally, for those who are less than pleased with capitalist society, in general, as well as by implication to the commercialization of HBPs, Leder offers a detailed picture of an alternate society where the supposed harms and inequities of capitalism would be disallowed (1999).

Our first point here is that once we "resuscitate" the public policy realm, all of the contributions of this volume become relevant to it. A given "representative of the people" may well subscribe to some particular cultural or religious view, but he or she is likely to be engaged, if he or she allows this, by many of our fellow authors, as they point out the implications or problems of certain ways of proceeding. This is not to say we expect such legislators to be philosophers, or to feel that philosophical points are decisive. We have already, with Engelhardt, abandoned the latter possibility. But, in the public policy forum, themes within a particular cultural or religious community will often echo basic philosophical traditions, and can be profitably tutored by them. In line with this, our sense is that many such cultural/religious perspectives, particularly as embodied in specific "representatives," do not, in fact, have clear, decisive views on many such issues, and thus may well profit doubly from the reflections of the philosophers, i.e., by also being tutored as to the relevant considerations in a given issue that their own tradition(s) do not speak to specifically. Finally, many representatives will owe their "seats" to the votes of those from multiple cultural/religious backgrounds and, along with assistance in sorting these out, may end up effectively "cosmopolitan" in some basic pragmatic, "weigh the pros and cons" sort of sense; such legislators should also profit from the reflections in this volume, including those regarding what reason would tend to license or abhor.

B. Regarding Whether the Commercialization of HBPs should be Legalized

Narrowing our focus to commercialization of HBPs solely, our second generic point here is that we do not see this issue as usefully approached as if it was all or nothing. Rather, we submit, it seems clear that some forms of commercialization will (and should) be allowed, some should absolutely be forbidden, and the real issue is which forms of commercialization, *between these extremes*, will be allowed or forbidden, and according to which principles or considerations. We offer the following heuristic regarding this generic point.

When one attempts a consequentialist weighing of the benefits and dangers of commercialization of body parts, it is instructive to make the comparison with a more prominent issue, physician assisted death. This exercise is helpful because the euthanasia/assisted suicide question is similar to the debate over a commercial trade in body organs and tissues for at least four reasons:

(1) They both concern themselves with an individual's right to control deeply personal aspects of one's body. Assisted suicide may be the paradigm case in this regard as it involves the most fundamental action to which the body can be subjected, namely its permanent destruction.

(2) There exist strong religious objections to a willful decision to destroy one's body. Furthermore, these objections are often based upon "ownership" grounds in that it is argued that persons do not have the right to destroy what is not their "property." We (and our bodies) belong to the deity and we may therefore not destroy them.

(3) This religious objection to self destruction is so pervasive in western society that it has blended into the secular system of law and societal norms in a manner that has obscured its religious origins. Thus, it has become difficult to isolate the discussion of assisted death in a purely secular sense.

(4) Concerns about the protection of vulnerable members of society form a central focus in both debates. There is appropriately a major fear that those who are already disadvantaged in society will be disproportionately drawn into the activity in question (assisted death or the sale of HBPs). It should be noted that this similarity with the most fundamental disposition of one's body (destruction) applies to other more mundane activities such as prostitution. In fact, such matters can never be taken out of their societal milieu and cleanly debated. It is all well and good to observe that most prostitutes have been driven by poverty and/or substance abuse, and that these are the root problems. This is, however, not productive in the public debate about permitting such behaviors. Linking sale of body parts to the elimination of poverty then appears simply ludicrous; it proceeds only with blinders as to the social pathologies that are operative (Richards, 1996). Remotely prudent and caring public policy should be addressing the latter, not accepting such pathologies and applying the bandaid of the latter. In sum, such bandaids eliminate nothing.

When we then attempt to strip away the religious arguments for the prohibition of physician assisted death, we find we are left with a group of arguments that concern themselves with the problem of avoidable harm to the individual. Arguably, the most persuasive arguments against legalization of physician assisted death deal with the danger of overlooking treatable forms of suffering such as unrelieved pain and major depression. Even among those who are reluctant formally to institutionalize the right to medical assistance in a suicide, there is still widespread support for such assistance in specific cases. Thus, the argument against assisted suicide in the medical context becomes a pragmatic dilemma for many since they do not object on moral grounds in each and every instance.

If the rightness of allowing commercial trade in tissues and organs (in a secular society) is thus, analogously, a matter of protection of citizens from unjust treatment, then one would assume that some practices would be permitted and some prohibited based upon certain parameters. The extreme cases anchor the spectrum. At one end, there is the voluntary donation of one's entire body in the absence of an otherwise lethal medical condition. Despite adequate consent, we would be loathe to permit an otherwise healthy person to be anaesthetized, intubated, ventilated, and then surgically disassembled.

At the other extreme is the practice of tissue donation that is already commercialized: sale of blood and plasma. It is easy to forget that blood is a human tissue and bone marrow is a human organ. When we break out of our preconceived stereotypes however, we must acknowledge that our society *already* permits the commercial sale of human tissue. Why have such practices not been problematic or highly controversial? It is because (1) blood is a rapidly renewable tissue, and (2) there is little risk and discomfort in its procurement. Now the contrast with donation of one's entire body for parts becomes clear; we already accept commercial trade in tissue when its procurement will not endanger or permanently disadvantage the donor (death here being the ultimate disadvantage). This is a simple consequentialist calculus and can form the basis of the regulation governing a just and fair policy. Public policy should thus not proceed as if it faced an all-or-nothing issue; rather it should separate the sheep from the goats between these extremes on the basis of such factors as immediate and long term harm, and the amount of disadvantage (or unfair advantage) that accrues to the contractees to the sale. Regarding such distinctions we will, in this space, only point out that the altruistic system has been grappling with such issues since its inception and direct the policy decisionmakers there, with the caveat that commercialization *per se* does not appear to add anything to such decision making. So our second point in this section is that the commercialization of HBPs, already an accepted fact in certain cases, should be legalized in general with public policy deferring to the mature decisions that already regulate the altruistic market regarding which actual transactions should be allowed as a function of the relative benefit, harm and disadvantage that the transactors are liable to.

C. Commercially Obtained HBPs within the Transplant System: A Few Caveats

If, as we suggest, HBPs should be allowed, to some degree, to be commercially obtained, a few basic issues arise regarding the effects of this on the transplant system. In sum, these considerations all go to the issue that although we may well allow people the negative right to enter commercial contracts for HBPs, this does not involve also giving them some positive rights resultingly to obtain transplant services for the organs thus obtained, and certain considerations regarding the integrity of the transplant system would have to remain operative.

(1) An analogy to medical research may first be usefully drawn here. Although consent is morally necessary in research and in the clinical setting, consent alone is not enough. Research procedures need to have a reasonable chance to generate the knowledge that is sought, with the favorable balance of benefits to the subject and the society, over risks to the subject, with fair selection of subjects. The same can be applied to the situation of organ donation/ transplantation as both are of similar financial burden on the society. Transplant technology (and research) is based on public funds. In addition, the society may need to look after patients long term if transplant outcomes are poor.

It may be desirable from the point of view of the society that the transplants are performed (it costs less to look after a successful transplantee than a disabled patient on the waiting list) but it is equally important that transplants do have a favorable outcome. This requires a certain amount of judgment of when and if to transplant. This judgment may be compromised in a setting of financial conflicts of interest. Now one may argue that decisions regarding organ procurement can be separated from decisions regarding a transplant itself. Yet, the organ itself is usually the triggering event of the entire process. It is thus especially important to insure that policy making attend to the potential outcomes of specific transplants, not just that one is available for a given individual.

(2) Another reason why society may have an interest in decisions regarding transplant prioritization is the fact that an imbalance of transplants in favor of the rich may be detrimental to the poor. The costs may be driven up by competing centers, the resources may be pushed towards the transplants that bring higher revenue, and the pool of organs available as a free donation may be reduced if the potential donors realize that (1) they could sell rather than give, and (2) they will have no control whether their organ will be given to someone where the therapeutic benefit will be likely as the choice will be determined predominately by the recipients' financial status. This, of course, raises issues of justice and fairness which we will address in our last section, but, at this juncture, we are only concerned to suggest that commercialization itself could radically alter the motivations and behaviors within the transplant system, and in ways that would be intrinsically disconcerting, both on the score of clinical appropriateness as well as that of justice.

(3) How such alterations might be seen by the public is not the least concern here. Financial influences on reasoning regarding organ allocation pose some risks to the patient-physician relationship. The

rising importance of fiscal factors in medical decisions has generated considerable anxiety in the society already, as demonstrated in the current concerns regarding managed care. Not certain whether their physicians are going to be guided by medical reasons or the financial pressures, patients may be reluctant to become donors with further implications on organ availability.

(4) Finally, excessive liberty of a recipient to control his transplant might not only reduce healthcare professionals to mere technicians, but by doing so might be detrimental to patients themselves. When market forces are applied to this process, the relation between the donor and recipient also faces new challenges. What right does a recipient have to demand the organ after the donor has changed his mind? In cases of bone marrow transplantation, a late withdrawal could prove lethal to the recipient, i.e., when their diseased bone marrow is obliterated by radiation and then the donor decides not to honor the contract. Equally, what rights does a recipient have if the organ is defective? Or, if the contract is for donation at some future date, what restrictions on donor behaviors can be contracted for, e.g., no alcohol if a liver is to be donated? We will here restrain ourselves, for purposes of space, from conducting a "negative scenario" review here, and merely suggest that the reader attend to the analogous practice of surrogate motherhood to realize that quite similar "distressing issues and scenarios" will tend to present themselves as commercially obtained organs enter the presently altruistic system, e.g., the surrogate who balks at turning over the child, fails to go to or follow agreed upon pre-natal care, or produces a "defective" offspring. Arguably these would present nothing new to what arises in the altruistic system except for the fact that what one might expect of a commercial donor might well be more stringent than an altruistic one (or at least the former might end up contracting to certain things that we would be loathe to enforce, e.g., forced donation of the contracted organ if they balk).

D. Some Final Concerns Regarding Justice and Fairness

A final generic point regarding the commercialization of HBPs relates to the fact that allowing commercialization *per se* is only one of a number of issues that must be settled as one attends to the practice of transplantation, and that other considerations regarding that practice will need to be decided, even if a given polity allows wholesale commercialization. Specifically, just because some individual purchases an organ for transplantation (under a broad commercialization policy), this in no way entitles them to actual transplantation.

It is one thing to opt for a libertarian approach here regarding commercial transactions. But modern medicine is pervasively supported and funded by public funds, so considerations of "justice and fairness" will surely come to the fore. That is, one may well have "secured" the organ one needs, but this surely in no way entitles one to move up on the cue ahead of those who can not afford such an advantage. We would suspect, in fact, that most "representative" polities would insist that those who commercially secure an organ remain in the same system of allocation for transplant surgery that everyone else is in. As all members of the public are equally entitled to access to publicly supported institutions, the purchase of an organ would *per se* only confer an advantage if the buyer was already at the top of the list for transplant. Further, though a fair number could afford organs, substantially fewer could afford to pay for the whole procedure, so the "inequity" created by wealth would tend to be quite circumscribed. Finally, for those who can afford to pay for it all, the public policy debate might well require, on the grounds of justice and fairness, that such people could move up on the cue only if they not only paid for their own transplant *in toto*, but also contributed substantial funds to allow others to get their transplants as well. We do not intend here to make any detailed recommendations as to how those who purchase their own organs should be treated in the transplant system. The preceding is simply meant to suggest that prudent, *and politically acceptable*, public policy here would probably tend to diminish substantially the advantage of wealth in all this. Our libertarian colleagues may well object here that such health care should not be publicly supported, and so forth, but assuming that it will be, as it currently is (legitimately by the action of the "representatives of the people assembled"), then the ways in which the advantages of wealth will be balanced against the entitlements of the many will be basic issues.

To depart from prior comments in this regard, when one considers questions of justice, the analogy with assisted death breaks down as euthanasia and assisted suicide do not concern allocation of a particular limited resource, i.e., transplant services. Because the person asserting control over his body by selling an organ must sell it to someone in particular, a just policy must take account of the use of that organ. One might argue that the problem of unjust distribution of commercially obtained organs and tissues is less important because it would simply provide an enlarged pool of goods for transplant. Rawls' "principle of efficiency" seems operative here, however:

> The principle holds that a configuration is efficient whenever it is impossible to change it so as to make some persons (at least one) better off without at the same time making other persons (at least one) worse off. Thus a distribution of a stock of commodities among certain individuals is efficient if there exists no redistribution of these goods that improves the circumstances of at least one of the individuals without another being disadvantaged. The organization of production is efficient if there is no way to alter inputs so as to produce more of a commodity without producing less of another. For if we could produce more of one good without having to give up some of another, the larger stock of goods could be used to better the circumstances of some persons without making that of others any worse (Rawls, 1971, p.67).

If one presumes that the commercially obtained transplantable tissue has no effect on the altruistically donated variety, then this would appear to be an instance in which the stock of commodities increases without causing any disadvantage to anyone. This presumption is flawed, however, since there is no way to ensure that most (much less "all") of those donating voluntarily will continue to do so. Hence, the commercially obtained organ pool will not only grow from the ranks of those who exercise their freedom to sell, but also from those who previously would have donated freely.

Does this injustice then prohibit (or at least render unworkable) commercialized trade in HBPs? Not if some means to ameliorate the injustice can be arranged in policy. Since the commercialization is not unacceptable in principle, then it is simply a matter of negotiating an acceptable policy which minimizes the injustice. This could be accomplished in a number of ways. One would be to impose a surcharge on the sale such that any commercial transaction would generate resources to supplement altruistic donations. This would minimize the disparity between those who obtain altruistically donated HBPs and those who purchase them. Alternatively, one could regulate the organ procurement totally from a bureaucratic perspective which could then manage the prevailing price and assure the base rate of altruistically donated tissue remains stable compared to pre-commercial levels.

VI. CONCLUDING REMARKS

We have presumed to suggest that although a given civil polity might morally forbid or substantially restrict the commercial trade in HBPs, there is no decisive reason in principle for one to do so, and that there are already well accepted instances of this practice, e.g., the sale of blood products, that make the issue one of when it is specifically appropriate for a given donor and recipient. Going further, we have suggested that it is one thing to allow people to obtain organs commercially, quite another regarding their access to the transplant system. In the latter forum, a significant number of complex public policy issues arise, including the scientific and professional integrity of the transplant system, how certain "disconcerting scenarios" will be dealt with (by analogy to the practice of surrogate motherhood), and how a just and fair system of transplantation can be designed and sustained where all citizens, equally entitled to its ministrations, are treated in the most fair (or least unfair) fashion.

Allowing the commercialization of HBPs, then, produces numerous complex and troublesome public policy issues beyond itself, and it is not clear what, if any, advantage the more well off may end up enjoying once the full public policy debate has been concluded. And to stretch this debate to its full, appropriate extent, one could imagine that the commercially generated injection of many more organs into this system (if such actually occurs), might well paradoxically lead the "representatives of the people assembled," to the extent such an influx requires substantial increases in public fiscal support, to conclude that public support of organ transplantation, whether altruistically or commercially obtained, should be sharply decreased or cease altogether, in favor of some utilitarian sort of calculation that much more good could be done for many more people by switching such funding to other areas, such as preventive health measures, social support of the disabled and dying, and so forth.

State University of New York
Buffalo, New York, U.S.A.

BIBLIOGRAPHY

Bole, T. J., III.: 1999, 'The sale of organs and obligations to one's body: Inferences from the history of ethics,' in this volume, pp. 331-351.

Boyle, J.: 1999, 'Personal responsibility and freedom in health care: A contemporary natural law perspective,' in this volume, pp. 111-142.

Byk,C.: 1999, 'The impact of biomedical developments on the legal theory of the mind-body relationship,' in this volume, pp. 265-276.

Donner, W.: 1999, 'A Millian perspective on the relationship between persons and their bodies,' in this volume, pp. 57-72.

Engelhardt, H. T., Jr.: 1996, The Foundations of Bioethics, second edition, Oxford University Press, New York and Oxford.

Engelhardt, H. T., Jr.: 1999, 'The body for fun, beneficence and profit: A variation on a post-modern theme', in this volume, pp. 277-302.

Eliot, T. S.: 1943, The Four Quartets, Harcourt, Brace and Company, New York.

Hankinson, J.: 1999, 'Body and soul in Greek philosophy,' in this volume, pp. 35-36.

Kline, D. C.: 1999, 'Despair, desire and decision: A fugal response to Engelhardt,' in this volume, pp. 303-330.

Leder, D.: 1999, 'Whose body? What body? The metaphysics of organ transplantation,' in this volume, pp. 233-264.

MacIntyre, A.: 1981, After Virtue: A Study in Moral Theory, University of Notre Dame Press, Notre Dame, IN.

Mack, E.: 1999, 'The alienability of Lockean natural rights,' in this volume, pp. 143-176.

Office of Technology Assessment.: 1987, New Developments in Biotechnology, Government Printing Office, Washington, D.C.

Powers, T. M.: 1999, 'The integrity of the body: Kantian moral constraints on the physical self,' in this volume, pp. 209-232.

Rawls, J.: 1971, A Theory of Justice, Harvard University Press, Cambridge, Mass.

Richards, J. R.: 1996, 'Nepharious goings on: Kidney sales and moral arguments,' The Journal of Medicine and Philosophy 21, 375-416.

Toombs, S. K.: 1999, 'What does it mean to be someBody? Phenomenological reflections and ethical quandries,' in this volume, pp. 73-94.

NOTES ON CONTRIBUTORS

Thomas J. Bole, III, Saint Anne Institute, Tulsa, Oklahoma, U.S.A.

Joseph Boyle, Office of the Principal, St. Michael's College, Toronto, Ontario.

Christian Byk, International Association of Law, Ethics and Science, Paris, France.

Mark J. Cherry, Department of Philosophy, Saint Edward's University, Austin, Texas, U.S.A.

Wendy Donner, Department of Philosophy, Carleton University, Ottawa, Ontario, Canada.

H. Tristram Engelhardt, Jr., Center for Medical Ethics and Health Policy, Baylor College of Medicine, Houston, Texas, U.S.A.; and Department of Philosophy, Rice University, Houston, Texas, U.S.A.

Jack Freer, Department of Medicine, and Associate Director, Center of Clinical Ethics and Humanities in Health Care, State University of New York at Buffalo, New York, U.S.A.

R.J. Hankinson, Department of Philosophy, University of Texas at Austin, Austin, Texas, U.S.A.

George Khushf, Department of Philosophy, University of South Carolina, Columbia, South Carolina, U.S.A.

Donna C. Kline, Attorney at Law, Houston, Texas, U.S.A.

Bogda Koczwara, Fellow in Medical Oncology, Department of Medicine, Roswell Park Cancer Institute, Buffalo, New York, U.S.A.

Drew Leder, Department of Philosophy, Loyola College in Maryland, Baltimore, Maryland, U.S.A.

Eric Mack, Department of Philosophy, Tulane University, New Orleans, Louisiana, U.S.A.

Gerald McKenny, Department of Religious Studies, Rice University, Houston, Texas, U.S.A.

Thomas A. Powers, Department of Philosophy, Santa Clara University, Santa Clara, California, U.S.A.

Allyne S. Smith, Jr., College of Health Sciences, University of Osteopathic Medicine and Health Sciences, Des Moines, Iowa, U.S.A.

S. Kay Toombs, Department of Philosophy, Baylor University, Waco, Texas, U.S.A.

Stephen Wear, Center for Clinical Ethics and Humanities in Health Care, Veteran's Affairs Medical Center, Buffalo, New York, U.S.A.; and Department of Philosophy, State University of New York, Buffalo, New York, U.S.A.

INDEX

Philosophy and Medicine

23. E.E. Shelp (ed.): *Sexuality and Medicine*. Vol. II: Ethical Viewpoints in Transition. 1987   ISBN 1-55608-013-1; Pb 1-55608-016-6
24. R.C. McMillan, H. Tristram Engelhardt, Jr., and S.F. Spicker (eds.): *Euthanasia and the Newborn*. Conflicts Regarding Saving Lives. 1987
                    ISBN 90-277-2299-4; Pb 1-55608-039-5
25. S.F. Spicker, S.R. Ingman and I.R. Lawson (eds.): *Ethical Dimensions of Geriatric Care*. Value Conflicts for the 21th Century. 1987
                                        ISBN 1-55608-027-1
26. L. Nordenfelt: *On the Nature of Health*. An Action-Theoretic Approach. 2nd, rev. ed. 1995   ISBN 0-7923-3369-1; Pb 0-7923-3470-1
27. S.F. Spicker, W.B. Bondeson and H. Tristram Engelhardt, Jr. (eds.): *The Contraceptive Ethos*. Reproductive Rights and Responsibilities. 1987
                                        ISBN 1-55608-035-2
28. S.F. Spicker, I. Alon, A. de Vries and H. Tristram Engelhardt, Jr. (eds.): *The Use of Human Beings in Research*. With Special Reference to Clinical Trials. 1988   ISBN 1-55608-043-3
29. N.M.P. King, L.R. Churchill and A.W. Cross (eds.): *The Physician as Captain of the Ship*. A Critical Reappraisal. 1988   ISBN 1-55608-044-1
30. H.-M. Sass and R.U. Massey (eds.): *Health Care Systems*. Moral Conflicts in European and American Public Policy. 1988   ISBN 1-55608-045-X
31. R.M. Zaner (ed.): *Death: Beyond Whole-Brain Criteria*. 1988
                                        ISBN 1-55608-053-0
32. B.Λ. Brody (ed.): *Moral Theory and Moral Judgments in Medical Ethics*. 1988
                                        ISBN 1-55608-060-3
33. L.M. Kopelman and J.C. Moskop (eds.): *Children and Health Care*. Moral and Social Issues. 1989   ISBN 1-55608-078-6
34. E.D. Pellegrino, J.P. Langan and J. Collins Harvey (eds.): *Catholic Perspectives on Medical Morals*. Foundational Issues. 1989   ISBN 1-55608-083-2
35. B.A. Brody (ed.): *Suicide and Euthanasia*. Historical and Contemporary Themes. 1989   ISBN 0-7923-0106-4
36. H.A.M.J. ten Have, G.K. Kimsma and S.F. Spicker (eds.): *The Growth of Medical Knowledge*. 1990   ISBN 0-7923-0736-4
37. I. Löwy (ed.): *The Polish School of Philosophy of Medicine*. From Tytus Chałubiński (1820–1889) to Ludwik Fleck (1896–1961). 1990
                                        ISBN 0-7923-0958-8
38. T.J. Bole III and W.B. Bondeson: *Rights to Health Care*. 1991
                                        ISBN 0-7923-1137-X
39. M.A.G. Cutter and E.E. Shelp (eds.): *Competency*. A Study of Informal Competency Determinations in Primary Care. 1991   ISBN 0-7923-1304-6
40. J.L. Peset and D. Gracia (eds.): *The Ethics of Diagnosis*. 1992
                                        ISBN 0-7923-1544-8

Philosophy and Medicine

41. K.W. Wildes, S.J., F. Abel, S.J. and J.C. Harvey (eds.): *Birth, Suffering, and Death*. Catholic Perspectives at the Edges of Life. 1992 [CSiB-1]
ISBN 0-7923-1547-2; Pb 0-7923-2545-1
42. S.K. Toombs: *The Meaning of Illness*. A Phenomenological Account of the Different Perspectives of Physician and Patient. 1992
ISBN 0-7923-1570-7; Pb 0-7923-2443-9
43. D. Leder (ed.): *The Body in Medical Thought and Practice*. 1992
ISBN 0-7923-1657-6
44. C. Delkeskamp-Hayes and M.A.G. Cutter (eds.): *Science, Technology, and the Art of Medicine*. European-American Dialogues. 1993   ISBN 0-7923-1869-2
45. R. Baker, D. Porter and R. Porter (eds.): *The Codification of Medical Morality*. Historical and Philosophical Studies of the Formalization of Western Medical Morality in the 18th and 19th Centuries, Volume One: Medical Ethics and Etiquette in the 18th Century. 1993   ISBN 0-7923-1921-4
46. K. Bayertz (ed.): *The Concept of Moral Consensus*. The Case of Technological Interventions in Human Reproduction. 1994   ISBN 0-7923-2615-6
47. L. Nordenfelt (ed.): *Concepts and Measurement of Quality of Life in Health Care*. 1994 [ESiP-1]   ISBN 0-7923-2824-8
48. R. Baker and M.A. Strosberg (eds.) with the assistance of J. Bynum: *Legislating Medical Ethics*. A Study of the New York State Do-Not-Resuscitate Law. 1995   ISBN 0-7923-2995-3
49. R. Baker (ed.): *The Codification of Medical Morality*. Historical and Philosophical Studies of the Formalization of Western Morality in the 18th and 19th Centuries, Volume Two: Anglo-American Medical Ethics and Medical Jurisprudence in the 19th Century. 1995   ISBN 0-7923-3528-7; Pb 0-7923-3529-5
50. R.A. Carson and C.R. Burns (eds.): *Philosophy of Medicine and Bioethics*. A Twenty-Year Retrospective and Critical Appraisal. 1997
ISBN 0-7923-3545-7
51. K.W. Wildes, S.J. (ed.): *Critical Choices and Critical Care*. Catholic Perspectives on Allocating Resources in Intensive Care Medicine. 1995 [CSiB-2]
ISBN 0-7923-3382-9
52. K. Bayertz (ed.): *Sanctity of Life and Human Dignity*. 1996
ISBN 0-7923-3739-5
53. Kevin Wm. Wildes, S.J. (ed.): *Infertility: A Crossroad of Faith, Medicine, and Technology*. 1996   ISBN 0-7923-4061-2
54. Kazumasa Hoshino (ed.): *Japanese and Western Bioethics*. Studies in Moral Diversity. 1996   ISBN 0-7923-4112-0
55. E. Agius and S. Busuttil (eds.): *Germ-Line Intervention and our Responsibilities to Future Generations*. 1998   ISBN 0-7923-4828-1
56. L.B. McCullough: *John Gregory and the Invention of Professional Medical Ethics and the Professional Medical Ethics and the Profession of Medicine*. 1998   ISBN 0-7923-4917-2
57. L.B. McCullough: *John Gregory's Writing on Medical Ethics and Philosophy of Medicine*. 1998 [CiME-1]   ISBN 0-7923-5000-6

58. H.A.M.J. ten Have and H.-M. Sass (eds.): *Consensus Formation in Healthcare Ethics.* 1998 [ESiP-2]   ISBN 0-7923-4944-X

59. H.A.M.J. ten Have and J.V.M. Welie (eds.): *Ownership of the Human Body.* Philosophical Considerations on the Use of the Human Body and its Parts in Healthcare. 1998 [ESiP-3]   ISBN 0-7923-5150-9

60. M.J. Cherry (ed.): *Persons and Their Bodies.* Rights, Responsibilities, Relationships. 1999   ISBN 0-7923-5701-9

61. R. Fan (ed.): *Confucian Bioethics.* 1999 [APSiB-1]   ISBN 0-7923-5723-X